To Dad
With our Love

From
June, Geoff & Maurice
Xmas '82

The Shearers

Patsy Adam-Smith

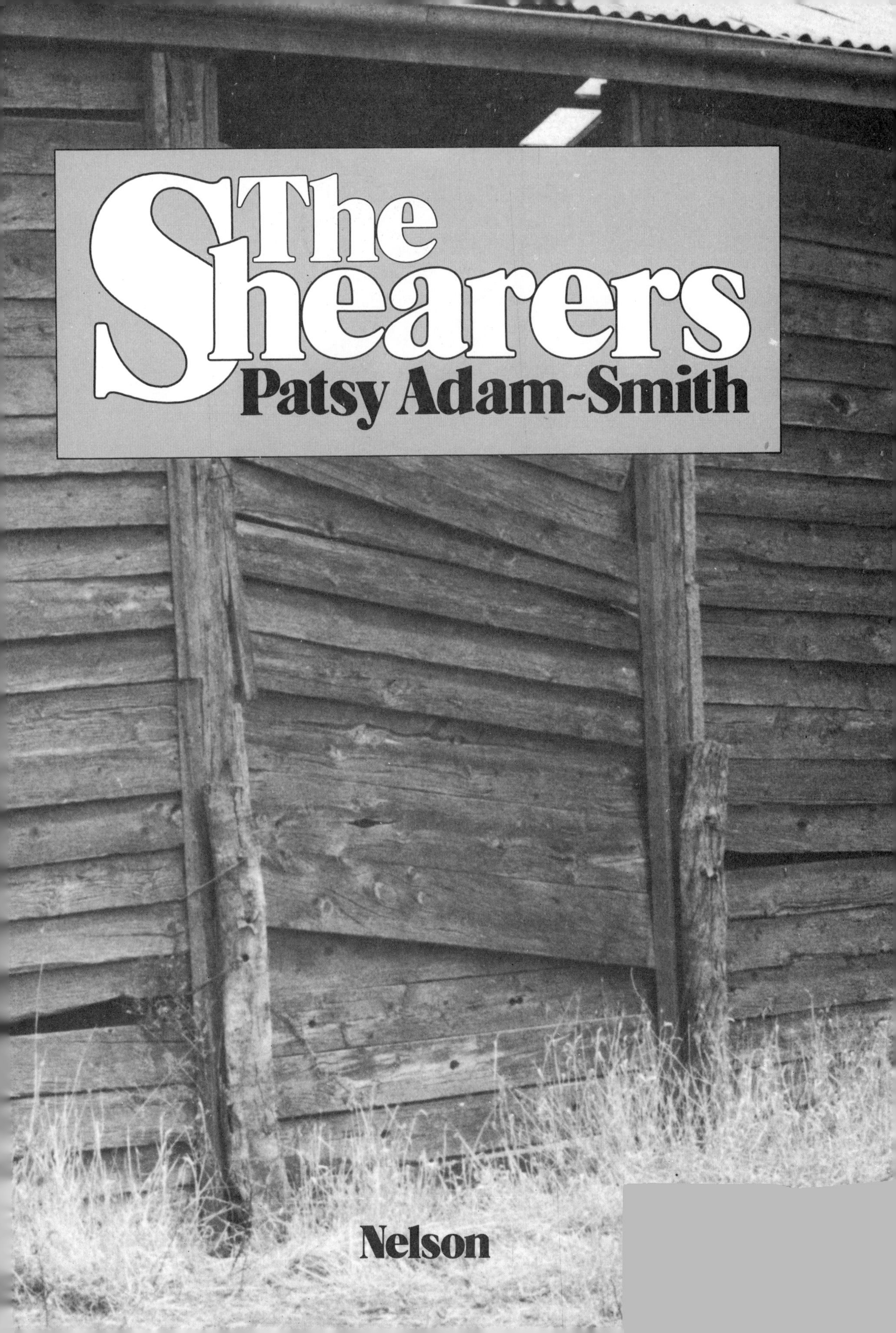

The Shearers

Patsy Adam-Smith

Nelson

Thomas Nelson Australia
480 Latrobe Street Melbourne Victoria 3000
First published 1982

National Library of Australia
Cataloguing in Publication data:

Adam-Smith, Patsy.
The shearers.

Bibliography.
Includes index.
ISBN 0 17 005884 0.

1. Wool trade and industry – Australia – History. 2. Sheep-shearing – History.
I. Title.

338.1'763145

Designer: Tom Kurema
Typeset in 11 on 12 point Plantin by Richard Clay (S.E. Asia) Pte Ltd
Printed in Singapore by Koon Wah Printing Pte Ltd

Contents

Introduction

There was a time not so long ago – many of us remember it well – when we loved this land as though our soul, the spirit of our body, were moulded by it, kneaded and shaped by the soil itself. We swaggered with pride because we believed we walked the greatest place on earth.

The song-makers sang to us of it and we learned the verses.

> You may sing of the shamrock, the thistle and rose,
> Of the three in a bunch if you will;
> But I know of a country that gathered all those,
> And I love the great land where the waratah grows,
> And the wattle blooms out on a hill.

Oh, it was innocent of us – call it naïve if you like. They say now that we were chauvinists; but by God, if we were, we did it with gusto!

Our poets were in the vanguard and we followed. To those of us who lived in the bush, the outback, much of what they wrote was part of the life around us. To city dwellers the poems were like a whiff of gum leaves, scented with the dust of this wide, barren country whose inland rebuffed us as if we were pirates trying to board a treasure ship. That men were droving mobs of sheep a thousand miles across that no-man's-land using the stars as a compass was thrilling. Those stars were new to men from north of the equator. 'Wake me when the Cross turns over,' the drover said to his mate on watch in the night-time stretches, when the sheep stirred uneasily and threatened to head for the wilderness, and the silvery cry of a dingo came down with the cold night air of the desert. We knew of the grinning skull of drought.

> Do you still receive the message that a ravished Border sends, of
> Water done all down the road, and starved stock coming through?
> Does Drought still ride by Hungerford and Death by the Paroo?

John Henry Adams, shearing at Boonoke (NSW) 1898, was typical of the small farmers who went shearing to support their own farms. His stories of the great treks of the shearers and the fabulous shearing sheds of Australia encouraged me, his granddaughter, to write this book

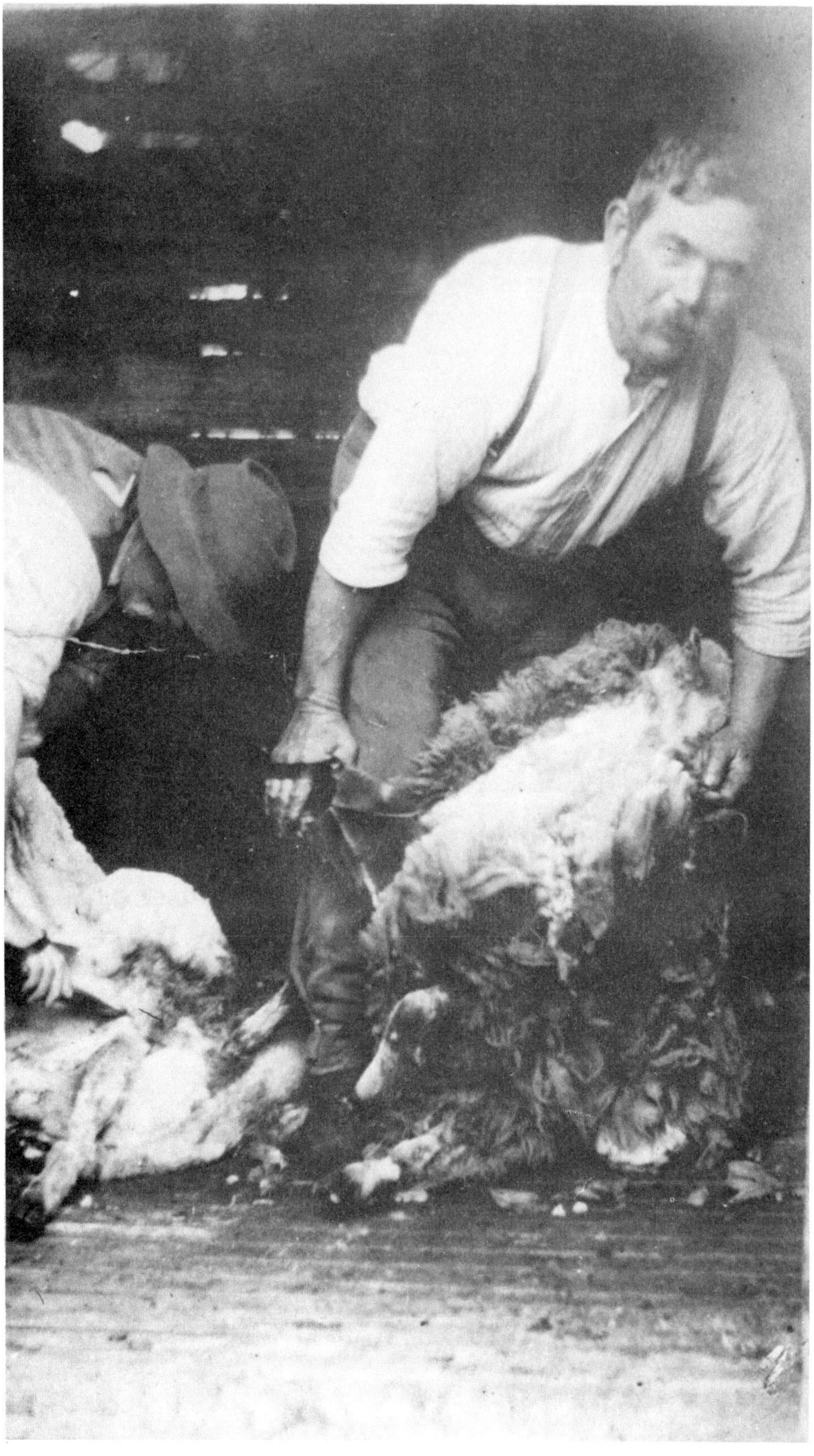

We spoke of the lonely homesteads where gaunt-faced women stood, waiting for those husbands we knew would never return:

Out on the wastes of the Never Never
That's where the dead men lie!
That's where the heat waves dance forever –
That's where the dead men lie!

Many a family remembered a kinsman whose bones were buried in an unmarked grave or bleaching out where the never-sleeping drover watched spectral figures pass by during his night watch.

Teamsters brought the wool down over Old Man Plain, bogging to their axles when the rare rain came on the black-soil plains, seeing magic lights called min-mins where no man walked. As late as 1936, Madigan and his party set off to explore the secrets of the inland. Swagmen humped their blueys.

For time means tucker, and tramp they must,
With stinted stomachs and blistered feet,
They must carry their swags outback.

Mailmen trying to get the mail through drowned when flash floods drove dry river beds five miles wide with water; shepherds and boundary riders were mad as hatters, and no one remembered if they were mad when they first went out to their humpies on the run or if they had gone mad with the dead loneliness of their unnatural lives. We knew them all – or thought we did.

The city was attuned to the bush and country news was of such interest to townspeople that until the beginning of World War II the staff of the *Sydney Morning Herald* moved out to the grounds for the Sydney Royal Easter Show and left a skeleton staff at the city office.

We rode on the sheep's back and didn't forget it for a day. Our economy revolved around wool. Geographical features were named after the greatest industry in the continent. We suggested Woolgold as a possible name for our new national capital in 1927. Before the introduction of decimal currency our

'Out there the boys are mustering, and I must be on the track,' sang the early shearers, and their folk-songs formed the largest part of the traditions and lore of the new land

The fabulous Tinnenburra shed of James Tyson (Qld), built 1895 with 101 stands (an arguable point, as are many in this industry). It was Australia's largest shed

shilling bore the head of a ram from the great Uardry station on One Tree Plain. The best of our stories, our poems and our romance concerned the bush; and the men caring for the sheep were the greater part of this lore. These men were the larrikin princes of our traditions and the shearer was the king. His peregrinations throughout the land, from coast to coast, cloaked him with a glamour no other man had. His daring and pluck in his long journeyings to these faraway places sounded like tales from fabled Araby, and the recital of places where he shore was a litany of the heartland of a new country: Augathella, Toganmain, Illilliwa, Cordilla, Widgeegoara, Jellalabad, Darraweit Guim, One Tree Plain, Alice Downs and Boonoke, Mulberrygong, Edeowie, Yanco, Burrawang, Momba, Mundabullangana, Terrinallum or Boodarie, Panshangar and Shene.

When shearing sheds cut out there was a Woolshed Ball; when the men came back to their home towns there was a Shearers' Ball. They, of all Australians, removed the parochialism from isolated settlements – often noisily, but they were the only army of men constantly on the move throughout this gigantic continent which stagnates if new ideas, new methods of work, new social impetus are not constantly wafted through it.

When they came home they were flush with money and they spent it like sailors, which was no bad thing and at times the only good thing for their home towns, when the recurring depressions that swung the pendulum from the highs were sitting like stagnant water on the land. Girls loved them – still do – with their aura of the spirit of the bush and the inland on their every word, on the way they walk; they were always 'dressers', always better dressed than other workers; but the over-riding factor was the way they went at life.

For the shearers, life was there to be lived. 'Grab it by the shirt front and shake it before it does the same to you!' was their philosophy.

Hundreds of thousands of them tramped this country throughout every state, sang their songs, fought their wars – for wars they certainly were – and all the time they worked the hardest of any man involved in bringing in the golden fleece that for a century and a half kept Australia solvent. In that, they were honest men. They had to be: they were and are paid only by the number of sheep they shear. They worked well, liked the life and loved their country.

There was one group of men who didn't admire the shearers: the squatters, those men who grabbed land early on in settlement, took what they could from it and seldom put anything back, not even affection.

1

The Squatters

One group of men did not admire the shearers – the squatters, those men who grabbed land early on in settlement, took what they could from it and put nothing back, not even affection. There is no denying the facts: this land was raped. The squatters ripped the guts out of it and moved on. In those first few years of settlement, sheep and their owners laid the foundations of vast disasters for Australia, setting off a chain reaction that has not ceased to this day.

They made no effort to deal with their labourers. In this unspoiled continent, isolated from outside influences or mores, they might have forged an harmonious union, master and man becoming partners. Instead, habits brought from the decaying old world were made worse in a land where servants were free: no payment was due, either in cash or responsibility.

Their North American contemporaries had great responsibilities. Slave owners not only bought their slaves, they had to feed and house them reasonably well if they wanted them to live and continue to work. They even had to keep them relatively happy, for unhappy men and women work less well. Families were kept together for the same reason, and within the confines of slavery their workers had a certain freedom. They could still sing, dance, and build up a strong religious fervour to sustain them.

In Australia, all this was missing. Mostly men without women, the convict workers, costless to the master, were worked until they could work no more and then returned to the depot from which fitter men were delivered to the property. As 'Anon' cried in the folk song, 'Van Diemen's Land':

> The first day that we landed here upon that fatal shore
> The squatters came around us, full forty score or more.
> They ranked us up like horses and they sold us out of hand
> And they yoked us up to ploughing frames to plough Van Diemen's Land.

Their accommodation was worse than any they had known in their own land:

> The hovels that we do live in are built of mud and clay
> And rotten straw for bedding and to that we daren't say nay.
> They fence us in with raging fire and we slumber as we can
> To keep away the wolves and dogs upon Van Diemen's Land.

In this country, as in the old, women were used by the masters as women workers have been used in all 'civilised' countries, but the new Australian women differed from the slave women of America's southern states: these women convicts were the same colour as their masters, therefore their cultural background was similar and, in many cases, a woman suited the position of partner to the master. In truth, she was his partner, working beside him, usually sharing his bed, until the man made enough money to import a 'suitable' bride or to travel 'home' to find one that would be acceptable to the new colonial society. Then it was back to the convict huts on the edge of the property clearing for the girl, or back to the convict depot.

Had the early land-grabbers been British 'gentry', matters might have been different. They would have been used to exercising the responsibility they had done at 'home', where a man who treated his servants ill was frowned on by society. 'One does not speak ill to one's servant for the fellow can't answer back.' This very British sporting behaviour was not transported to Australia. Not for nothing were the ruling class here dubbed the 'Bunyip Aristocracy'.

From the beginning, the distinction between master and man was greater in this country than it had been in Britain. (The exceptions to this rule are recorded, sometimes merely in the behaviour inherited through generations of landowners and reflected in the independent faces of the generations who have worked on the family properties.) It was most natural for masters to take a political stance almost entirely involved with ensuring that their own new-found wealth would never decrease, and to evolve a system whereby those less grasping than they would serve them. The 'Anons' of the countryside were singing of this a century ago.

The Death of Ned Kelly

> 'Come out of that, Ned Kelly, you done a lawless thing;
> You robbed and fought the squatters, Ned Kelly, you must swing.'
> 'If those who rob', says Kelly, 'are all condemned to die,
> You had better hang the squatters; for they've stolen more than I.'

'Go tell your boss', says Kelly, 'who lets the rich go free, that your bloody rich man's government will never govern me.' The song finished, after Kelly's death, with a warning and an invitation.

> And so they took Ned Kelly and hanged him in the jail,
> For he fought single-handed although in iron-mail.
> And no man single-handed can hope to break the bars;
> It's a thousand like Ned Kelly who'll hoist the flag of stars.

The land came as easily and as freely as did the labourers. Land was granted to men who brought a letter of introduction, who had a sum of money in the bank, who brought servants or animals to the colony, or 'for services rendered',

'These lands are to be a sheep-walk forever!' wrote Robert Lowe in 1847 in response to the too lenient land grants to squatters. Flock of 23 000 sheep at Wingadee station (NSW)

which could and did mean anything, including corruption. They moved inland to where there were acres enough to make a man lord over greater realms than any landholder in England, no matter what his rank. No wonder they assumed the airs and graces of an aristocracy that had taken centuries to refine its stance in Europe. Impelled by a common impulse, the pioneers, socially prominent and otherwise, had headed for the boundary, the 'Limits of Location' of 1829, beyond which land was neither sold nor let, nor settlers allowed. (On the south of what is now Victoria, the Henty brothers had been in occupation of the Portland Bay District for six years before their presence there, as squatters, became known to the Surveyor-General in Sydney.)

George Ranken, Commissioner of Crown Lands, stated that by 1831, 'they were pouring across the frontiers in scores, north, south, and west – the Governor could not have prevented this, all the police and military in Australia could not have guarded an open frontier of 500 miles – the trespassers had found a new name for themselves – squatters.'[1]

The word 'squatter' seems first to have been used in Van Diemen's Land in the 1820s for those who gathered flocks by both legal and illegal means. Their holdings, convenient depots for stolen sheep, were called 'runs'. The inland frontiers of the United States of America were at that time infested by outlaws and vagrants who called themselves 'squatters', and this name was adopted by the Australian adventurers.

Lord Stanley, in a dispatch of 30 January 1845, wrote: 'I know that the great source of wealth of New South Wales has been mainly the work of those who are termed squatters. I am also aware that though they are described familiarly by the term of American origin, they differ greatly and in most important particulars from the class whose name has been transferred to them; that instead of consisting mainly of the least educated of the population as is often the case with pioneers of civilisation in America, they include many of the most educated, intelligent and wealthiest of the inhabitants of the colony. I believe, moreover, that they constitute a body whose influence in the colony, out of the Legislature is very great, and in it, at this moment, paramount to every other.'[2]

Not all squatters had peerless public backgrounds. Judge Burton, closing proceedings of the Supreme Court in 1835, said, 'Another source of crime, the occupation of the waste lands of the colony by unauthorised and improper persons both bond and free, who, commencing with nothing, or a very small capital, soon after acquired a degree of wealth which must lead every reasonable man to the conclusion they do not get it honestly.' Many terms were used to describe the later squatters, and their 'nefarious practices', but they were 'only following in the steps of all the most influential and exceptional colonists whose cattle and sheep stations are everywhere to be found side by side with [them], and held land by no better title'.[3]

In spite of its origins, the word gradually acquired a complimentary meaning in Australia as the squatters' way of life became more opulent.

Benjamin Boyd in 1844 held fourteen stations in the Monara District, paying licence for four only, and four stations in the district of Port Phillip, paying licence for two. He had come to the colonies only two years previously.

Men holding these vast runs used pressure on all governors. John Macarthur claimed to have secured the recall of the first four governors of the colony of New South Wales and he used his influence to further his wool empire. Governor Gipps, in a dispatch of 19 August 1843, wrote of the depressed state of New South Wales: 'insolvency amongst all classes of the community, even amongst men who were, a short time ago, considered in wealthy circumstances, wages are falling . . .' Sheep were selling at a half to one-third the price they had realised in 1840. A new process was introduced in the colony, 'that of boiling down sheep for their tallow . . . several establishments have been fitted up with cauldrons capable of boiling three hundred sheep at a time'.

The number of sheep had increased from 1 334 593 in 1839 to 3 023 408 in the New South Wales of 1844. Gipps attempted to regularise the land tenure act and the number of animals and became as hated as were King and Bligh before him.

By 1850 the pastoralists' undisputed sway was such that much of the inland was either absorbed by runs or tied up in pastoral leases. 'It has been said that the runholders used "bribery, corruption and all forms of roguery" in their anxiety to secure their leases and to forestall those only too anxious to take their place . . . Vast areas were privately explored and squatted upon long before the Government had any definite knowledge of such proceedings'.[4]

Pastoralists annexed whole provinces. The great squatting areas that were taken over were virtually principalities. (The homes of some stand to this day: Como, once the Melbourne townhouse of the Armytage family, and Werribee Park, the impressive mansion built by the Chirnsides that later became a seminary for priests and now belongs to the Victorian State Government.) Sir Thomas Elder owned a tract of land in South Australia larger than the whole of Scotland where he was born. It was Elder who imported camels from India

'The boss of the board with his eyes everywhere', was referred to in song and verse even after the days when machines had superseded blade shears

with their attendants and accoutrements. Using Port Augusta as the supply pick-up because of its railway, his 'desert schooners' ranged over an area with a radius of hundreds of miles.

The new land had become a sheep walk and wool was king. That may sound like a cliché, but Australia indeed 'rode on the sheeps back', as it was claimed, from the 1830s until after the end of World War II.

These vast estates, measured in square miles or by millions of acres, were stocked, overstocked and eaten out, their native grasslands lacerated by the sharp little cloven hoofs of the foreign animal with the turf-destroying teeth.

The absolute political power of these first wool kings ensured that the judiciary was also in their hands, and neither men who had come free to the new country nor those born here could believe in equality and justice before the law when they were tried by their master for crimes said to have been committed against him.

This total power that wealth and politics gave the squatter over other men both increased his social affectations and set the labourers on edge. What could have been a new world, a place free from humbug and hypocrisy, was spoiled for all time by the millions of tiny black feet carrying enough wool for a king's ransom. That the schism at times appears wider than the gulf between master and man in Britain is the fault of the man in Australia, not the master. These wide acres may have been good for sheep, but they also bred a man dissimilar to others, one who fitted the place, a unique fellow. There was nothing hackneyed about him: he plagiarised no one, because he saw himself as superior to all who had gone before him. While the landlords took their pleasures in the cities and 'touring Europe', this man not only worked the wide dry acres, he thrived on them. Many of the men who became known as squatters spent little time outback, and that only during the rush to grab land. The men who tamed their runs, as those first sheep properties were known, knew the country as well as the Aborigines and were proud of it.

As for the Aborigines, they suffered worse than the land itself from the coming of the woolly mint that squeezed out money whenever money-hungry owners fleeced them. The patter of tiny hoofs crushed the last of the laughter from the tribes who had husbanded this land carefully for 40 000 years. They had walked barefoot over its delicate surface, had dug only with sticks for food; they had burnt it regularly to help regenerate growth: its fragile balance had been a pattern into which they fitted. Once the sheep came these people were doomed: as the squatters said, the Aborigines just would not see that their killing and eating a sheep was in no way comparable with that of the white invader killing and eating a kangaroo. The Aborigines were angry that their best food-gathering lands were now the best sheep-growing lands, and they were exiled from their old homes so that more sheep could graze safely.

Wool was God, and in their greed the men who reaped the wealth resented and treated with disdain any who were not of their kind. They gave no concessions to smaller farmers, some of whom still work the holdings their grandfathers took up, families whose toil today is almost as hard as when their forebears owned the land.

With this greed and, more particularly, his social antics, the squatter ostracised the battling small-holder. In 1977 this disdain was still apparent: it was claimed that 'only seven of Victoria's properties were still held by the original land-holders' families', ignoring those score of farming families in Gippsland and elsewhere who are descended from men who did not get their holdings free, but bought land passed over by the squatters because it was not as easy to work as the acres they took up. These small farmers never had servants to help carve their farms out of the bush. I am indebted to one of the finest, most public-spirited 'landed' families in the country for that point. The Crooke family took up The Holey Plain (Vic.) in 1841 when there was no land route to Melbourne and they must at first sail their ships laden with sheep to Tasmania to market. The fifth generation now runs this splendid old property in Gippsland and I mention them because they represent those who early took up land and who, unlike some of their kind, earned the respect of their neighbours – a priceless commodity in a rough, new society.

Since England and the English were amused by anything the colonials did 'down-under', we set out from the beginning to depict ourselves as hard-doers. In one sense this was no great fault: we could tell a tale of hardship or of woe without appearing to whinge, whine or complain. In his poem, 'Hanrahan', John O'Brien has told what it really was – and is – like for the small farmer. 'We'll all be rooned!' said Hanrahan, whether it be flood, fire, drought, heat or cold that the settlers spoke of. 'We'll all be rooned', said Hanrahan, 'before the year is out'. They have been 'rooned' regularly, ever since, but they've stuck it out, often going shearing to make ends meet. These small farms were settled

by the working men who emigrated from England, Scotland, Ireland and Wales.

For generations these men had been proud of their various trades and callings, their workmanship, their trustworthiness and loyalty to a master or firm. They were the founding fathers of Australia. The industrial revolution had done much to rob the working man of his dignity and these free men hoped to return to the days when masters had an obligation to their servants. Many came because they heard there was land to be had for all who wanted it; they saw opportunities and security for them and their families. Many came because the spirit of adventure was in the air and they were ready for anything. Some came from starving Ireland; others fled from the Highlands of Scotland as the result of the Clearances. Welshmen wished to escape the death-dealing life below ground in the mines; Londoners came because they heard of fresh air, trees and clean rivers that were unknown in their great city with pollution at that time worse than at any time since. Even those who 'left their country for their country's good' hoped to find a rainbow's end eventually, some sort of freedom. All these hopes built muscles on the spirit, developed men of such originality and defiance of precedent that this strong, vigorous strain showed up again and again in their sons – or in their spiritual sons, the working men of a later period, the farmers, navvies, road menders, farm hands, wheat lumpers, engine drivers, boundary riders – all the men who toiled.

Leading the men who swaggered with pride in their independence was the shearer. He was not so much a nonconformist as an unconformist. There was a novelty, a freshness of approach to life about him from the earliest days. He was the genuine, unadulterated article from the beginning. He tramped the land from north to south, east to west, but he was no true swagman: he moved with a purpose, even though he made the movement a thing of adventure in itself. He saw from the earliest days how things were in the outback, saw the struggling selector with his few acres and fewer sheep, and had an admirable distaste for the absentee landlord and his often obnoxious manager of the vast, unloved but rich runs that had been grabbed with so little effort. He developed a skill that is unique, finer to this day than anywhere else in the world. Because of this skill, and his roaming, rambling-seeming life and his stories of the Great Outback on returning home, he became the romantic, freedom-loving hero of a people. Songs and stories about him were on everyone's lips, and writers saw in him the inimitable qualities that made Australians an autochthonous people: a race that sprang from the soil itself.

2

The Master – the Servant

It was the day when the boss was truly the master. When convicts were still being sent to this country and not all men were 'free', we come across evidence of the power of the employer. H. S. Wills is writing in his diary at Lexington, Mt William (Vic.), on 8 November 1843: 'Under necessity of inflicting correction with a stick upon one of the shepherds. Also had angry words with shearer (but proved I was master).'

At that time there were 8000 sheep on his property. By 22 November 1847 he wrote, 'Shore 20 400 sheep, including lambs'. And in 1852 he had '28 000 to 29 000 more-or-less sheep on Lexington'.

The Russells began to arrive in Van Diemen's Land in 1822 and quickly began to do well. Descended from tenant-farmers in Fife, Scotland, they lacked both wealthy friends and influential relations, but within a few years they were settled at Dennistoun near Bothwell (Tas.). In 1836 they spread to the newly settled Port Phillip District under the umbrella of the Scottish-formed pastoral Clyde Company. There they secured Golfhill out of Geelong and Terrinallum near Mt Elephant. Later properties owned by the Russell family were Mawallock, Carngham, Barunah Plains and Native Creek. The correspondence between Van Diemen's Land and Port Phillip gives a valid picture of shearers of the day and how those ex-convict ticket-of-leave men influenced the way squatters would use and treat their workmen in the future: 'Andrew Smith, year ending February 1838 @ £28 shearing, washing and labouring by the day; William Ross, 7 months @ £28 absconded; Thomas Hill, October 1837 to 15.3.38 by the week @ 12s 0d, by the day @ 2s 6d, washing, dressing and shearing at current rates. Settled per order on Eddie, Walsh & Co. £22 12s 5d. George Price – a year ending 29.11.1838 @ £28; 6 months ending 31.5.1839 @ £16. May 25, 1839, to paid expenses at Corrayo during the time the bullocks were lost, 7s 0d. To carting wool in consequence of getting drunk £1 13s 0d; F. W. Cobb Dec. 1837 Shearing @ £1 per 100, 4½ days @ 5s p.d.; Richard Jones Dec. 1837, shearing @ £1 per 100, 4½ days @ 5s p.d.'[1]

Lyndoch Valley (SA), 1844

Phillip Russell wrote in 1838 from Tasmania to his brother George at Geelong: 'I have purchased bagging, wire etc which I will forward soon so that you may have no delay with the shearing. The men will get clothing in town. You will deduct it from their wages. I have likewise given them 10s 0d each which is to be taken from their wages. Tilson is a good shearer. Chapman has not had so much practise as yet.' The new settlement in what is now Victoria was then only three years old and Phillip was able to send stores as well as convicts or ticket-of-leave labour across to the new station.

Two years later he wrote to George at Golfhill, Port Phillip: 'I . . . was glad to learn that the Sheep Shearing was progressing, tho' not quite so rapid as you could wish; I hope by this time it is drawing to a conclusion. I am sorry to find the men lately sent have turned out such blacquards, but a determination on your part to make them do their duty will I have no doubt shew them the necessity of complying with the terms of the agreement; it might certainly prove advantageous to hold out some inducement of an additional pecuniary nature to those who behave well, but the evil disposed I would by no means give into.'

By 1841 George Russell's books show that the rate for shearing had decreased from the 1837 rate to 10s 0d per 100 sheep, and Bells and Buchanan, agents, wrote to George about the depressed state of business in New South Wales: 'The Sheriff (Macquoid) has shot himself on becoming bankrupt, and the House of Gore & Son has stopped payments, and several other failures are expected. It is said that there is a very large amount due Gore & Son by Settlers, and it is feared that the price of stock will fall there.'

Shearing took some weeks, and of the 14 070 sheep shorn in 1841, shearer 'Moffatt' had a tally of 2406, followed by Walford, Kimbers and Muirhead, with Weir having the lowest tally (2153) for these five men. Thomson shore 1467 and Baxter 1071, while a Robert Moffatt clipped 89. The fifteen shepherds, employed for yearly wages of £32 10s 0d per man, each had their own flock ranging from 1245 sheep down to 663. Nine hutkeepers were paid the same wages. The cost of 'woolbagging – 1240 yards' and 'twine 44 lbs' was £57. This was before wool packs were being sent to the property.

That some shearers were determined to be just that – shearers – is evident from correspondence between J. H. Patterson and George Russell the following year. 'I could send you an excellent shearer, but he will not take a job at other work: let me know.' Patterson speaks of shearing in Melbourne to save cartage, and says: 'I have got all your hands, but I had to promise to give them a passage by the Steamer down, which is no great thing after all.' Enclosed with this letter was 'a list of men's names that leave by the steamer today for you. I have 40 wanting the job.'

> We agree to engage with Mr Russell of the Leigh for 2 months, at the wage of 20s 0d per week, to wash sheep, press wool, and anything else required. Dated this 13 October, '42.
> John Branagan, Joseph Walker, John Aicken, Thomas Calcutt, John Culinan. Witness: J. H. Patterson.
> [Walker and Culinan used a cross for signatures.]

Two women squatters, the Misses Anne Drysdale and Caroline Newcombe, were partners in Boronggoop station in the same area as the Russells. Both kept diaries and Miss Drysdale's diary is quoted in the *Clyde Company Papers*. '*Tuesday, 12 October 1843:* All the shearers arrived in the morning, and were engaged the whole day making a washing place, which consists of logs or spars put out in the river between 2 trees; the spars form 2 pens, into the first of which the sheep are flung and allowed to swim awhile; they are then pushed by a forked stick under the middle spar, and men with flat sticks rub off the dirt, after which they swim out . . . and go dripping and exhausted to join their companions. We went to see the operation, which seemed rather cruel, as the day was cold and windy; the river close to the bank is very deep. One of the sheep got out of the Penn and swam across the river, but one of the shearers followed, and brought it back; the river here is fully as broad as the Clyde at Hamilton. They washed 700 to day.'

At Boronggoop, they were still 'abuilding', but finally she announced at the beginning of November: 'Yesterday the wool shed was finished. It looks very well: 30 feet in length, by 14 wide; a skilling or verandah along one side, under which some sheep are to be placed all night, that they may be ready to shear in the morning, otherwise they would be so wet.'

For the next fortnight Miss Anne's diary notes details of sheepwashing, making hurdles, weaning lambs and a 'Mr Dredge, who preached, the shearers attending'. Then, in mid-November, she writes: 'Yesterday our shearing was concluded: the number shorn was 2060. We had all along been afraid that Vere, Armstrong's wife, would be confined before the shearing was over, which would have been unfortunate, as she is cook for the whole establishment; but

she continued quite well and active the whole time until yesterday afternoon, when she took ill. Armstrong rode to Coria, and brought Dr. Shaw, who arrived to tea. At 5 o'clock this morning, a fine boy, the 6th boy and 7th child, was added to the family, and she (Vere) is doing nicely.

'The weather continued fine all the time of our shearing, but today it has rained the whole day, and is now very stormy; we have therefore reason to be thankful that everything has happened so well for us...

'On Monday the shepherd at the out station was discharged because he would not take less wages than £35; we got another the same day for £25. The sheep are now divided into three flocks, two of which are to go to the out station; these with the hutkeeper will take away from the house three of the men. Mr Dredge to tea, and remained – preached to a congregation, with shearers, of 20.'

In October, 1844, shearing rates had increased and Alexander Russell could write to George Russell from Hopkins Hill (Vic.) that: 'I have engaged my shearers, and intended to commence shearing on the 24th Inst.: the terms are twelve shillings pr. 100, and to find themselves in rations: I engage to find them a cook, and supply them with rations at the following prices pr. lb – Flour and Meat 2d., Sugar 4d., Tea 3/-.'

He tells of the trial of a shepherd (foreshadowing what was to happen in fifty years time to shearers when they complained): 'You will see that they are fifty-three sheep short of the number I sent down on the 12th August, principally Lambs: there were two flocks rushed at night when I was down with the Drays – there was 12 Old Sheep missing, and several lambs smothered; in another flock the shepherd lost or could not account for 20 lambs. I refused to pay him all his wages, when he summons me down to Port Fairy; the case came on yesterday, and was decided in my favour.'

Mid-November brings news of shearing and a subject which preoccupied most early property owners and managers: washing sheep. 'I am getting pretty well on with the shearing: we will finish all the Old Sheep to day, and will commence to the Lambs tomorrow; the wool is pretty clean, and with the exception of one flock is as clean as yours was last year. I made a new washpool in a large water hole off the River: it is very good water, much superior to the River, and if I could only keep the sheep clean after they are washed I could get the wool up very clean; but the wool shed being so near the huts, and much exposed, I find it very difficult to keep the sheep clean outside, although very clean next the skin.'

These letters are full of rates of pay, costs of materials, labour problems and each year's shearing. The Van Diemen's Land shearers (Derwenters) are pronounced slow, but 'good, clean shearers'; we read of drays going up to Geelong with 'thirty-five bales of wool' and returning with provisions and essentials like twine.

George Russell's manager of Terrinallum (Vic.) often wrote daily letters reporting on the running of the large, isolated property. Sometimes shearers were scarce, other times 'weekly men' were hard to hire. There are requests for sorters, shepherds and hutkeepers, and wages are constantly mentioned, as are unsatisfactory employees – 'he is a careless, lazy old rogue'; 'I have turned off one crawler'. One employee, John Galbraith, demanded higher wages. 'John Galbraith leaves tonight. I offered him £27, but he wanted £30; so I let him go . . . John Jones has bolted; if you should not hear anything of him I will require another hutkeeper, and also a shepherd in Galbraith's place . . . I neglected to mention a Tar pawlin in the list. I have two but think it would be better to have one for the bullock drivers to sleep under, as they strip the wool to cover themselves in wet weather.'

In late September they began to wash and shearing got under way early in October after some of the men had been 'employed paring amongst the sheep: the lame ones are recovering fast, but I find the best way to be to pare all, sound and lame together . . .' (Footrot was a perennial problem and even in the 1980s one comes on afflicted flocks.)

That year (1848) the manager of Terrinallum, A. C. Cameron, proposed to cut out the issue of rum in an attempt to prevent drunkenness amongst the men. All the early diaries complain of this, and Cameron thought 'it would be better not to allow the washers grog this year, but give them more tea and sugar instead . . . I have not fixed any price with my shearers, but am allowing them at the rate of 10s 0d per week for jobing until shearing commence; they are a very decent looking lot. I will want about a dozen weekly men, but will be able to get some here, so you had better send as many as can be got before the Dray.'

The tea and sugar of 1848 was obviously not as popular with the hands as rum, because by 1850 A. C. Cameron was writing from Terrinallum: 'I suppose I will have to give the washers grog this year as usual: it will take eight gallons of rum to do them: probably you will have a chance of getting it sent up; if not, they must do without until the wool dray returns.' The rum had not arrived by 27 September, the date of the next letter: 'I am happy that you have been able to get 2 men for me, for it is a difficult matter to get men here . . . I told my washers that I would give them a glass of grog a day; could you manage to send the Rum up . . . With regard to the stores required, I think 90 bags of flour, 25 cwt. sugar, and 6 chests teas will be quite sufficient as I have a considerable supply on hand. I should like one chest of Black Tea and two bags fine sugar; and a good supply of tobacco will also be wanted. The number of sheep shorn will be about 24 000.

'I hope you will be able to come up next week, as I feel quite lonely not having any one to talk over the affairs of the state with. I will commence shearing on the first or second proximo. I might have begun a week ago if I had had shearers: the most of mine are forest men, and I told them they would not be required before the 1st Oct. I allowed Bostock to wash a flock yesterday at the wash while I was getting the tubs ready: he washed 1000 very clean without the spouts . . .'

The Clyde Company Papers were prepared for publication by Phillip Brown and an appendix contains details of some payments made to shearers, washers, hutkeepers (and their wives and sons) at Golfhill. From 1837 to 1840, shearers were paid £1 per 100 sheep shorn and washers earned 5s 0d per day with keep. Shearers were charged for their hutkeepers' wages from 1834 to 1844, 15s 0d per week for 39 days and 10s 0d for 51 days, respectively, and for their rations from 1842 to 1844. From 1845 to 1850 the station paid the cook but debited the rations to the shearers. In 1847, the rations for 59 days cost each man 10½d per day. In 1849 the shearers were debited £5 for the cook's wages, but in 1851 shearers paid for neither cook nor rations: gold had been found that year and workmen were very scarce.

In the records for 1849, carters and pressers *on contract* appear, carters at 2s 0d per bale from the station to Geelong. Pressers and washers were 'kept'.

Current rates were:

	1842	*1843*	*1844*	*1845*	*1846*	*1847*	*1848–50*
Shearers	20s	13s	12s	10s	13s–14s	14s 6d	13s
Washers	3s 4d	2s 6d	12s	2s	12s?	14s	12s–15s

In 1850, eleven shearers were paid 13s 0d per 100 as were nine station employees doubling as shearers for the season.

George Rhodes, wool sorter, earned 29s 0d per week and found; the pressers, Daniel Curran and Jeremiah Riley, received 10d per bale and found; John Robinson, Bullock Driver was paid 12s 0d per week.

Three cooks, two for the men's hut and one for the woolshed, were Thomas Gothart (12s 0d per week), John Hames (£25 per annum), and Mrs Smith, who cooked for the woolshed for 12s 0d per week. (She and her husband earned £20 per annum until shearing then '9s 0d and 12s 0d respectively p.w., found'.)

Other wives, mostly of shepherds, were hutkeepers, as were Mrs Murrell, Mrs McDonald and Mrs Gillies. The two boys on the board, Francis and Richard Jollie, each received 8s 0d per week.

Two thousand miles to the north of Tasmania, at Gigoomgan station (Qld) the 1854 shearing was got though with the maximum number of problems. One feels sympathetic to the shearers but even more so for Anderson and Leslie, pioneer squatters who lived 25 miles from the nearest neighbour and a thousand miles from civilisation. They spent the last of that year's typically cold winter trying to wash sheep. These extracts from their diary need little comment. Day after day the entries continue and none mention a day of rest. If shearing is held up by rain other tasks are given to the men, with never a break in the monotony of work.[2]

'*12th* [*October*] Humphries, Carberry, Heffernan and Gregory at wash pool. Got the dry ewes washed, about 260. Singer with cattle. Black gins let the lambing ewes mix with the ewes and lambs.

13th Humphries, Carberry, Heffernan and Gregory at the wash pool. Washed today half of the ewes and lambs. Water cold – 8 deaths, 2 from drowning 6 from exhaustion. Delaney refused to go out with sheep but afterwards thought better of it and went. Blacks assisting at washing.

14th Not so much spirit shown among the men at the wash pool and they seemed inclined to be quarrelsome. Carberry especially behaved in an unmannerly way. The cold weather threw a dampener on them. Owing to the state of the weather we could not get the ewes as clean as the others.

15th J. Dell did not come. He never left Gayndale. Has been drunk all the time.

16th Gregory putting shed in order for shearing. Billy looking for horses but receiving a bite from snake was unable to get them.

18th Humphrey, Carberry and Singer shearing. Gregory stripping hurdles and pressing wool.

19th Gregory stripping hurdles and rolling up fleeces. Lea pressing. [The manager arrived back this day after going by dray for stores and found 'the man I left here has bolted after having drunk all the grog – 10 gallons'.]

20th Humphries, Carberry and Singer shearing. Ogden not well. Gregory drunk. Heffernan drunk, Thackeray drunk. Dray came from Wide Bay last night accounting for men being drunk.
21st Humphries, Carberry and Singer shearing. Gregory rolling up fleeces.
24th Humphries, Carberry and Singer shearing. Gregory pressing wool and shifting hurdles. Delaney and Ogden with ewes and lambs. Heffernan and Charlie with cattle. Blackfellows stripping bark. Paddy came back. P. Lea keeping tally, shearing etc. One wether killed for rations.
25th Carberry and Singer putting bark on the hut as the rain in the morning prevented the shearing.
26th Humphrey, Carberry and Singer shearing. Singer improving rapidly at shearing. P. Lea in shed. Gregory shifting hurdles and pressing wool and rolling fleeces.
30th Rain prevented shearing. All the men salting the meat from the killed bullock in the morning. Humphreys pressed a bale of wool and then with Singer brought in a load of bark. Carberry pressed a bale of wool assisted by Heffernan. Ogden not well. Gregory drafted sheep in the morning and sewed the wool bales. Delany with sheep.
31st Humphrey, Carberry and Singer shearing. Gregory rolling up fleeces. Heffernan barking hut and doing nothing. Gregory and Heffernan had a quarrel today, the former was discharged in the evening.
1 Nov Humphrey, Carberry and Singer shearing in the morning and pressing and loading in the afternoon. Heffernan assisting. Gregory left this morning.
2nd Carberry, Singer and Heffernan washing the rams. Humphreys started with the dray.
5th Carberry, Singer and Heffernan soaking sheep. Blacks did not come up so could not wash them.'

They washed rams until 7 November, then went back to the shed to shear them.

'*8th* Carberry and Singer shearing remainder of lambs and barking hut. Heffernan rolling up fleeces and assisting to bark hut.
9th Humphreys arrived with dray. Some blacks came up to take away some of the bull which died in the paddock. Carberry and Singer shearing. 97 sheep sheared. Wool got away by the *Herculean*.'

On 11 November the diarist writes that 'Two shearers arrived' but he does not mention them again and continues to record the names of Humphries, Carberry and Singer as shearing – when he doesn't have them doing a score of other jobs, including 'felling timber.' On 16 November he records the shearing tally for the day as: Humphreys 57, Carberry 47 and Singer 47.

The owner had many other matters to occupy him and shearing was but one of his problems. Each day men are out looking for cattle, and almost daily they search for blacks, attempting to make them come and work. Sometimes the men sleep overnight in the bush on their expeditions after blacks and report back shamefaced at having sighted none. The entry for Sunday, 19 November, might explain the blacks' reluctance to come in.

'*19th* P. Lea looking after cattle, brought in about 400 but they got out of the paddock again. Gave Mickie a stockwhipping for letting them go.'

On 22 November shearing was finished and the following day's entry states: 'Humphrey and Singer loading wool. Carberry and Heffernan pressing wool and putting to rights.'

'Shearing in Australia', from *Australasian Sketcher*, 3 October 1874

Throughout the whole time they were shearing, entries mention men being sick. Now he records: 'Ogden sick. Carter sick. P. Lea not well today. Water very bad and apt to give dysentery.' Humphrey went off immediately with the dray for supplies. By December the men were 'grumbling and rather troublesome' and each day the thermometer registered higher temperatures until:
'*15 Dec Sun*:- Thermometer on verandah 121, Bay filly got bogged in creek. Had to pull her out with ropes. Carberry found himself robbed of £8 suspects Heffernan but can prove nothing, being done when he was asleep.
20 Dec Carberry and Heffernan not at work at all. Drinking.
25 Dec . . . All drunk. Carberry took up an axe to knock Heffernan down, accusing him of stealing his cheque on the 20th. Also hit Beilby with a spade on the head. Heffernan then came up to the kitchen and drew his knife.
26 Dec [men] not at work. Not recovered from their drinking yet.
1 Jan 1855 . . . Charley brought the mail and news of the Battle of Telma and the capture of Sebastopol.
4 Jan Carberry still very ill. [he had not yet returned from his drinking fit.]
6 Jan Sally the black gin went Bong and was buried. Humphreys half day only, says there was no meat.
Sun 7 Jan Humphrey to Wide Bay to be married.
8 Jan Advanced Carberry £2.16.0. Heffernan owes Carberry 30/- Carberry owes Cosgrove 18/-
12 Jan Humphreys came from Wide Bay with his *to be* wife – slept in the men's hut.

The diarist does not tell us of Humphreys' end matrimonially, but Singer carries on splitting palings, chopping down trees and so on. (It is interesting that there was no more trouble in the men's hut after shearing ended.) He notes that wages were high and a man could not be got for less than £2 a week, then goes on to list the workmen on his station:

'Cosgrove and wife, Carberry and Heffernan – splitters etc
Humphreys, Bullock Driver – very good man
Delaney and McLachlin, Shepherd and Hutkeeper
Singer [he doesn't say what Singer does].'

All the shearers at Winninburn (Vic.) in 1858 were men who worked on the property. 'Bawdon', the gun of the shed, shore up to 129 per day, quite a record for that era, but in between times he milked the cows, turned tobacco and carried out various chores around the place. He consistently had a higher tally than the other shearers. With seven men shearing the average per day was 582. Like many other owners and managers, Adam Turnbull, the owner of Winninburn, faithfully kept a diary: 'Commenced shearing 18 October, 1858 @ 15s 0d per 100. Showery, hailstones . . .'[3]

'Tommy' brought in the sheep daily and did general work on the property during shearing until Monday 14 November, when Turnbull's diary records, in the midst of a page of notes about the comings and goings of neighbours and work on the property, 'Blacks killed Tommy, a young black who has been working for us the last month, by taking his fat.'

In 1862, 17 930 sheep were shorn on his property, with sometimes nine shearers on the board. With Sundays off, it took them 30 days to finish the flock. The sheep were washed by blacks, who were not listed on wages sheets but given rations only. Two men were 'foot-rotting' as shearers finished each sheep.

3

Laban

If shearers were to have a patron saint they might well choose Old Testament Laban: 'And Laban went to shear his sheep.' Laban was the brother of Rebekah, father of Rachel who was a drover. '. . . And behold, Rachel cometh with the sheep. And while he yet spoke, Rachel came with the sheep: for she kept them.'

Laban's method of shearing some 2000 odd years ago had not altered when the men at the new settlement of New South Wales began to clip the wool off the sheep brought out on the First Fleet in 1788 for meat. The earliest Australian shearers clipped away like Laban with 'glorified scissors', usually holding the sheep on their laps, sometimes sitting on a tree stump or on the ground as they worked. Sometimes several worked on the one sheep, one shearing while the others held the animal.

It took the vast explosion in sheep numbers for the old method of clipping to evolve to the advanced state of the art that shearing is today. It has been brought to its highest degree of development in Australia.

At first, sheep numbers increased very little because settlement and exploration remained near the coastline. But after that great barrier to the inland, the Blue Mountains, was crossed, the sheep population exploded.

While flocks were still small, the owner and his servants shore the sheep; as properties grew and the hungry woollen mills of England needed more raw fibre, flocks grew larger. Soon shepherds and 'hands' were employed, mostly ex-convict ticket-of-leave men (sometimes called 'Derwenters' and Sydneys) and assigned convicts from the hulks and over-crowded convict prisons, together with a few free labourers. These did the shearing along with their other tasks.

Aborigines were found to be adept with sheep and up until the 1930s many properties not only had a full 'board' of Aboriginal shearers but all the mustering, yarding, penning-up, piece-picking and cleaning was done by tribesmen and women living on the properties.

German women were noted as shearers in some areas, and Chinese were so proficient that they were sought after as teachers of the art – even though the White Australia policy made them objects of fear.

As flocks increased, a station's regular employees could no longer handle the work, as they were needed for mustering on the huge properties, washing the sheep, and re-locating after shearing. Individual men began offering themselves for work as shearers. Often they were forced to agree to wash the sheep beforehand, but as time went on they became more skilful and refused to do other than shear. Invariably they were obliged to sign a contract to remain on the property until all sheep were shorn and to abide by certain rules.

Skilled men were always in short supply. As early as 1826 shearers were paid 17s 6d a hundred and before 1840 they were getting £1 a hundred, a price that was not eclipsed until the twentieth century.

The building of fences replaced shepherds and, by the 1880s, far fewer men were employed full-time on properties than there had been during the early years of settlement. Shearing now became the province of travelling 'experts', who travelled from shed to shed during the shearing season. The 'old hand', the ex-convict, was now not part of this band of skilled men. Too old, lacking in incentive because of his being managed by others for most of his life and by the lack of women to take as wife, he dropped out of our history. His only legacy to the shearing industry was the way he influenced the squatters to see and treat their workmen in the future. These men usually travelled in groups on foot, horseback, camels, bicycles and later in cars, and camped and lived off the land in between 'runs', as shearing on a property was called.

A system of sending £1 ahead to an owner in the hope of getting a pen was not always successful. Up to 200 men might gather on a property, waiting for shearing to begin, and only on the starting day would a man know if he was one of the chosen band. If he was not, he shouldered his swag and tramped off to the next shed in the hope of finding better luck there.

Each property had its own agreement. Until late in the nineteenth century men had to be 'on the board' at 6 a.m. and work ceased at sunset (in the northern latitudes this could be very late). On Saturdays it finished at 4 p.m. Usually there were three short smoke-ohs, an hour for breakfast and another for lunch.

The men were to buy their rations from the station store at the price set down and were not to leave the property until shearing was ended. If a shearer broke the fleece he would not be paid for shearing such sheep, nor would he be paid if he made second cuts or in any other way dissatisfied the owner. Of the score of clauses on the various 'agreements' none were concerned with the rights of the employees.

The sheds were primitive, low roofed, dark. The living quarters were the same, but neither was part of the contract.

At Golfhill, (Vic.) on 11 September 1849, twelve shearers signed an agreement which contained, among other clauses, the following: 'We the undersigned Sheep-Shearers hereby agree to shear the sheep of the Clyde Co. at the Leigh this ensuing season of 1849. On the following conditions, namely. The shearing will commence about the last week of September and continue until finished. The price per 100 sheep is agreed to be 13s 0d and we agree to find our own provisions which are to be supplied at the following rates namely Flour 2d per lb, Meat 2d per lb, Sugar 4d per lb, Tea 2d per lb; we also agree to find a Cook . . .'[1]

Like biblical Laban, early shearers went about their task in a leisurely way

Although it was not mentioned in the agreement – there was no need, it being understood – any sheep not shorn 'to the satisfaction of the superintendant' would not be paid for.

Margaret Kiddle, herself from an old 'squatting' family, wrote in *Men of Yesterday** of the squatters' views and opinions of shearers: '... The independent workers already most characteristic of the Australian colonies were the shearers. They travelled the country taking work as they needed it, and "carrying their kit along with them".

'The squatter regarded shearers as foul-mouthed, drunken reprobates who maintained an unbroken solidarity among themselves and inevitably stirred up trouble among his regular hands.

*The 'men' of the book's title were, of course, the squatters.

'There were two classes: the Derwenters from Van Diemen's Land, and the Sydneys. The former were the better shearers but the latter were the faster. The Derwenters were distinguished by their tall hats and kangaroo knapsacks, or Derwent drums as they were called, and the Sydneyites by their quart pots and blanket swags.

'Throughout the Western District the Derwenters seem to have been more in favour than the men from Sydney, but both were difficult to handle.

'The squatters' low opinion of shearers in particular and station hands in general was based on solid evidence as many a station diary bears despairing witness. The Learmonths had a reputation for treating their men well. In 1841 their Ercildoun run extended for 50 000 acres, twenty of which were cultivated; they had over 13 000 sheep, 325 cattle and 16 horses. To assist them they employed thirty-five men and four women...

March 2, 1839 Fine day. John Mooney bolted...

March 15, 1839 John Mooney has been assisting with the wool at Geelong having consented the morning of the third to return to our service...

December 25, 1839 ... Thomas Green came down before sunrise to say that Henry Macdonald's lambs had got out of the fold during a severe storm yesterday morning – that about 900 were missing – and that no traces of them had yet been discovered. Mooney and George Gabb started to look for the lost sheep. The Shearers procured Rum from a dray of Mr Birch's and got as drunk as fiddlers.

December 26, 1839 ... Old Sowesby in bed all day from the effects of last night's debauch – as also all the shearers...

January 22, 1840 Mrs Good [the cook-housekeeper] did nothing yesterday and today being unwell from the effects of drink brought up by the dray...

March 31, 1840 ... Drays returned from Geelong, George Taylor having been put in the Stocks for drunkenness.

April 22, 1840 Beautiful bright day – commenced ploughing. S.L. [Somerville Learmonth] went up to see the splitters and when he came back found the men raving drunk – wine had been brought up by Lang yesterday and stolen out of his box...

February 15, 1841 All the men drunk and unruly except James Coleman and Edward Ferriss. James Harper said he intended to bolt but was persuaded to remain on being given a bottle of wine...'

And so it goes on. No doubt the restrained writer of the diary would have preferred to crack James Harper over the head with a wine bottle, but as a valuable commodity James Harper had to be cajoled. As the terse entries indicate, Christmas was a time of special celebration. Then each man was allowed a bottle of wine or spirits, and, as one young squatter wrote: 'There is a continual scene of dissipation and drinking among the lower orders. And we poor swells, instead of partaking in any sort of enjoyment, are kept continually galloping from station to station in mortal terror that Long Jim, Lanky Dick, with Bobby the Bull are all drunk with the sheep looking after themselves on the tops of the mountains.

'All the station hands and most independent workers were liable to drink themselves into delirium whenever there was a sufficient supply of grog, but as a general rule their serious drinking was reserved for the intervals between working. They were often aptly compared with sailors who after abstemious months indulged themselves to revolting excess. The practice of "drinking their cheques" was such a well-recognised feature of colonial life that it was taken for granted.

'With three or Six Months Wages in their pockets they gird their Kangaroo Skin Knapsacks upon their backs containing their whole stock on hand – say a blanket, 2 shirts and stockings. Thus equipt they journey often 150 miles to the next Grog Shop where they have a "bright flare up" for a few days.

The early board (*c.* 1864), when men could shear wherever they pleased in the open shed

'When all is spent they are again ready to engage with the first that applies or again return to the bush indifferent whether they are employed or not as they live comfortably at the expense of the settler. A shearer was known to take £16 to Melbourne and spend it in three nights.'[2]

Some squatters who could afford it brought out labourers under the bounty immigration scheme in the hope that hand-picked men might prove less drunken and more satisfactory than the old hands. But Niel Black's troubles with his Highlanders are an indication that the hope was illusory. They were engaged for him by his partner, T. S. Gladstone, who took the greatest care to choose sober and trustworthy characters, but was hampered in this laudable object by his even greater care to pay them the lowest possible wage. In 1841 he expected to procure ten or twelve men, but eventually could persuade only seven to engage, and they insisted on engaging at £30 a year with an extra £3 for their passage.

In 'both sorrow and anger', Black reported the results of his partner's pains several months later: 'Four of the Hired Servants sent from home have already bolted . . . I fear the Moral Character of most of them is of a very low order . . . They left me in the busiest season of the year (shearing). Not together, but one after another while here I am 140 miles from where I could get their places supplied.'

In 1850 shearers were earning 14s 0d per hundred but a few years later, because of the gold-rushes, it rose sharply, only to drop to 12s 6d a hundred in the 1860s. This rate remained fairly constant, and in the 1880s, when men began to object, many were still getting only 12s 6d per hundred. Wool winders (rollers) were earning £1 a week and pressers, 1s 6d a bale – on agreeing to put 'not less than 400 lbs in each bale of greasy wool and not less than 300 lbs in every bale of washed wool'. The agreement between squatters was that 'the work may, if possible, be got done at any lower rate'.

'On properties where sheep had to be washed before shearing (to remove sand, dust or seeds from the wool)', said Margaret Kiddle, 'the washers became shearers after the last sheep dried off and was ushered into the shed. On some properties a ceremony of sorts took place when the shed was ready to begin, a whistle was blown, rum handed round to the men and then they were heads down and tails up until the shearing ended.'

Alexander Mollison took up land sixty miles out of Pyalong (Vic.) in 1838, only three years after the first settler arrived in the colony. On 30 April 1841 he wrote: 'It would not be easy to give you a description of our mode of life as you request. I may, however, simply say that the Colliban establishment numbers about fifty men, women and children, and more than eleven thousand sheep, cattle and horses. That the eleven thousand quadrupeds are spread over

From the earliest days, sons followed fathers into the shed. In this early Queensland shed, Shelford, Jim Butler Senior works with his son and grandsons Sam, Eddie and Charles Butler

Birrie Creek (NSW) and brushwood fences typical of many early stations

forty-five miles of unenclosed country east and west, that it is the duty of the fifty men, women and children to keep the eleven thousand animals in order, that both the bipeds and the quadrupeds are continually going wrong; and finally that it is my duty and never ending care to put them right again. Our servants are now mostly emigrants, and others are "old hands" as they are called, that is men who have been convicts but whose term of sentence had expired. These are quite useful men whom I have selected from the many scores of them who have been in our employment during the last five years.'

He gives an interesting account of shearing in a letter from Colliban station dated 28 November 1847: 'I returned about three days ago from my station called Pyalong, about thirty miles hence, my sheep being all stationed there now. I do not bring them to Colliban to shear as formerly, but have built a shearing shed at Pyalong. I began rather earlier in the season this year, and the weather being fine, I finished this somewhat heavy task in four weeks. About 500 or 550 sheep were shorn every day, and the wool packed in bales ready for shipment. It is classed by a wool sorter. That is, each fleece when shorn is carried by a lad to the sorter who removes any stained or inferior parts, rolls it up and sends it to a bin in which are other fleeces of the same quality only. Three men are employed at the press. Two men to bring sheep to the shed, take away the shorn sheep, sweep, and water the yards in which they are kept.

'Ten shearers were employed. We began at six o'clock and generally worked until half past six, when the shearers were dismissed to their huts. After that a good deal remained to be done by the master and overseer, and I was glad to

Canowie (SA), 1870

smoke a quiet pipe at nine and tumble into bed. Another party of eight men under an overseer were employed at the same time in washing the sheep. They begin about a week before the shearers. The shed at Pyalong holds six hundred sheep in pens or enclosures, separated from the shearing floor, wool table, bins, etc. When the weather is unsettled the shed is filled at sunset, so that the men are employed, even if it should rain. In the fine weather a few sheep only are housed to preserve their fleece from dew, and these are shorn before breakfast.

'When the Pyalong flocks are washed the washing party was sent to St Agnes (a station adjoining the Colliban, and which I manage) and in due course the shearers followed. We are now shearing at St Agnes and making hay at Colliban.'[3]

4

Magic Soap and Ammonia

In the early 1840s the price of sheep and wool fell and the industry was in crisis. As they had paid little or nothing for their land, the obvious saving the squatters could make was through wages. They built up their flocks and lowered the wages of shepherds and hut keepers.

Some men got jobs at the boiling down works, where sheep were boiled in vats for their tallow which was exported to Britain for the making of soap and candles. (Some estimates of the number of sheep boiled down between 1843 and 1851 are given as four million.) A further effort to weather the crisis was to wash the sheep before shearing, because cleaner wool brought higher prices.

With a flat board on the end of a long stick, shearers scrubbed the sand out of the wool on the sheep's back on the Edwards River (NSW)

'Thou art fearing expense, but stop a bit,' wrote Francis Cotton to Joseph Mather in 1861. His speech, with its 'thou's' and 'thee's' identifies both him and his friend as Quakers. Kelvedon on the east coast of Tasmania has been identified with the Society of Friends since Cotton and his wife Anna built there in 1830. The letter of 1861 was to advise Mather of a new method Cotton was using of 'hot water washing' to clean wool while it was still on the sheep's back.

'Get the try-pot and heat water, get the tubs, mix water to a proper temperature, dissolve 1 lb or so of soap in each tub, put 3 or 4 sheep (if the tubs will hold them) in each cask, wash one lot while the other soaks . . . two persons lift the sheep out of the water on to the grating, squeeze the soap and water out roughly and turn them onto an incline – and down they slip into the River. Here they swim themselves clean. Four persons may wash 500 sheep in a few hours; the washing will be less labour to the men than at present. Mention this to Captain Crosby. Nothing like diffusing information, others may profit by it. Crawford thinks they may gain £1000 the first year by using Hot Water.'

The Cottons may have gained too. Certainly Kelvedon prospered, and like many of the early Tasmanian homes it is still owned by descendants of the original builders.

So much was different from what they had seen in Britain. The lovely white sheep, for instance, where were they? Here they had all turned brown, their

Weilmorangle (NSW), like many western stations, had its own wool scour where the wool was washed, in this case in warm water to lessen the weight for cartage

Rivers were utilised for sheep washing; the Yarra, Melbourne, 1872

wool clotted with sand and dust. But it wasn't aesthetics that bothered the growers; the sand and dirt-filled wool slowed down shearing, blunting blades (and later cutters) which often cut the sheep because of the pressure needed to push through the extra thickness and weight. And there was worse. The cost of transport was much greater because of the weight of sand.

On Illilliwa (NSW) a 'lagoon' was made specially for washing sheep. It was in reality a huge tank with high clay banks around it. The manager, Donald Grant, a self-taught surveyor, put a pump on the river and pumped water into the excavation. The 'lagoon' was large enough for Grant to teach young lads to scull and, it is said, they made splendid oarsmen. The property was owned by an English company which wrote to Grant, telling him to wash the sand and dirt out of the wool to lessen the cost of transport. Grant installed a steam engine and, using bark piping to carry water down the lagoon to it, the water was heated. He employed forty men with their own camp and cook at the sheep-wash; 250 000 sheep were washed twice in hot, soapy water, then rinsed in clear water, then driven across five miles of sandy, dusty plain to be shorn at the shearing shed. This went on for three years until, exasperated, Grant wrote to the company and the scheme was abandoned.

On some stations spouts were a popular refinement, with two men holding each sheep beneath the spout while it was soaped (until soap was ruled injurious) and rinsed. The animal was then swum, being ducked occasionally by means of long poles, until it reached draining, then drying pens.

At Warbreccan on the Edwards River (NSW) 1000 sheep could be washed daily 'by proper treatment and management. Twelve or thirteen blacks, one white overseer and a shepherd were employed at this work'.

There were many sheep-washes, all different, but all of them awful for the men who worked at them. Often the men worked in water all day to the waist – sometimes to their armpits; some stood on planks, some on barrels, but it made no difference. 'It was a fair cow of a job,' Charlie Gough remembered his father saying.

Scouring the wool after it was taken from the sheep was not much better, as far as the men were concerned. Many stations had railway tracks running out from the shed to the scour. Large stations such as Yancannia (NSW) used their scour until 1935. At Cordillo Downs, isolated out in the desert, the greasy wool was trucked to a scour. After washing and drying, it was piled up in heaps, like huge haystacks, awaiting pressing.

In later times, huge scours were erected in towns such as Winton and Longreach (Qld), an industry where steam and chutes and perforated columns made the work quite unlike that on properties.

Certainly the men were not required to urinate into buckets as they did earlier. In 1897 J. L. Corrigan, a wool scourer of Hay, wrote of wool scouring to I. B. MacLeay of Booligal. 'Hay, October 12th 1897. J. B. MacLeay Esq, Wool Works, Booligal. Dear Sir, I loaded yesterday, per carrier John Brannock, some

On Cecil Plains (Qld) a trough was made for the task

Clayton, Shuttleworth & Co. proudly advertised their steam-powered sheep wash

magic soap and 1 tin of ammonia, 14 lbs, to treat the sandy wool with. This and plenty of hot water, I think it will come up tolerably well; of course it might show a reddish tinge. Use some of your ordinary soap with Magic, and some ammonia, as much urine as you can collect in old buckets. I prefer it to ammonia. In fact, when I scoured the "Paddington" clip I used as much urine as the men could make, and found it thoroughly cleaned the wool, making it nice and open. The ammonia and magic soap are also two *very strong cleansers*.

'I should make three tubs of a bale. We are doing that with some wool we are now treating. I do hope you will push the wool through quickly, we are doing it this year at a farthing less... Brannock has one 3 bags of flour, 1 sugar, 1 tin ammonia (14 lbs), 2 boxes magic soap, 4 bales – 200 wool packs... He will deliver to you at scour.'

5

Nemiwaddinana

The effect of sheep occupying the Aborigines' hunting grounds was nowhere as acute as in the settlement of Tasmania. This island is so ruggedly mountainous that one-fifth of the state, the south-west, is still unoccupied by man in 1982; but the valleys and flats of the grazing lands of the midlands were always rich, and for 10 000 years they had supported numbers of Aborigines. Within seventy-three years of white settlement, the black people were all dead. It was, Charles Darwin said, the speediest extermination of a native people by foreign invaders in recorded history.

This race that disappeared so quickly after the coming of the flocks of sheep gave the name nemiwaddinana to the animal they had not seen before. And it was nemiwaddinana that became the cloven-hoofed seed of destruction of the Tasmanian Aborigines.

No one is sure, but it is thought that there were about 4000 Aborigines on the island in 1803 when white men came from Norfolk Island with their sheep. The last native died in 1876, although it could be said that the race had died out by 1845, because by then no babies were being born and the score or so of dejected remnants of this most ancient and isolated race were herded to a muddy beach to wait for death.

The Bengal and Cape sheep that arrived with the first settlers were followed in 1806 by merinos. Captain Paterson had obtained 300 from Macarthur's flock in Sydney but 100 had died during the passage to Launceston. The flocks spreading thickly up the rich valleys of the midlands were strengthened by 181 pure merino rams that Lieutenant-Governor Sorell distributed to settlers in 1820. By this time there were some hundreds of thousands of sheep on the island (1 700 000 by 1860) and the end of the Aborigines was in sight. On 13 March 1819, from Government House, Hobart, the Lieutenant-Governor dictated a proclamation addressed to the settlers, '. . . Aware that many of the settlers and stock-keepers consider the natives as a hostile people, seeking without provocation opportunities to destroy them and their stock.' He continues that it is most certainly not the case, because if they were intent on destruction they could have increased their attacks a hundred fold on the livestock dispersed for grazing over so large a part of the interior country. Instead, their meetings with herdsmen were generally incidental. The blacks were seldom the assailants, and when they were it was retaliation for recent injuries inflicted on them by white people. 'It is undeniable that many cruelties have been perpetrated repugnant to humanity and disgraceful to British Character, while few attempts can be traced on the part of the colonists to conciliate the native people.'

That was in 1819. In less than ten years time, proclamations would be greatly changed.

In 1825, 250 000 acres of land in north-western Tasmania were granted to the Van Diemen's Land Company. This company had been formed in England by a group of English capitalists interested in producing fine wools for the largest woollen manufacturing country in the world. In March 1826, their secretary, Edward Curr, arrived in the Derwent with surveyors to take up land 'beyond the ramparts of the unknown'. They took the valuable lands of the lovely north-west coast and adjacent islands, and imported Saxony merinos from Germany and rams and ewes from the stud of George III of England. By 1830 they had 6129 very well bred merinos and had brought out the *Caroline*. On board were servants and sheep.

This large holding was at first almost totally cut off by land from the rest of the state. One of their surveyors, Henry Hellyer, had once travelled for eight hours and covered only a half mile through the 'impenetrable west coast vegetation'. Its shepherds and hut keepers in their brutalising isolation were in constant conflict with natives. But help was at hand. Shortly, George Augustus Robinson, an evangelical Englishman, was about to be launched onto the island to round up every native, bring them in to 'civilisation' and transport them to an island in Bass Strait.

Those rich, grazing midlands were the powder keg.

For thousands of years the Tasmanian Aborigines had migrated from the central areas to the warmer east coast each year to feast on shellfish, crayfish and the eggs and flesh of the sand-burrowing mutton birds. When the invading whites came in numbers to the eastern area with their flocks, the natives' migratory route to the coast from the central areas was almost immediately cut off by sheep runs. They were no longer able to get to the coast for food because their forays (and the dogs they had got from the white man) frightened and dispersed the sheep and certainly alarmed the sheepmen. Thus, half their annual food supply was lost to them. They were angry, of course. All over the island their kinsfolk had been murdered or treated brutally and pushed back; white kangaroo hunters had depleted the food they could once obtain.

Now they retaliated. They made savage sorties down from the hills to the sheep-filled valleys and flat lands. Lonely shepherds were easy prey, isolated and sleeping either in 'watch-boxes' in paddocks or in lonely huts; their muzzle-loading guns were not as reliable nor as accurate as the silent spear. Allied to the ten-foot spear was the skill of great guerilla fighters, secret, swift, silent. On the first year of their attacks on this corridor through their migratory route, the blacks killed thirteen shepherds. In 1825 nine whites were fatally speared in ten days; in 1929 Montecute homestead itself was attacked six times in six weeks and its neighbouring homestead burnt to the ground. For five years they terrorised the settlers who lived in dread of them and generally grossly exaggerated the numbers of blacks who attacked. These colonisers were hard pressed.

In his editing of the papers of G. A. Robinson, N. J. B. Plomley has written: 'The story from 1825–31 is one of killing, injury, plunder and burning, at its peak in 1829–31. There is list upon list of such happenings.' He is speaking of blacks' superiority over whites.

And the clamourings of the settlers for the extermination or removal of the Aborigines held the weight of flocks of sheep and white faces. The government was powerless because it depended for its support and existence upon the settler. The press was in full cry: the natives must be removed, 'if not, they will be hunted down like wild beasts and destroyed'. As Plomley says, 'To the distant colonising power, Britain, the Aborigines were a natural feature capable of being pushed aside in much the same way as the bush was cleared for ploughing'.

By 1828, the Secretary of State was writing to Governor Arthur: 'I am aware of the extremely difficult task of inducing ignorant human beings . . . to acknowledge any authority short of absolute force, particularly when they are obsessed with the idea which they appear to entertain with regard to their own rights over the country in comparison with those of the colonists.'

The blacks were fighting with the only weapons they knew, the waddy and spear. The boomerang or woomera had not reached Tasmania before the rising waters of the later-named Bass Strait cut the people off from the mainland of Australia. But the limited choice of weapons was no drawback; they used both brilliantly. They were fighting for everything, for the right to cross what they considered to be their own land, for the return of the children who had been abducted – not for care or affection but for their usefulness – and kept in a state of slavery and prostitution ('a widespread practice' as G. A. Robinson wrote), they fought for the right to hunt kangaroo . . . they fought for all the lost causes.

There is no doubt that they cried havoc and at times went berserk with bewilderment and frustration. They killed women and children – as whites had done to them. By 1825, fear of the natives caused shepherds to desert their flocks and settlers to threaten to leave their sheep properties if the government didn't 'do something'. They had used their free convict labourers to build woolsheds of thick blocks of stone, with slits in the walls through which muskets could be fired on Aborigines. Though they may have huffed and puffed, those men who received grants of 6000 acres of rich land and free servants were not going to leave. Already shepherds and other isolated workers were being spoken of as the perpetrators of murders and cruelties on blacks, and, indeed, blamed for the outbreak by blacks who were said to be retaliating against white men's cruelties. Shepherds certainly killed, but their masters in no way dissuaded them: settlement was being hampered by the presence of the blacks and they must go. Landholders passively supported their employees 'constant sport' of hunting, maiming and killing blacks and stealing their children.

Settlers were given military protection in November 1827, and stock-keepers too in the remote areas. Semi-official roving parties rampaged through the forests seeking victims. When rewards were offered for each black brought in by a white man, these roving parties were successful in capturing some Aborigines – but, as Robinson cried, 'How many are killed for each one captured?'

Martial law was proclaimed on 1 November 1828, and the Black Line was drawn across the island on 7 October 1830. This last effort, with much of the manpower of the colony supposedly drawing the blacks into a net as the line pressed southwards, trapped one old man and one boy.

Not only the rich sheep country of the midlands was under seige by blacks: the whole state trembled. The Van Dieman's Land Company had 6129 well bred merinos by 1830, and the shepherds were panicking because of stealthy attacks by blacks. George Augustus Robinson was sent out to 'bring the natives in' – for their own benefit.

Robinson, while travelling in the Company's area on 21 June 1830, wrote that some Aboriginal women told him that the Company's shepherds had got the women into their huts 'and tried to take liberties with them'; that the men resented it and speared one white man in the thigh; that one man was shot dead. Then some of the natives killed some of the Company's sheep and drove them off the rocks into the sea; but sometime afterwards the whites caught, by surprise, the whole tribe which had come for a supply of mutton birds at the Doughboys and massacred thirty of them, then threw them two hundred feet over a cliff. (Robinson later interviewed Chamberlain, one of the four men who shot the natives. 'This man related this atrocious act with such perfect indifference my blood chilled.' Chamberlain told him that yes, they had shot 'about thirty'. 'We threw them down the rocks where they had thrown the sheep.')

'Since that day the white people at Cape Grim are called nowhummoe – devil – and when the blacks hear the report of a gun they say, "the nowhummoe have shot another tribe of natives".'

Three days later, Robinson viewed the rocks where the natives had been massacred. 'The road lay through excellent feed where the sheep was grazing.' He went another five miles to the point of rock opposite the two humps sticking out of Bass Strait – the Doughboys – and was told that the natives used to swim to the islands for mutton birds. A steep path led down the 200-foot cliff to the rocks lining the sea where there was a 'beautiful spring of water the natives used to quench their thirst when going to the islands'. There was 'a large cave which had often sheltered them during a storm. On the occasion of the massacre a tribe of natives, mostly women and children, had come to Cape Grim. They swam to the islands leaving their children at the rocks in the care of the elderly people.' Then they swam back with their birds tied together with grass and cooked and began to eat the hard-won fare, 'When down rushed the band of fierce barbarians thirsty for the blood of these unprotected and unoffending people. They fled. Some rushed into the sea, what remained the monsters put to death. Those poor creatures who sought shelter in the cleft of rock they forced to the brink of this awful precipice, massacred them and threw their bodies down the precipice, many of them perhaps but slightly wounded.' Robinson was aghast. 'Whilst I stood gazing on this bloody cliff methought I heard the shriek of the mothers, the cries of the children and the agony of the husband who saw his wife, his child, torn forever from his fond embrace. I was shown a rock where an old man had endeavoured to conceal himself was shot through the head by one of the murderers – who mentioned these circumstances as deeds of heroism.' He clambered down the cliff where the bodies had been thrown and saw several human bones, 'some of which I brought with me, and a piece of the bloody cliff. As the tide was flowing I hastened from this Golgotha.'

In June 1833, dogs 'belonging to the natives' (there were no dingoes in Tasmania and no dogs at all until the settlers arrived) destroyed numbers of sheep at the Van Diemen's Land holdings near Surrey Hills. By 24 January 1834, Robinson 'The Conciliator' was passing 1200 sheep 'clipped at Surrey Hills' and the police magistrate at Circular Head was reporting to the Colonial Secretary that 'extensive devastations have lately been committed among the V.D.L. Co. sheep flocks by a migratory tribe of aborigines whose bearing is such as not to afford any hope for a friendly communication with them by persons unacquainted with their manners'. And there was the rub. Those few 'migratory' natives were almost the only blacks at large in Tasmania. The rest of the black population had been rounded up by Robinson and exiled to Flinders Island in Bass Strait, and few were 'acquainted with their manners'.

By 1836 all had been exiled to that island in Bass Strait. And the sheep grazed safely.

While the owners of 'the walking goldmines' in Tasmania were winning their short war of land ownership, their compatriots in New South Wales were in open conflict with the blacks. Because settlement had been primarily a gaol for convicts there was no encouragement for men to attempt to tame the inland; prisoners spread over a large area would be too hard to control. But when the benefits of growing wool became manifest, there was a rush to be rich and the Tasmanian solution to the black 'problem' began to be repeated.

In the wide open spaces of the mainland of Australia things were different from Tasmania: there were large areas to which a defeated people could escape, even though naturally it would have to be land that was useless to the woolgrower. By the mid 1820s the battle was well and truly joined, and here, as in Tasmania, the blacks fought well.

In August, 1824, the Government declared martial law in the Bathurst district and this was in force for four months. Blacks were killed, whites were killed. At Mudgee sixteen blacks were shot. Bloodshed, violence, mistrust ruled.

Forced from their hunting grounds by sheep and cattle, the blacks rallied to the attack. First they attacked by spearing sheep for food, then the white owners attacked them; they in turn counter-attacked by spearing whites – any whites – thus gaining a reputation often for killing those who had most helped them. And the whites killed just as wantonly.

The new settlers who pushed inland from the coast were young, vigorous, plundering men without women of their own. They took their women from the blacks. Sometimes the blacks, greedy for sugar, flour, tobacco and alcohol, bartered their women away – and wanted them back as soon as the trade goods had disappeared. It is a confused area, a story no Australian writer has attempted to understand, the blacks with their cavalier treatment of women, the whites with their equally cavalier treatment of black women and total indifference to the children born of the thraldom. (A century later the once-removed indifference of the white wife to the white husband's behaviour is exemplified by the many half-white children seen running on the property and the disappearance for a time of the housemaid, or the wife of the black stockman.) The lofty brows in Whitehall were no more to blame for the tragedy than the lusty, buccaneering young men on the frontier lands. It is the way the world works, even in our own time. When fuelled by the drive for great profits, the fierce trio of lust, adventure and greed are almost unbeatable.

In Victoria no black tribes resisted for long. The squatters' power at this later stage was bolstered by the new double-barrelled gun, for they were no longer at the disadvantage of having to ram powder home while the blacks loosed volley upon volley of spears at them. Also the terrain itself was unsuited to the guerilla warfare the blacks waged so well elsewhere.

The battles in Western Australia were fierce. In October 1834 the Governor, Sir James Stirling, rode at the head of an armed body of settlers who killed fourteen blacks in retaliation for the murder of whites at an earlier date.

In South Australia in 1840, all the survivors (except a little girl) of the wrecked ship *Maria* were murdered by blacks. The following year a large punitive party 'led by the Protector of Aboriginals' murdered 'over thirty' blacks in revenge. That same year, when Aborigines at the Glenelg River stole a small flock of sheep from newly-arrived squatters, fifty-one of the blacks were murdered by an avenging expedition at what is now known as the Fighting Hills.

W. H. Bossence, writing of the blacks in the Numurkah (Vic.) area, says of the Bangerang tribe that 'the tribal way of life came to an end well before selectors populated the country and it is to the early squatters that we owe the little we know of the Bangerang in their natural state'. (He takes care not to leave it at that.)

Here was a tribe who lived on the rich lands between the Murray and Goulburn rivers and 'the early squatters' reported on their physique. Albert Le Souef of Tallygaroopna, William Locke of Kotoopna and Edward Curr of Tongala noted that they were 'fine specimens, upward of six feet in stature, exceedingly active and warlike in appearance. The river, or fish eating tribes are the most numerous and the most robust, many six foot, well built and stout in proportion.'

These white men all wrote that they had 'good relations' with the blacks. Doubtless they had, although at what cost and to whom may be gauged when we remember that, of the Bangerang, no full-blood was left by the end of the century. No full-blood Aboriginal is alive in all Victoria today.

The sort of relationship it was is described by William Locke. 'A gentleman made his appearance on the back part of the run, intending to "sit down" there with a flock of sheep; but my sable friends would not hear of it at any price. They said to him, "What for you comballee alonga this one country? This one country all about belonga Mr Locke." Eventually he had to depart, the blacks at my request cutting bark canoes to enable him to cross his sheep.' That was in 1869 when there were 144 Bangerangs living. In 1841 there had been 1200 of the tribe between the rivers.

'During the shearing season of 1869 about one hundred Aborigines were employed at the Wyuna Station (near Tongala Run). These shearers all worked well but drank very hard and had no difficulty in obtaining drink at the shanties and even at many of the public houses.'

John McKenzie, Manager of Wyuna Station, employed ten blacks that year; 1s 0d per hundred shearing sheep and 'found' (food, tobacco, etc.), and 10s 0d per week and 'found' for sheep washing. 'On Saturday, 23 October 1869 at

5 p.m. paid them all off. They went direct to a shanty about three miles off and on Sunday 24th one of them died drunk, he was one of the best shearers I've had in the shed in the last four years.'

It would appear that McKenzie tried to combine his responsibilities to the squatter owner with a private view on the black/white situation, for in May 1872 he wrote: 'To my view no act of Parliament can put a stop to any immorality or disease resulting from the intemperate habits encouraged by the whites.'

To write of Australia's black citizens and their relationship with whites is to enter a mine-field. There are no clear-cut issues, no side which is totally right; all edges are blurred. Cullenlaringo is an example.

Horatio Spencer Wills, the son of a convict who had been transported for highway robbery (and, incidentally, the first to introduce wire netting fencing to keep dingoes out, as well as to put down strychnine baits to poison wild dogs) was said to have been kind to blacks. In 1839, Wills overlanded south from Sydney to Victoria's Western District, crossing the Murray by floating his goods and servants across. He imported fine merino sheep and his property, Lexington, shore 20 300 sheep in 1846 and 30 000 by 1851.

In 1861, with a large party of servants and station hands and 10 000 sheep, he trekked to Cullenlaringo, a property in the Dawson River country (Qld). They first sailed up to Brisbane, bought stores, stock and wagons at Ipswich, and set off on the ten-month journey to the new country.

A fortnight after their arrival at Cullenlaringo, on 17 October 1861, Horatio Wills and eighteen other members of his family and station hands were massacred by Aborigines. This was the worst episode in the history of Australian land settlement. (Perhaps the only good to remember of that event was that Horatio's son, Thomas, who survived the slaughter, throughout his life carried no grudge against the natives. He taught them cricket, and his team – with Thomas playing with the blacks – played at the Melbourne Cricket Ground on Boxing Day, 1866: Australian Natives versus The World.)

Seven days later, on 24 October 1861, that survivor, twenty-five-year-old Thomas Wentworth Wills, is writing to a relative by marriage and friend of his father, H. C. A. Harrison. 'I have not time to go in to particulars. I can only say that all our party except six have been slaughtered by the blacks on the 17th.' Nineteen men, women and children were dead. 'I am in a great fix, *no men*. If we had used common precautions all would have been well.' His father, Horatio Spencer Wills, had advocated trust and friendship with the Aborigines. 'My poor father, and Baker were most brutally murdered. I want a *good* man up here that thoroughly understands sheep, a good overseer. Do all that can possibly be done as soon as possible in the shape of shepherds, good resolute men that will shoot every black they see . . .' He gives the names of the dead. H. S. Wills (his father), D. Baker Snr, D Baker Jnr, Mrs Baker, Elizabeth Baker, Ken Baker (a boy) and an infant eight months old; G. Elliott, P. Mannion, Mrs Mannion and three children, McConnell, C. Weeden, Jas. Scott, Henry Pickering, George King and a man called 'Tom' who had been engaged at Rockhampton. Alive: T. W. Wills, Jas. Salter, C. Kenny, P. Malony, John Moore, Wlm Alberg.

The real tragedy of Cullenlaringo was that Horatio Wills was the sort of man a new country could do with, yet he died, aged 49, a few paces from his tent door, 'evidently killed by a tremendous blow'. He had lived a courageous life; having run away to sea at fifteen, seen the ship's company put to death by savages in the Pacific, and saved his own life by throwing himself at the feet of the chief's daughter, later escaping to sea again to go whaling. Later again, with his friend Harrison, he invented Australian Rules Football. Henry G. Lamond, writing in *The Bulletin* of 28 January 1953, reported that Wills's 'mistaken idea of black-brotherhood with the whites' was foolish. 'Wills fostered the idea that the blacks were to be trusted. The massacre killed that idea. Murray, Commanding Officer of the Native Mounted Police, quickly avenged the treachery.'

This police corps was formed in 1848 by squatters of the Burnett and Condamine districts. They levied themselves and employed Frederick 'Filibuster' Walker to command it. For the next half century, until 1900, this corps rode the west and centre of Queensland. Lamond insisted that 'They did their duty'. He says, 'old maids of both sexes may pretend to shudder in horror over the methods used by the N.M.P. If they have Australian lineage long enough they may realise that they are alive now because the early N.M.P. protected their ancestors.' Some of the terms used at the time survive: 'Punitive Expeditions; Summary Justice; Salutory Lessons.' As Lamond says, 'the blacks had been taught a lesson – by their own kind.'

When the squatters wearied of paying the levy Filibuster Walker was given a vote of thanks and the sack. The native police were sent packing. 'That was an error of judgement. The blacks had added to their native cunning much knowledge they had gained from the whites.'

Unlike the American Indians, the Aborigines had no trade goods to barter for guns, and the few they stole were of little use to them without training in the care of the gun and the gunpowder. Had they been supplied with guns by traders, the war would have taken much longer than it did.

The natives of Australia were not servile, and they were not easily subjugated. In December 1854, Sub-Lieutenant Robert Walker and three of his Native Mounted Troopers were riding on the Ranne's pastoral property to which they had been posted when 200 Aborigines armed with spears blocked their way. Walker and his men killed (according to Walker's report) two men. That night the Aborigines attacked again and were driven away by the horsemen.

On the same property in September 1855 the Aborigines crept up on the police camp and relieved them of carbines and pistols before attacking and killing three troopers and wounding four others. They were determined to try to force the sheepmen off their lands. Before Christmas that year fifty warriors with nulla-nullas and spears killed three men and one woman at Mt Larcom sheep run, seventy miles from the Ranne's. The blacks took as booty sheep, sugar, flour, pipes and tobacco, and were easily tracked down by the 'punitive party'. At least eleven blacks were said to have been killed, but statistics of the battles of this war are totally unreliable – particularly when it comes to 'punitive' deaths.

If settlement was to be achieved, the 'salutory lesson' must continue. By 1857 Filibuster was back with the corps, at the behest of government, because 'things had become so hot'.

Indeed they had. In that year, at the Hornet Bank Massacre, the Fraser family was murdered. Mrs Fraser, a forty-three-year-old widow, and her children worked the property. The blacks attacked the station in June 1857 and were driven off but, angered by the theft of their land and their women, they returned on 26 October and killed the widow, seven of her children and three workmen.

It was Filibuster Walker who led the punitive expedition 'to avenge the Fraser family', although it is widely reported that one of the two sons who survived the massacre did his own avenging to the end of his days.

By 1861, the year Horatio Wills was murdered, Filibuster was dead of 'Gulf Fever', a form of malaria, and a new commanding officer was appointed. 'The officers of the N.M.P. did not rise from the ranks,' wrote Lamond – his father had been one such officer and he himself had been born in an N.M.P. camp – 'they were chosen as men possessing qualities of education, breeding, knowledge of drill and firearms and ability to handle natives.' A few months after his appointment in 1861, G. P. M. Murray 'quickly avenged the treachery' of the blacks who had crept on Cullenlaringo and massacred the Wills family. From such bald historical facts do we glimpse the double-edged sword of settlement in a new land.

6

Ten, Ten, Two and a Quarter

When New South Wales and Tasmania were the only colonies in Australia the largest group of workers was the shepherds. They guarded the wealth of the land. In March 1846 there were almost 13 500 shepherds in New South Wales. In outer districts three of every four employees were watchers of sheep.

Still as a statue he stood there, a tall, gaunt figure in a long, frayed overcoat, a shabby felt hat half hiding the sunken cheeks and thin grey beard, and hard lines drawn as with a ruler across his face. So stood the shepherd, watching his flocks. Alone for too long, he was regarded as being a hatter. Some out near the Lachlan and 'the other side' of the Darling River saw another man only when the ration cart came every five or six weeks. Their nights were interrupted by the cries of dingoes. 'Curious thing is, they'd always sit down outside the pen and howl whilst they were thinking of it,' an old shepherd recalled. 'All would be quiet as death for an hour then, without any warning, comes a scamper, a rush of sheep in the pen. That means he is over the fence and amongst them. The dog is out of it like the wind – he won't come back for an hour or two. It was generally enough to sing out, sometimes I'd have to pick up a fire stick and shake it at them.'

Before paddocks were fenced, folds were made, often of brushwood or a ring of fire if even that was lacking, and the sheep were herded in for the night. The shepherd slept in the fold in a watch-box, a construction no bigger than a coffin, a sort of bush-wacked palanquin made of odds and ends of wood. Dingoes made this necessary; the object was that the rush of sheep in the fold would warn or awaken the shepherd and he would rise and hunt the dingoes off.

After paddocks were fenced there was no further use for the watch-boxes and the rheumaticky old shepherds must have sighed with relief.

For many years after the shepherds had disappeared from the scene a small memorial remained to them. In the midst of scrub in the western lands, encircled by leopard-wood trees and emu bush, one might come on a patch about the size of a tennis court, as brilliant and soft and close as a square of rich green carpet. This was where some lonely shepherd had penned his sheep night after night to protect them from dingoes and blacks in the days when the land across the Darling was no-man's-land. Nearby, a rusty sheet of iron is all that remains of his old water tank.

No one asked shepherds what their rations were. Everyone knew. Ten pounds of flour, ten pounds of meat, two of sugar and a quarter of a pound of tea: Ten, Ten, Two and a Quarter.

When his shepherds ran away to the diggings in the 1850s, Horatio Spencer Wills of Lexington (Vic.) employed Chinese to shepherd his sheep, 'even black boys', said the *Argus* of 12 March 1921. At one time nine Chinese, angry at their treatment by an overseer, marched on Lexington homestead, 'a gesticulating infuriated mob armed with shear blades fixed to poles, evidently with the intention of running amok'. It was said that they were eventually soothed and their services dispensed with.

Early in 1861, when Wills set off from Ipswich (Qld) with 8000 sheep, bullock drays and carts for the Springsure district on the Nogoa River, his shepherds shore the sheep en route to the new station, Cullenlaringo.

Folklorists and Australian balladeers have fostered the belief in the Great Australian Outback as a land where working men were plentiful (as well as matey and happy in a larrikin way). This was never the case, and from the first settlement it was always difficult to get workers in the bush. Often those who stayed longest in isolated posts were lonely to the point of despair, and the outback has always shown a higher *per capita* suicide rate than the cities.

An old shepherd told James Bonwick in 1880, 'It's a lonely life. I never see

Set in a fold made by a surround of brushwood, the shepherd slept in his watchbox at night. If dingoes got in his flock he was wakened from his sleep by the rush of sheep into the enclosure. (Villamatta station was an early 'run' near the You Yangs (Vic.)

Western Australia, 1882

anyone but the ration carrier once a week. I follow the sheep and camp where they camp . . . I've been pretty well broken into not going to sleep: I've been made to pay for lost sheep, so that for three years I hadn't a cent of wages to take.'

Although Australia is not a great place for ghosts and therefore ghost stories languish, there was a Haunted Gully between Lismore and Derrinallum (Vic.). On an outpost of Gala station a shepherd lived in a hut, caring for the sheep. A cook was also living there, as was often the case, but when the weekly supply waggon called one day the murdered body of the shepherd was found – and the cook had disappeared.

A new shepherd was sent to the hut, but when the supply waggon next called he told the driver he was going back with him because he was being haunted by the ghost of the murdered shepherd. The new shepherd was induced to stay on if the hut was moved across the stream from a nearby spring, the theory being that a ghost cannot cross running water. Apparently the move was a success and the new shepherd settled down happily.

Julian Stuart knew the 'monkey-dodgers', as he called the shepherds. The fold they protected the sheep in at night was 'a yard built of stakes driven into the ground and high enough to keep the dingoes out. They led lonely lives, saw few human faces, and were often out on the run when the weekly ration carrier came and missed seeing him. The only human voice they heard was their own, for nearly all shepherds and other lone bushmen talked to themselves.

'They lived on damper and mutton year in, year out. The nearest approach they had to vegetables was pig face, horehound and marsh-mallow that grew around sheep camps. They reckoned any edible vegetation tended to keep off scurvey, of which they were always in dread.

'A shepherd's dog always slept inside the hut at the foot of his master's bunk – any that stayed out at night was certain to be "chewed up" by the dingoes before morning.

'As well as talking to themselves, bushmen got into the habit of holding long conversations with their dogs.

'In the old days [pre-1870s] huts were built of much stouter timber, iron-bark slabs two or three inches thick on account of the danger of attack by the blacks. The walls had loopholes so the shepherds could shoot the blacks. On the Upper Dawson in the eighties I camped at a big water hole near an old-time hut spear-marked and loopholed, and the old chap who occupied it told me two shepherds had been killed here by blacks before his time. There was hardly a hut about the Dawson or McKenzie of which similar yarns were not told. If the blacks killed chinamen they ate them too!

'The men who lived in the old huts led lonely lives and many of them died lonely deaths.

'One shepherd on the Dawson had been dead nearly a week before the ration carrier found him. The sheep were locked up and it was summer, they were in a bad way for want of water.

'On another run the shepherd died when out with the flock, bitten by a snake most likely, and the wild dogs played havoc with the sheep for a night or two.

'As time went by the huts were connected by telephone with the head station and the blacks became tame and harmless through the influence of rum and opium. So the armoured, loopholed huts were unnecessary.'

7

Raddle

When visitors enter the drawing room of a certain fine homestead on a large station between the Murrumbidgee and the Darling rivers, they 'invariably note a peculiar-looking article under a glass globe on the mantelpiece. On inquiry they learn that it is a piece of "raddle", and the station manager will tell them it is kept there as a reminder that it was a liberal use of "raddle" which paid for the construction of the woolshed. "Raddle" is a harmless substance used for putting a mark on the sheep's wool. Its too liberal use during shearing time was one of several causes which led to the formation of the Amalgamated Shearers' Union in 1886,' wrote W. G. Spence, the first president of the union.

As at all stations, Mount Eba (SA) station store was the only source of supplies for shearers. Here, they were 750 miles from Adelaide and must pay whatever was asked

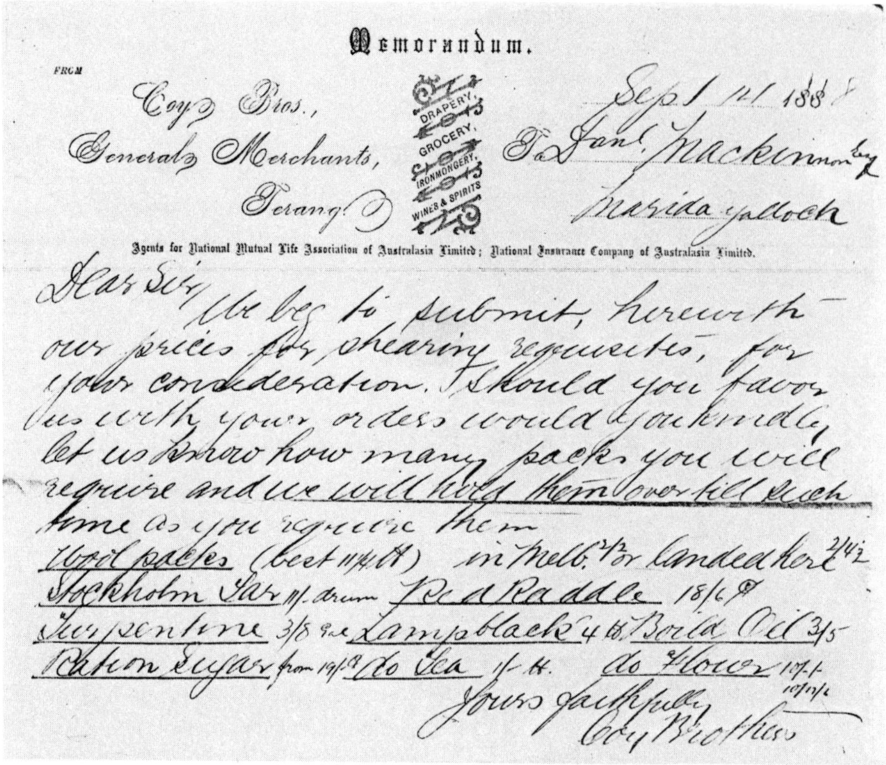
Memorandum.

FROM Coy Bros., General Merchants, Terang.

DRAPERY, GROCERY, IRONMONGERY, WINES & SPIRITS

Sep 1 14 1888

To Jan. Mackinnon Esq Marida Yallock

Agents for National Mutual Life Association of Australasia Limited; National Insurance Company of Australasia Limited.

Dear Sir,
We beg to submit, herewith our prices for shearing requisites, for your consideration. Should you favor us with your orders would you kindly let us know how many packs you will require and we will hold them over till such time as you require them
Wool packs (best 11 lb) in Melb. 2/1½ or landed here 2/4½
Stockholm Tar 11/- drum Red Raddle 18/6 [illegible]
Turpentine 3/8 gal Lampblack 4 lb Boiled Oil 3/5
Refined Sugar from 19/- Do Tea 1/- lb. Do Flour 10/- 10/10/-
Yours faithfully
Coy Brothers

Red raddle was used to mark sheep for several reasons; until the end of the nineteenth century its greatest value for some squatters was to withhold payment to shearers for supposedly poorly shorn sheep

'Shearing being paid for at so much per hundred sheep shorn, each shearer has a pen which holds a considerable number, into which he turns out each sheep after he has taken the wool off. As the quality of the work affects the result very materially, both in quantity and value, the employer or his representative naturally demands good workmanship. This is right and proper, but in many of the agreements which men had to sign in pre-union days, a clause gave power to the employer to refuse to pay anything at all for a sheep which he considered had not been properly shorn. He therefore "raddled" such sheep by putting a mark on it, and it would not be counted. Some of the squatters went one better than this, and if they found one sheep badly done they condemned the whole penful, and would not pay for any of them.

'The manager of the station referred to boasted of having raddled 40 000 sheep in his time, so one can understand his reverence for "raddle". Another nice little game they had was . . . to deduct half-a-crown and in many cases three shillings per hundred for all sheep shorn before paying a man off in cases where they discharged him prior to completion of shearing. Seeing that the "boss of the board", as the man in charge of operations was termed, was sole judge, it was to the interest of one not over-burdened with honesty to find fault with a man in order to get rid of him and thereby save money, perhaps £5 by doing so.'

The agreement signed by the shearer may be termed the specifications for the work, as well as being a binding contract under which a man was a prisoner until he finished the shearing. If he left before cut-out, he was liable to be fined or imprisoned or to forfeit all his earnings. 'In pre-union days he was little better than a slave. The employer dictated the terms and was sole judge as to their fulfillment, and if an appeal was made to law it would be a squatter who would sit as magistrate to decide the matter in dispute,' explained Spence.

An agreement of twelve clauses, printed in large type and posted at Mount Eba shed (SA) in 1887, stated that if a shearer was discharged he must forfeit six days' shearing money and return shears and 'any articles got from the store'. The man could be sacked at any time by the employer. Clause 6 reads: 'Singing, whistling, swearing and noise prohibited', and in Clause 12: 'Payment to be made by money order payable in Adelaide.' Adelaide was 750 miles away.

An agreement from Prince Royal shed states that there will be no money until cut-out (that was usual, even in sheds when the season lasted ten weeks). The price would be 12s 0d per 100 'with a bonus of 3s 0d per hundred if shearer remains until cut-out'. If the shearer averaged less than fifty per day he was charged for his tucker. If 'signs of drink' were on him, he was fined 10s 0d per day or part of a day.

Under an agreement used in Moralina shed in 1888, shearers were liable to a fine of 2s 6d and also to be discharged for 'bad or improper language'. The price was 14s 0d with a bonus of 3s 0d if the shearer cut out the shed. Rail fare was deducted if the shearer was discharged before he finished the contract, and all other costs to the station were also deducted from his earnings. If a man failed to average over fifty sheep per day he was charged 5s 0d per week for his 'grub' (this latter was the word actually printed in the agreement).

In each of these sheds engagements had been made in Adelaide – in the former case in Snell's, and in the Moralina case at Hittman's Labour Agency. Varying fees had to be paid, and it was notorious that in many cases from one to five pounds would have been paid by a shearer to make sure of certain sheds, thus giving the man without cash no chance at all. (The abolition of engagements through labour agents was one of the first steps taken by the Amalgamated Shearers' Union when it was founded.)

Conditions in an 1887 shearing contract for Meteor Downs were 'that the said shearers agree to shear all or any of the sheep now depasturing on Meteor Downs, in a clear, careful and workmanlike manner and not in any way to injure or maltreat the said sheep. The said shearers also agree to conform to the rules of the shed, viz:

1 Shearing to begin at time appointed by person in charge of the shed and continue till sunset if necessary.
2 Any shearer absent from shed during working hours without permission will be fined a score of sheep.
3 Any shearer using obscene language or swearing will be fined a score of sheep.
4 Any shearer cutting a teat, button or pizzle will be fined the value of said sheep.
5 Any shearer treading on his fleece or tearing it will forfeit the shearing price of said sheep.
6 All sheep to be lifted and carried from pen to their position on shearing board or not to be paid for.
7 All sheep to be properly dressed with tar on any cuts made by shearer.
8 All sheep not shorn to satisfaction of person in charge of shed not to be paid for.

9 No shearer to allow his sheep to carry wool outside the shed without bringing it back.
10 Shears will be supplied to each shearer at 5s 0d a pair, no selection allowed, the first pair to be taken. One pair shears allowed to each five hundred sheep.
11 No shearer allowed in wool room or on the wool table.

William Gough, an old drover from Hay, saw the conditions the shearers worked in. He wrote of the 'no union or no shearing contract' days. 'In the 1890s the housing conditions was very crude, unlined iron huts, some with earth floors, two tiers of bunks and chaff bags to fill with dry grass or whatever was around the shed.

'The station provided each man with a tin pannikin, two tin plates, a knife, fork and spoon and a kerosene tin cut down for a wash basin. No baths or showers but an iron boiler or sometimes a copper was provided to heat water.

'The shearers had to buy kerosene from the station store and make their own lamps for light. Those lamps were made from pickle bottles with an old pen or some small metal tube fixed through the cork, half filled with water and kerosene to hold a wick.

'Most of the stations kept a store where the shearers could buy the bulk of their rations, tea, sugar, flour, yeast, onions, potatoes, jam, tinned bully beef, eggs, vegetables and tinned milk.

'For other things shearers had to depend on hawkers (mostly Chinese) that travelled round from shed to shed – hawkers of clothing would travel round and it was the custom for the shearers to pay the hawkers by an order on the shearing boss. This money would be deducted from the shearer's cheque at cut-out.

'The shearers had to pay for everything – and they could have everything they liked in food providing it was available.'

The status of the shearer is evident in the old drover's words. He had lived on damper and a bit of mutton, but 'everything they liked' was what the shearers should have. In his thirty pages of memoirs, William Gough, who is still spoken of with respect in the Riverina, only mentions a squatter, selector or manager as evidence of his having done the droves he writes about. Instead, he describes the shearers in a way that sets these men apart, as they deserved to be.

'Shearers was completely separated from the shed hands (or rouseabouts as they was called in those days).

'It was very hard on a learner . . . I have seen a young chap with his hands bandaged and his back aching battling to try to make his tucker in his first shed and I have seen the old shearers take the hat around for a trier at his first shed to help him on his way to his next shed. That was blade shearers.

'The shearers used to come to the roll-call – mostly riding and leading a packhorse; some on foot carrying Matilda and some in sulkies and four-wheeled vehicles with a hood to keep the sun off them – some of those men would camp in their waggonette during the shearing. It was the custom then for the station to provide a horse paddock to agist the horses at 1s 0d to 1s 6d per head per week and in some cases they had to pay a lad who may have missed out on a job at the roll-call to act as horse tailer at 6 pence or 1s 0d per horse per week. There could be sixty or seventy horses at a shed.

'The shearers had complete control over a cook. If he didn't suit them they

Quarters for permanent station hands were primitive enough, but no artist attempted to portray the airless, window-less huts for shearers' accommodation. (Doogalook, Vic.)

could sack him and put another man in but they were liable if any trouble arose from changing the cook or if any dispute among themselves held up the shearing.

'I think it was Tarwong Shed that one of the first shearers' meetings was held over the raddling of the sheep, and there was the formation of the Shearers' Union – but this is a picture of what I saw of the work and conditions the blade shearers had before the unions were formed. The result of the first strike led to the abolishing of the Raddling Penalty.

'This information is from what I saw myself, not from hearsay.

'Now for the shedhands or rouseabouts [he wrote], the wool rollers, piece pickers, pickers-up, penners up, pressers and wool classers.

'The shedhands' lighting system was jam tins half filled with clay or sand and filled up with mutton fat with a bit of worsted wrapped around a stick for a wick. Those were slush lamps, gave a bit of light and a lot of smoke and smell. But the shedhands could afford no more, they couldn't buy kerosene on £1 per week.

'Pickers up (lads) was paid 15s 0d to 17s 6d per week. Penners £1, piece pickers £1 to 22s 6d, wool rollers 25s 0d. Pressing was done by contract, all hand presses, screw, lever and rack and pinion presses. It was very hard. Some screw presses required three men to operate them. The pressers would board with the shearers or batch on their own.

'The wool classers would board with the Boss. Their pay was £1 to 25s 0d per hundred fleeces by mutual arrangements with the station.

'The rest of the shedhands was housed separately – completely away from the shearers – and found with the ordinary station rations cooked by the station cook . . . mutton, bread, potatoes, plum jam, treacle or golden-syrup, tea, sugar, salt, pepper and a rice pudding now and again with a few currants for cakes about twice a week. If they wanted milk, butter, vegetables or fruit, they would have to buy those things from the passing hawkers and pay for them out of their wages. They was housed in galvanised iron huts, mostly earth floors, two-tiered wooden bunks. Like the shearers they had bags to fill with whatever they could find as [filling for a mattress], tin mugs, etc. and a kerosene bucket to wash in, and make their tea in and all these things had to be accounted for at cut-out.

'The rouseabouts were expected to be on duty at 7 a.m. and could be allocated to any job that the boss might order them to do – wool rolling, piece picking or if shearing was held up by rain or any other cause the boss could use the shedhands to work outside the shed. I have seen them working in gangs cutting wood, cutting rushes and digging out rabbits – only a few miles from Hay.

'There was nearly always an overflow of men at the roll-call at the big sheds and they came there somewhat like the shearers, some with horses and vehicles and some on foot with swags up. Some of them could shear but had missed out on getting a pen . . . no money, so they'd take a job as rouseabout. Of course, food and clothing was cheap then or a lot would have starved. The good old days – so called by some.

'I myself have done a little of all shed work, picking up, pressing and classing, but although I can shear a sheep I never took on shearing because I wasn't strong enough to do that work.

'So I took on droving.'[1]

Some of the sheds William Gough worked included Till Till, Tarwong, Ulonga, Toganmain, Burrabogie, Illillawa, Mungadal, Toberly and Gunbar.

The high price charged for rations caused great dissatisfaction among the shearers. Most squatters refused to allow men working for them to purchase anything elsewhere than at the station store, many not even allowing a hawker to come on the ground. Pre-union prices are indicated by those charged by Lachie McBean at Moulamein station in the Riverina. The town was less than a mile away, and goods were not cheap there. Flour in town was £1 7s 6d per bag; at the station it was £3 2s 6d; tobacco: town 6s 0d, station 12s 6d; matches: 6d in town, and 1s 0d on the station. All other requirements were charged for at proportionate rates. On the Murrumbidgee 50s 0d per bag for flour was charged at a station on one side of the river, whilst in the town just across the stream it could be obtained at 20s 0d.

Practically all stations bought their goods wholesale in the cities and had them transported wholesale so they were in no way disadvantaged. The shearers knew this and thus knew what great profits the station store made from the unfortunate men who travelled hundreds of miles at their own expense to take the wool off the sheep.

In all such 'contracts', everything was weighted in favour of the proprietor. Except for the three 'smoke-ohs' he had complete control. If a squatter was immoral – and such a document assumes some were – a shearer might be sacked as incompetent in the final week of shearing and receive no payment at all for his weeks of work. (Some sheds shore for up to twelve weeks.) As for the fines, up to quite recent times they were a matter for jokes among shearers: 'The big squatter making a big man of himself by presenting a cheque with his name on it to the hospital committee. With our money!'

The contract refusing payment until cut-out made no allowance for a man with wife and children to send money home, and by so doing kept shearing with single men, thus losing the leaven of the many men who may have been sober and industrious through having a house to keep.

Counting-out pens, Newinga, Talwood

Shearing rates varied. As much as 15s 0d and 16s 8d was paid in the Riverina and as low as 12s 6d elsewhere, with men finding their own rations 'and all requisites', which means finding their own tools.

In Victoria, the rate at Eumerella was a penny per sheep and rations.* Wando Vale, near Hamilton (Vic.), also shore as low as 8s 0d per hundred. The general price was from ten to thirteen shillings and the average about eleven. All rate districts, such as Monaro and New England (NSW) and Darling Downs (Qld), shore at very low rates. The north-east of South Australia had rates similar to Victoria. The hut accommodation was universally bad, and most shearers believed that conditions and wages were worst where employers were best off.

In their greed the squatters over-did their role of master and the shearers revolted.

*£1 = 240 pennies.

8

The Breath of Revolt

Revolution is an arbitrary thing; it allows choice of neither time nor place, not even the manner in which it will be conducted. Had those three choices been available to the shearers when their moment was at hand, even with the wisdom of hindsight we can only imagine what would have happened. All we know is that they were pressed too long, that the day of unpaid convict servants was well past, as was the time when working men humbly accepted oppression with a doffing of their caps in homage to the magnanimous giver of labour.

We do know that, like all groups who have had unquestioned command over the lives of others, the squatters did not notice or correctly appreciate these changes, nor would they accept that workmen could successfully direct their own lives. They were taken unawares by the intensity of events. They reacted violently, as does anyone who blindly believes that their wealth, position or way of life is challenged, and they called in all the mighty power at their command. All this brought Australia closer to total anarchy than anything before or since.

The shearers may have been itinerant bands, but in their midst were many men who had been miners and knew of the miners' struggle for better conditions and pay. Many of the shearers were well read; some had recently arrived from Europe and were aware of the agitation among working men overseas against the unfair usage of their labour.

It began in 1874. A shearer, Benjamin Graham, who travelled and worked in the sheds along the Condamine, wrote a letter to his home paper, the *Toowoomba Chronicle*, and placed an advertisement in its pages for which he paid £9. He was beginning to agitate for some kind of organisation among the bladesmen. Editor of the *Chronicle*, Mr W. H. Groom (later speaker of the Legislative Assembly in Queensland and MHR in the first Federal Parliament), became interested in the shearers' problems and, after Graham had collected 1s 0d per man from a number of shearers, Groom acted as treasurer.

Kilfera shearing shed

By 3 November 1874 they had 53 names from Jondaryan station with £2 13s 0d cash; by 12 December the treasurer held £22. The idea had been taken up with great spirit.

A meeting held in Toowomba fixed 20s 0d per hundred as the price of shearing; the hours were to be from 6 a.m. to 5 p.m. with one hour each for breakfast and dinner, and 15 minutes each smoko. Saturday knock-off time was fixed for 4 p.m.

During the same period a committee was formed among the forty shearers at Peak Downs station (Qld), and a meeting held in March 1875 resolved to adopt rules and elect officers etc. Subscriptions were fixed at 20s 0d per annum, a storekeeper at Clermont was appointed treasurer, and a Mr Blair was secretary.

Wolfang station started non-union, and three members of the new union in the district went out and 'interviewed' the men in the hut. The men agreed to stop work; no doubt inflammatory remarks were passed – and overheard by a permanent station hand (a 'boss's flunkey' as W. G. Spence was apt to write). This man gave evidence that caused warrants to be issued for the arrest of the leaders. The men escaped, it was said, even though 'the constable was offered a reward if he caught them, but he did not make any special effort'. The attempt to set up a union fizzled out.

Other sporadic attempts were made but men could not overcome the difficulties of the distances and lack of united effort until Mr A. J. Sullivan put his active spirit behind the movement in 1884. He was a member of the Amalgamated Miners' Association of Creswick (Vic.), and on 12 January 1884 he wrote a letter to the *Australian* advocating the formation of a union on the lines of the AMA. In August that year he wrote to the *Town and Country Journal* in Sydney, and to the *Leader* in Melbourne. Sullivan was the principal speaker at a meeting held in Kerr's Farmers' Hotel, Wagga Wagga, on Wednesday 28 October 1885, with a view to forming a union.

Earlier that year, in February 1885, he formed a union with others in Longford, Tasmania, and branches at Campbell Town, Oatlands and Hamilton

(Tas). In Victoria, Mr W. E. Lawless started a union and enrolled a number of shearers, but was unable to continue and returned the money paid as membership dues. A Mr Priestly started a Shearers' Union in the south-east of South Australia, but he also failed to bring it into practical use.

At Midkin in 1885 there was anger when the rate was reduced from 17s 6d to 16s 8d. Early in 1886 the shearers determined to resist the reduction. They bought rations and formed a camp. Very decided action was taken and they blocked any men going in at the low rates. 'So determined were they', later wrote W. G. Spence, 'that any man trying to get in on the squatters' rate was promptly dumped into the station water tank. From these men's endeavour a union was formed at Moree (NSW).'

The board at Kilfera

That year there was a big strike at Dunlop, on the Darling, west of Bourke. Here the shearers stood out for nine weeks against a rate reduction. The Darling was in flood. The shearers had a tarpaulin muster in the camp and each man threw in all the money he had. The station manager had cut off supplies, expecting that lack of food would force them to submit, but shearers elsewhere subscribed money to send flour by river steamer to the striking men. After nine weeks the manager agreed to the old rate and they went back to the shed.

But with all their earnest intentions and idealistic mateship, the organisers lacked experience; they were not widely known enough to inspire confidence, and the rules they had drafted were crude and incomplete, and indicated a lack of stability or permanence. Men had become habituated to a particular state of affairs and, as each station had its own agreement and all were not unjust, it required some universal attack in the shape of an additional infringement upon men's rights to bring about the right psychological moment. It came in 1886. The following advertisement appeared in the *Australian* on 2 April:

TO SHEARERS

'The owners and managers of the undermentioned stations hereby give notice that the price for shearing during the current year will be 17s 6d per 100 sheep, but if any shearer is discharged for wilful bad shearing or other misconduct he shall accept payment at 15s 0d per 100 for all sheep shorn to date of discharge. That rations will be charged at a fixed rate, viz., flour £2 per bag, sugar 4d per lb, tea 2s 0d per lb, and other articles at rates equally reasonable. This advertisement is inserted in order that shearers coming from a distance may have due notice of the reduction, and from this date it shall be understood that any applicants for pens agree to the terms specified therein. For Talyalyea, E. D. Campbell; Wallara, O'Donnell Bros. and Ryan; Tinapagee, M. J. Feehan; Wanaaring, Heddon Bros; Nocholeche, Walter Gamson; Goorimpa, David Brown; Urisino, A. S. Fowler; Elsinora, W. W. Killeon; Thurloo Downs, Rowan and McTeir; Berriwinnia, Davidson and Co.; Woongamana, J. Breaton; Kallara, David Brown; Mara, A. W. Paxton; Curranyalpa, Mr H. Seal; Winbar, J. Thomas; Fort Bourke, M. P. Fitzgerald and Co., Tonga, W. H. L. Rankin.'

The seventeen stations on the list all started early, hence they were to take the lead in the reduction of rates.

It was the half-crown cut which enabled the organisers to arouse and enrol men and lay the foundations of the Amalgamated Shearers' Union. A shearer from Creswick (Vic.), David Temple, contacted W. G. Spence, at that time General Secretary of the Amalgamated Miners' Association, which he had led to become the AMA of Australasia in 1884. David was a member of this association, for many shearers went mining in the off-season.

Later Spence was to write: 'He came and consulted me as to the wisdom of trying to organise the shearers, and we made a start at once. I wrote a letter which was published in the *Ballarat Courier* on the 29th May, 1886, urging combination and offering to assist. This was followed up by other letters, and rules, elected officers, etc. I was elected chairman, and Mr D. Temple secretary . . . On the grievances of the shearers being placed before the late Mr Edward Bateman, editor and one of the proprietors of the *Courier*, that gentleman wrote a slashing article, which he reprinted and distributed.

'Before we began I laid down the principle that the Union must be intercolonial – must ignore political boundaries – and every member must carry his rights and privileges as a member with him, no matter what colony he was in. Mr Temple went shearing, and on reaching Riverina wrote to me giving a report that the men were eager for a stand being made. By arrangement, Messrs Temple, J. A. Cook and James Slattery undertook to travel amongst the sheds and organise. They were to take the risk of non-payment for their services if their mission failed, but to have £3 per week if success followed. I acted as treasurer and directed the work generally, as well as organising from my office. We followed up the Riverina and Victorian shearing by sending the three men named to New Zealand, where they enrolled over 2000 members and founded a branch. We enrolled many Maoris and had the rules translated into their language. The result of our efforts on the continent was an enrolment of about 9000 members. I enrolled about 800 by communication from the office, but the success of the [Union] has been due mainly to the system of sending out paid travelling organisers.'[1]

The Jondaryan team

For the remainder of 1886 they organised, addressing weekly meetings at Camperdown, Kyneton and Winchelsea, and formed unions and branches in Bourke, Wagga Wagga, Cobar (NSW), Casterton (Vic.), eastern South Australia and Port Augusta.

At the first conference held in Ballarat on 19 January 1887, it was resolved that the union rules and agreement should be put into force during the shearing season of that year. They had already approached the squatters, but few had agreed to come 'to mutual agreement' as Spence put it. They had £730 to start what was to become a major battle with 'the powerful wool kings of Australia'.

There were by now several Bush Unions that covered bush workers, and the Queensland Shearers' Union, which began in 1886 with shearers at Isis Downs. These unions remained separate although in contact and sympathy with the Amalgamated Shearers' Union. (The Queenslanders had had a conference in January 1887 in Blackall and the manager of Alice Downs 'and a few others' came and a union was begun. By the end of the year they had £833 10s 9d and on 3 August 1888 they registered as a Trade Union. At the end of 1888 they had a cash balance of £1817 0s 7d, and by the end of 1889 they had almost 3000 members.)

The policy of the union was conciliation, but the breath of revolt was in the air. Copies of the proposed agreement were sent a season ahead to pastoralists in all the new country organised, with a request to meet or suggest amendments. There were no replies.

And the unionists were as yet unsure of their strength. The season had hardly begun when the rule in regard to engagement of union men only had to be suspended. Instead, travelling organisers were sent out to be at every roll-call if possible and act as the mouthpiece for the men.

It was the beginning of the great horse rides for which the union became famous. There were 33 ASU agents out in 1887 and most of them had a very lively time racing from one property to a roll-call at another. Frequent changes of horses were necessary as one after another was knocked up.

These organisers saved the union from defeat that year. Where they failed to reach a shed it was invariably lost: the men signed non-union at the behest of the sheep owner. When men stood out to work union or not at all, and were told to leave the property, camps had to be formed for them, rations supplied by the union and the true location of reserves and roads discovered. Maps were obtained, and in more than one place when the squatter ordered men off his run, he was astonished to find them defying him by camping within a few yards of the woolshed. They had located a road – or maybe a reserve – which he had forgotten existed.

The union claimed 500 sheds that year, costing them £1200. They came out with 16 000 members and under £600 in funds.

The first stand is thought to have been made at Kilfera station, (NSW) owned by the Kilfera Pastoral Company. The roll was called on 23 July 1887, Union terms were ignored by the management so the men left and formed a camp at Ivanhoe, about fifteen miles away.

Word came to the camp that Kilfera had obtained ten non-union men, but a night raid to the shearers' hut brought the ten into the union camp. In two weeks the company agreed to the union's terms and the 82 men required were balloted for out of the 200 then in camp.

This greater number in a camp than seems reasonable was one of the problems union shearers would have to contend with for the next six years while camps were necessary. The difficulty was this: when word got down and up the rivers and tracks by bush telegraph that the shearers' camps had tucker for union shearers, any bagman, swagman or sundowner could move in. If the camp organiser tried to put them out they could – and did – say they were shearers and, if thrown out, they would go to the shed and work non-union, thus winning the war for the squatter.

The owners were again asked to a conference the following year, 1888. The reply to one such appeal was written on the back of the circular and returned in the sender's envelope.

Woorooma,
February 25th, 1888.

To the Shearers' Union,
Victoria or elsewhere.
President and Secretary.

Gentlemen, – This, your circular, is addressed to manager Woorooma Station. I beg to inform you that I the writer is the manager and owner. I have others assisting me for 16 years, 12 years, and down to 6 years. They are content and satisfied all with their positions and, I do not wish that their minds should be disturbed, poisoned or corrupted by any Union, any bondage society, or any corrupt league, and I give each notice when they hold any communication with any league or union that purpose evil they will be paid off as soon as I find this out, and I have to request that you do not write or send any of your print to anyone here. You have sent parcels of print and writings here, but they were not read by anyone, nor will they be heeded no matter how often they come. I do not want to interfere with your Union in any way but let them leave me alone, keep clear of me and my property, and if they do trespass after a caution they will get the hardest treatment that law can inflict, and as often as they trespass on any part of my land, and to be plain with you not a Union man will get a day's work from me. Last year 69 of the lawless ruffians camped in 9 camps opposite my shed for a term till the PM gave them notice to break up. I had a watch armed on my shed with guns loaded with sparrow shot, not to kill anyone, but to mark their legs so as to be known next morning. For all their bounce my shearing went as well and peaceable as ever I saw. The men took high wages out of the shed – some £32 in six weeks 3 days after paying for their stores and rations bill, 5 or 6 meals a day, as they liked the best of provisions, and the lowest average was £3 a man per week clear. If this is not enough I do not know what to say. If it comes to this that I cannot get my sheep shorn with free not union or bond men I will sell the sheep, put cattle on the run as I had before. I know the evil of Societies, clicks and corruption since 1833 in Kircaldy [Scotland], when Russels' foundry was closed with the villainies of loafers, agitators sucking the wages of working men out of their pockets, then setting idle, exciting, ignorant labourers on to evil. This caused a panic in Kircaldy at the time. But it is now time to press Parliament to bring in a bill like the crimes bill in Ireland to punish first the leaders and agitators and then those that are led on and take part in evils that disturb the peace of the nation and the prosperity of the country. Gentlemen, I have explained myself.

Hoping that I will never hear or see any of your society after this. Gentlemen, I remain,

Yours respectfully,
L. McBEAN.*

The agreement drawn up by the management for the shearers to sign at Manus shed indicated the feeling in 1888. 'We, the undersigned, do severally and each for himself declare that I am not a member of any society of Shearers' Union, and that we have not pledged ourselves not to shear with non-union shearers.'

Another clause read: 'No person other than the shearers employed allowed to be in the hut unless by permission of the overseer in charge.' This was to prevent union organisers from speaking with the non-union men.

According to W. G. Spence, 'Riverina squatters contained some strong characters. Many of them were so highly individualised as to oppose union of any kind, and fought their own battles. This suited us, and we came out with about eighty sheds employing Union labour.'

The hardest struggle for mastery the fledgling union had in 1888 was in Victoria. Through inter-marriage, there were strong links between families and properties; the Western District squatters were organised and the shorter distances to travel meant that non-union labour was more readily available in

*The redoubtable Lachie McBean was looking not too far into the future when he advocated the use of the Bill that had been brought into being by the British Parliament as the Irish Coercion Bill. Although this bill had been rescinded in England it was resurrected in the 1890s in Queensland to convict unionists.

Victoria than in other colonies. As the year rolled on union shearer after union shearer was arrested, tried and gaoled.

The strike camp in that district was at Penshurst and from here men were taken to Hamilton and charged with 'conspiring to raise wages, forcible entry, deporting shearers and assault and riot'.

The first such case, against Thomas Walters and John McCormack, broke down on all counts except that of conspiring to raise wages, and each man was sentenced to twelve months hard labour. (Unionists have always resented this type of charge because, as members of a union registered under the Trades Union Act, they believe themselves to be exempt, the Act having been designed to protect unionists from the law against conspiracy.)

In 'pure merino country', as the shearers waggishly refer to the class-conscious squattocracy of that area, pastoralists' associations were as militant as the shearers, but not all pastoralists joined – as not all shearers joined shearers' unions.

By 1890 they were reading the Riot Act on Circular Quay, Sydney, also. *Jumna* (left) waits for Jondaryan wool to be loaded

The Millear family of Edgarley decided not to join the sheep farmers' association. It was 1888 and 'nearly all the owners here have joined. They are bound to ignore the Shearers' Union in every way even to the giving of 15s 0d per 100 except in cases of heavy sheep where 15s 0d per 100 had to be given before 1887. Sheep owners to the north of here have all bound themselves to accept their own rules and to give from 12s 0d to 14s 0d per 100.'

Because he had not joined the Sheepmen's Association, Millear could not get non-union labour and by 1890 he was scraping together eight shearers for his

It was the beginning of almost a decade of revolt. And two years later, in 1890, the Maritime Strike drew the other fledgling unions into the confrontation

twenty-stand shed: 'One very slow, a blackfellow also very slow who pinked [cut] them, the rest nondescript.'

They were a poor lot but they were in demand. When Millear finished with them they were to go on to Murndal to the Winter Cookes but: 'I hardly know on what terms to express my disgust,' he writes suddenly. The seven men had been 'talked over' while on the way to Murndal. 'Several of them have shorn for several seasons past for us and always non-union but their present conduct tends to impress one that no man can be trusted.'

The revolt was beginning in all colonies. At Woodlands station (Qld) on the night of 9 November 1888, union men were in the men's hut trying to persuade non-union men to join them when the cook, Charles Hill, fired a gun up the chimney as a signal to the owner, John McMaster. McMaster raced into the hut brandishing a revolver and was disarmed. Walter Ball, Edward Wythe, Joseph Pitman, James Brown and Pat Power were arrested and charged with having assaulted, imprisoned and detained John McMaster of Woodlands and received sentences ranging from six to twelve months.

At Alice Downs (Qld), Julian Stuart was engaged in the fight for union conditions. The shearers had managed to get their agreement signed but the rouseabouts had difficulty getting the management to agree to employ 'Union Labour Only' throughout the station. 'We got a list of the names of all the men in the musterers' camps, at the wool scour, the teamsters, boundary riders, as well as the shed hands and, supported by the shearers, requested the manager to give the labourers' delegate a cheque covering the membership for the lot. The manager agreed – he was anxious to get on with the shearing. The amount of each man's ticket was to be deducted from his earnings. But Narky Ned, an old bloke, objected.

'"For what and for why should I have to pay ten shillings into your union? I get twenty-five shillings a week." We said, "What about when the union gets it raised to thirty shillings?" and he said, "I'll be getting the same as they get."

'Rather than have shearing delayed, the manager sent Ned out to the paddock to cut wood for the cooks where he wouldn't cause any friction.

'Alice Downs was a good shed for rouseabouts that year. Any amount of rain and some of them had cheques as big as the shearers. Old Ned lined up with the rest and got the shock of his life to receive only 25s 0d when the others were all getting 30s 0d – the union men that is.

'"Take it and get!", said the manager.

'That manager was Charley Gray, a well-known athlete, an exception to the general run of managers, and a fine, manly fellow.'

Julian Stuart had a bad year in Queensland in 1888. 'The only shed I got that year, Daandine on the Darling Downs, we had to knock off because of the drought. The sheep were too weak to be mustered and a lot of us were sent to wire-net the creeks and waterholes to keep them from being bogged. Sheep were dying everywhere.

'Shearers there got 15s 0d a hundred, hours from daylight to dark, conditions in the hut and shed beyond description. On the board we had a mixture of black, brown, brindle and piebald as well as Chinese cooks.

'On our way out west we travelled through very dismal surroundings, sheep dying everywhere, and not for one second, day or night, did their bleating cease – surely the most plaintive prayer ever sent to heaven for rain. The cruelties were made still more horrible by the kites and crows who plucked the eyes and other titbits out of the animals before they were dead. We sickened of the desolation. The dingoes had fine times that year and were the biggest and fattest and cheekiest wild dogs I ever saw anywhere.

'We travelled west to be in time for the early sheds on the Barcoo. The first to call the roll, Barcaldine Downs, was owned by Fairbairns, Victorian absentees – and when we talked unionism they got their backs up and postponed the shearing rather than depart from the old conditions.

'The shearers were only partly organised, the labourers not at all; but we camped around and kept in touch with one another until the next shed called the roll.

'This was Northampton Downs, owned by Murphy and Sons, also of Victoria where Sir Francis was President of the Upper House, the son being Member for Barcoo in the Queensland Parliament.

'They were installing the Wolseley machines and were very anxious to get a start, but put up a bitter scrap for retention of the old conditions. There were no union officers available but we had a pretty effective solidarity.

'One incident made a life-long impression. I was pitching my tent some miles from the shed and a stranger came to explain things. He finished up by producing a roll of notes and saying, "Don't go short of anything till we beat them. There's thirty pounds there and you're welcome to half of it." Afterwards I got to know him well – Frank Latreille, the life and soul of every shed that ever he was in. Of course we won with men like that about and the shed started late in the season with full recognition of both unions – shearers and labourers – the first to do so, as far as I know in Queensland.'

Arms and ammunition were beginning to appear. In August, the affair known as the 'Brookong Riot' was enacted. The *Riverina Herald* of 17 August 1888 printed the contents of a telegram sent by W. Halliday MLC, Owner of Brookong, to the Union Mortgage and Agency Company.

'Please purchase 40 Colts revolvers, with 100 rounds of ammunition for each; send them to Hanging Rock railway station by to-night's mail. Brookong is in a state of siege. Situation intolerable. Will you kindly inform law authorities that police authorities here are powerless. My servants have been dragged out of their beds and terribly maltreated.'

The Union declared that this so-called riot was trumped up. 'All this row because a small camp of Union shearers was waiting for Brookong management to give Union terms. Not one servant of Halliday's had been in any way interfered with, not even so much as to awaken them out of sleep. Some of the men from the camp had visited the shearers' hut and had brought away some of the scabs. They did this under the eyes of the fifteen police who were in the camp. Presumably Halliday termed the few scabs his servants, hence the misleading wire, which gave the idea that the homestead had been attacked, especially when he said it was in a state of siege.'

Fifteen police were sent the 52 miles from Wagga Wagga on the Urana road to Brookong station and the local cavalry volunteered to assist, if necessary. The Police Magistrate, H. Bayliss, went out to read the Riot Act, 'Among the said persons or as near to them as he can safely come', as the Act demands, 'and with a loud voice, command silence . . . and openly proclaim "Our Sovereign Lord the King chargeth and commandeth every person here assembled . . . who is not entitled to attend this meeting immediately to depart from this meeting to his lawful business. God save the King!"'

The absurdity of the Act had perhaps never been more evident than on this day. In the harsh sunlight of the new land, 12 000 miles from the seat of government that had framed the Act to control an older colony, green Ireland, the shearers stood and stared. 'Our Sovereign Lord the King chargeth and commandeth all persons here assembled immediately to disperse themselves and peacefully depart to their habitations or to their lawful business. God save the King!'

Had the Act itself been read out in its totality, the bewilderment of the listeners would have been complete. This Act invoked by Government during the strikes of the 1890s included Statute VI, George IV (1819), known as the Irish Coercion Act. 'An act for more effectually preventing seditious meetings and assemblies to the great terror and danger of His Majesties' Loyal and Peaceable subjects and in a manner manifestly tending to produce confusion and calamities in the Nation.' In short, before any meeting could be held, notice must be given in writing six days previously, such notice to be signed by seven persons usually resident within the parish or township'. This, of course, disqualified all shearers from ever calling a meeting. The shearers listened intently – and from that day knew that the lines were drawn.

'If twelve or more persons continue together by the space of half an hour after such proclamation is made, every such person being legally convicted shall be adjudged guilty of felony and be liable to be transported for any term not exceeding seven years.'

Bayliss, the local Police Magistrate, continued: 'It shall be lawful for any person lawfully attending such meeting, to seize, and apprehend any person not entitled [to attend] such meeting, and bring him before the law.' Clause 15 further enacts that should any person or persons, being part of 'an unlawful assembly . . . be killed, maimed or hurt, in the dispersing, arresting and apprehending etc., those causing the death or injury, including those aiding and assisting them, 'shall be free, discharged and indemnified for, or concerning the killing, maiming or hurting of any persons so continuing together aforesaid'. After that, the further injunction that, 'Attending meetings with flags, banners and other ensigns or emblems can merit a fine and imprisonment for a term not exceeding two years' hardly seemed to matter to those earliest of 'rioting' shearers at Brookong.

Extra police were sent out; W. Halliday asked that they be armed with swords as well as revolvers. The union camp was about half a mile from the shed, and the number of men there soon increased to over 200. Two union organisers, John Parker and Brian Lee, were in charge. These two, with Alfred Collins, Joseph Casey, Wm Blair, Wm Tyson, Alexander Armstrong and J. J. Woodcock were arrested. When the trial came on in the Circuit Court at Wagga Wagga on 19 October, 1888, Secretary W. E. Head was arrested in court. Parker and Lee were sentenced to three years' imprisonment; Tyson, Armstrong and Woodcock to two years each; Collins, Blair, and Casey to one year each.

The Judgment delivered by Sir William Windeyer at Wagga Wagga on 22 October 1888, in the case of Parker, Lee and others for the Brookong Shearing Riots, was the first of a number of historic findings brought down that decade. His Honour addressed the prisoners and said: 'You have been found guilty on the clearest evidence of a most dangerous attack on society. If organisations of this kind were permitted – if lawless bodies of men, organising themselves with the discipline of a military force, were permitted to go about the country interfering with honest working men, there would be no safety for any man in the community. I am glad to find that the jury have arrived at the only conclusion rational men could arrive at in such a case, and with no more deliberation than was necessary to consider the case of each prisoner separately. I do not suppose any except the residents of this district know of the state of

One of the major causes of complaint – the Vermont sheep. Bred to great size, excessive body wrinkles gave more skin area from which wool could be cut. Shearers resented being paid the same rate for these as for normal flock sheep

things going on here. I have had a tolerable knowledge of the country, but I had no idea until I came to try this case what state of things existed here, a state of things probably unparalleled in the history of the country. I should not have thought it possible that two or three hundred men should camp on a road for the purpose of preventing honest men going to work, and to capture and bind them prisoners and hold them for days. Let anyone contemplate what may follow if this kind of thing were permitted. There would be an end of liberty and safety, but the law exists for the protection of every man, high or low, in the community, and those who took part in proceedings of this kind must expect to have the hand of every man armed against them. If a man's liberty were interfered with, if his life were threatened by overwhelming numbers, he

Beginning of the Manifesto sent in 1890

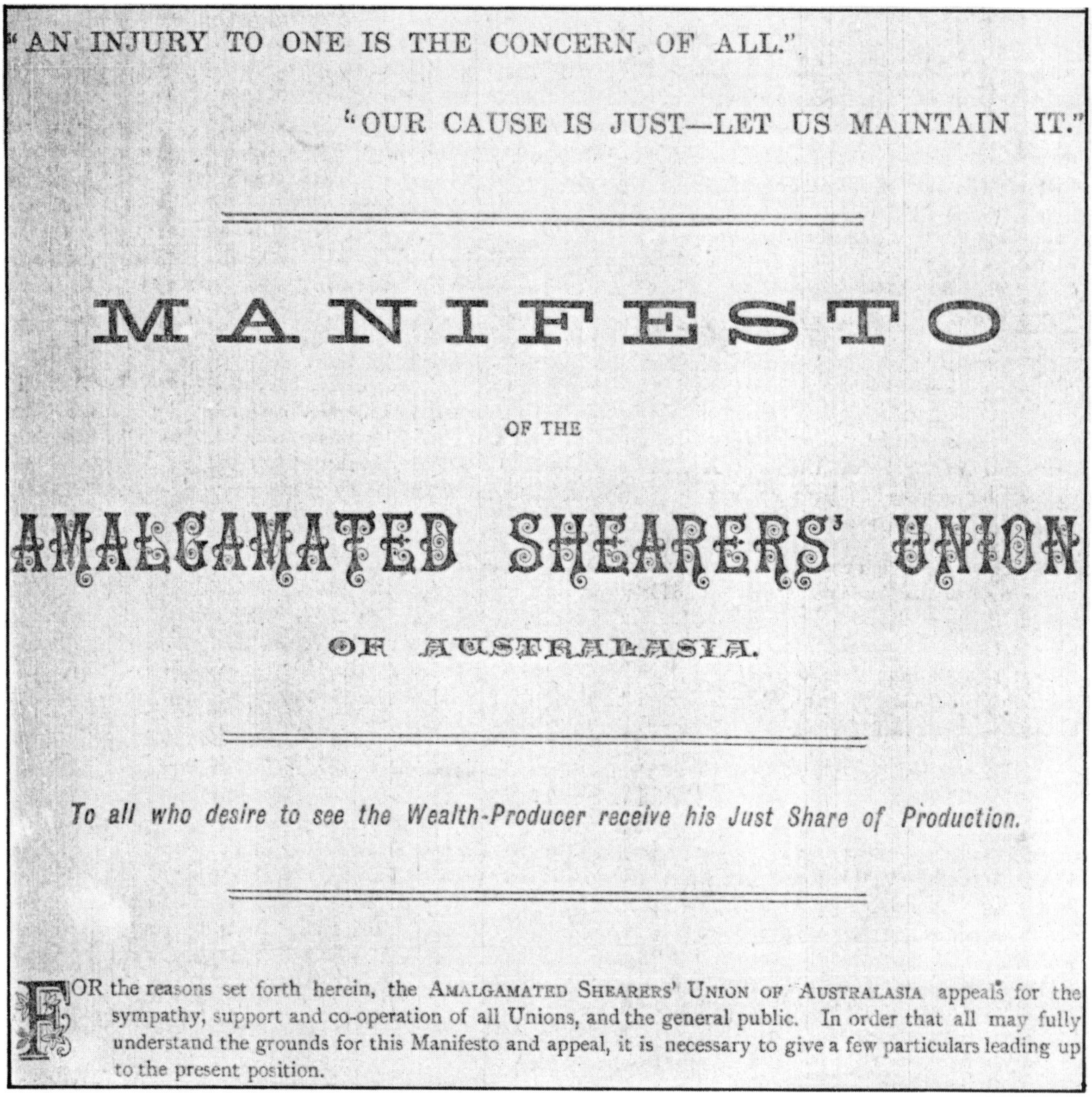

"AN INJURY TO ONE IS THE CONCERN OF ALL."

"OUR CAUSE IS JUST—LET US MAINTAIN IT."

MANIFESTO

OF THE

AMALGAMATED SHEARERS' UNION

OF AUSTRALASIA.

To all who desire to see the Wealth-Producer receive his Just Share of Production.

FOR the reasons set forth herein, the AMALGAMATED SHEARERS' UNION OF AUSTRALASIA appeals for the sympathy, support and co-operation of all Unions, and the general public. In order that all may fully understand the grounds for this Manifesto and appeal, it is necessary to give a few particulars leading up to the present position.

and every other honest man is entitled to protect himself by taking the lives of those who come upon him. This in law is deemed justifiable homicide. On the other hand, if lawless persons took life, they were guilty of murder.'

Matters were hotting up. At a shed twenty-three miles out of Adelaide the union declared that the owner, brandishing a revolver, had attempted to ride his horse over a union rep.

On 4 September, J. Wilson, a union organiser, was fined £10 at Deniliquin for 'obstructing' shearers at Zara and Wanganella. And at Woorooma, Lachie McBean's run (which shearers had declared they would see shear union, and eventually did), when the tough and unyielding manager was driving his buggy through the crowd, they asked him to stop for an interview and his restive horse bolted – after the winkers 'came off' and the manager was injured.

Two unionists were arrested, but before the trial came on another, G. Smith, was recognised as the man blamed for 'interfering' with the winkers at the time of the accident and was given nine months imprisonment.

Men who had 'come out' on strike after signing agreements on squatters' terms were brought up and fined heavily under the Masters and Servants Act.

The year had cost the union £3000 in strike camps, but they came out of 1888 with a revenue of £11 000 and 19 000 members with a credit of £1446. By the end of 1889 (after more strikes and camps and costs) there were 2400 union sheds in the three colonies (NSW, Vic. and SA), an increase in membership of 800 and a credit balance of £3284 10s 0d. South Australia had been particularly successful, with the whole colony being brought under the union. It seemed that the battles were almost won.

But excitement and skirmishes had fanned the breath of revolt into a wind that blew to hurricane force the next year, and 1890 was to see a turning point in the history not only of the shearers and squatters, but of Australia itself.

9

Part of the Glory

A detailed account of the shearers' strike of 1890 was written by one of the convicted men, J. A. S. (Julian) Stuart, and edited by his daughter, Lyndall, into the book *Part of the Glory*. The title was taken from the judge's injunction to the jury at the Rockhampton conspiracy trial of shearers: 'It does not matter that large crowds of men are outside this court shouting for these men when they leave. It is no part of the glory if that should be shared by you.'

John Alexander Salmon Stuart was born on 18 December 1866 on the Williams River in northern New South Wales. The 'Salmon' was to honor a captain of one of the river boats that played such a prominent part in the lives of the families who dwelt on the northern rivers.

Some of the Stuarts, a comfortable, middle-class family, believed their forebear, who emigrated from Ardnamurcha on the Scottish island of Tobermory, to have been bound up with grazing interests in western New South Wales and industrial projects throughout the Commonwealth. J. A. S. Stuart believed no such thing. This widely read young man, who so wanted to be a poet that he changed his name to Julian, believed that his maternal and paternal grandparents had been crofters, small tenant farmers. During the 1850s these people migrated to Australia following the 'clearances' of cottars and crofters from the large estates. (A. G. McDonald, in *My Scotland* writes, 'On an island near Tobermory, the proprietor, a Mr Clark, closed up with stones the only good well, and so cleared the island entirely of its numerous inhabitants.')

Stuart's early life was comfortable for that period. His father's farming and grazing properties on the Hunter and Clarence rivers produced maize, a sugar plantation, vineyards, lucerne, sheep, stock, a sawmill and horses. But the chains and leg irons left behind by the convicts who worked the place for the original owner seemed to Stuart's young mind 'vile and to cloud the day'.

'I remember, I remember,' he later wrote, 'the place where I was born.' Transportation of convicts had ended before his father had bought the property, but 'Old Jacob', who had finished his sentence there, stayed on and young Julian

'The Labour Crisis!' *Sydney Mail* headlines read (1890), when teamsters went on strike in sympathy with shearers – the first time such determination had been shown. Wool-growers, supported by mounted troopers, drove their own teams to the ship from the railway

heard him tell of it. 'He said that more than once he had been 'strung up' to the ironbark flogging post that stood in the garden opposite the dining-room window, to get the 25 strokes of the cat, and floggings mostly took place at breakfast time, so that the previous owner could watch them while he ate. It saved time.

He studied in Sydney for three years and became a school teacher, but by the time he was twenty-one he had returned to the bush. On the Upper Dawson he took a fencing contract, then, living and working hard on the miles of fence posts and wires for the large property owners, he met a 'wanted' American strike-maker from Pittsburgh, Pennsylvania. It was 1887 and the effect on the young bushman was lasting. He had never questioned a way of life in which his family prospered, but while he was studying he had seen hungry unemployed in Sydney working for their keep, and the Pennsylvanian's talk of going on strike for better wages and conditions reminded him of the rebellious poetry he had read by the Scottish crofter, Robert Burns. The middle-class prejudice which governed his family's thinking and their circle was leavened. He got his hands in wool.

By 1888 he was working with immigrants from Britain, men who had been brought out as a pool of cheap labour 'for those who tied up the brown land for their own private use and gain. These immigrants were the natural followers of the convicts, ticket-of-leave men, the time-expired men, as a source of practically free labour for the squatter.'

Time and travel drew him inexorably into the vortex that was about to sweep an old order away. The next year was 1889 and the six-foot horseman was riding down the Warrego to join a cattle drove making for Bourke. Arriving at Burenda he saw for the first time the conflict and some of the reasons behind it.

Having lived an intinerant life himself for a few years, he knew that men who moved from place to place looking for work – such as seamen on ships, men on river boats, shearers – were not entitled to vote in their country's elections. He had thrown in his lot with these men; he knew they were interested in getting a voice in the management of their country and that the only way they could do it was through their own organisations, their trade unions.

These unions had attempted to begin early enough in the new Australian colonies. After the arrival of the outlawed British chartists there was no going back to accepting oppression humbly. The Chartist, James Stephens, led a march through Melbourne streets in April 1856; he was followed by stonemasons who had downed tools on the building of the new university in support of the claim for an eight-hour day (a 'perfectly suicidal' thing, 'stupid, mischievous', was the thought of an eight-hour day, said the employers). But that great leveller, the goldrushes, had thrown working men together and they were no longer isolated. And they talked.

An almost insurmountable obstacle to gaining advancement was the difficulty of having one of their own kind elected to parliament. The so-called two-party system of Westminster tradition was no more than two separate groups of employers divided by niceties of rhetoric and financial interests. The system had been transported to Australia with as little change as there had been in the usage of labour for the benefit only of the rich or privileged in the old lands.

But the truly unscaleable wall to parliament for the working men was that members of colonial parliaments were not paid: a working man, if elected, would have to do his day's labour and still sit in parliament that day.

Trades and Labour Councils were formed in the capital city of each colony, but the rambling great hinterland was almost without organisation and its itinerant labourers were forced to travel from place to place to seek work. Yet in 1879, when Julian Stuart was only thirteen years of age, an attempt had been made to bring the various groups into a body that would lend strength to their aims. In that year the first Inter-Colonial Trades Union Congress was held in Sydney.

The following year, 1880, the great financial boom began for all Australia. It was the greatest boom yet known in the colonies, particularly in Melbourne, which drew as much foreign capital, population and expansion as any city in the world at the time. It was the Golden Age! The great land benefited all, it was cried. All Australia was glowing.

Then, like all artificial booms, the wealth that had poured in from abroad dried up as foreign investment dried up on the other side of the ocean. The illusion ended. Among the first to feel the pinch were the wool-growers and other men on the land. In February 1892 a run started on the Savings Bank of New South Wales. For two days crowds pushed and shoved to get their money out.

Police rescued those who did manage to withdraw their deposits, but many were trampled, screaming at the fearful knowledge (in many cases) that their life's savings were lost. Similar scenes took place in other cities and by 1893 the collapse of the banks was in full flood.

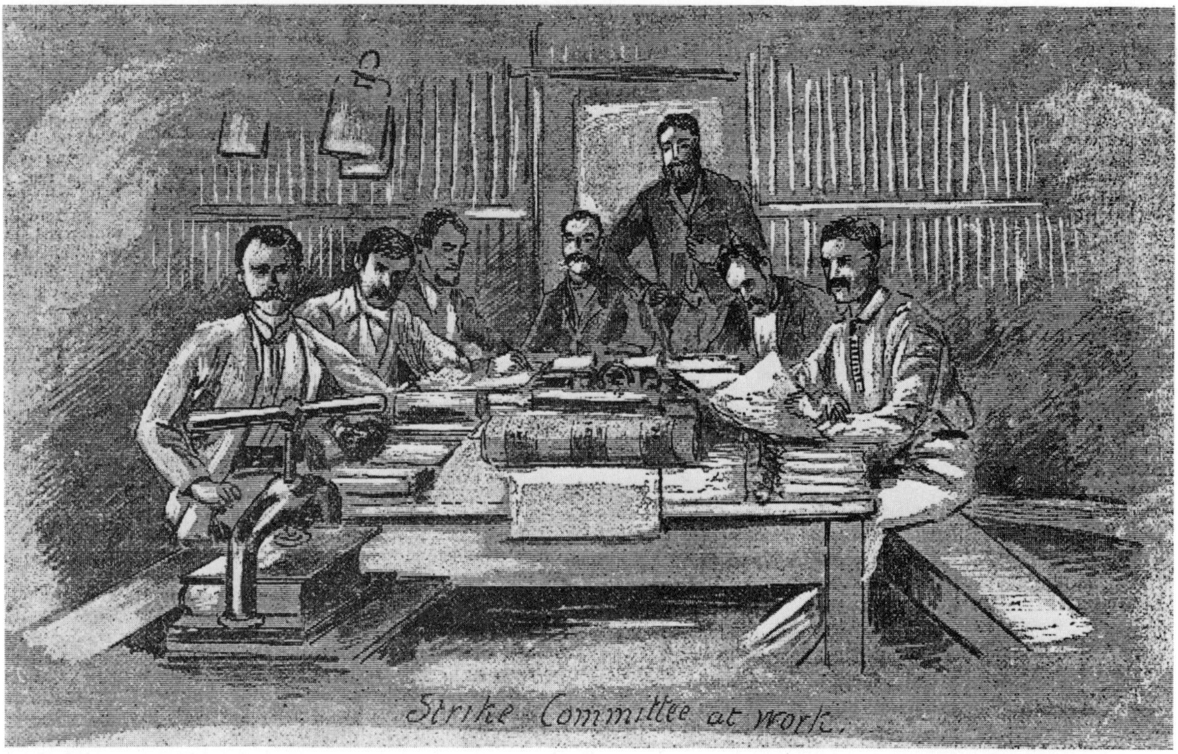

Throughout the back country, men had organised. By 1891 groups such as this at Barcaldine were planning strategy. Committee from left: L. Crane-Daly, J. Murphy, H. Jackson, J. Payne, J. Campbell and R. Woods. Their letterpress (left) would leave evidence of their industry

Julian Stuart, better educated than many of the men he now travelled with, would have known much of this, but his greater interest was in those men who were most hard hit by the catastrophe, those itinerants who had been forced to travel the countryside taking whatever work and pay was offered them by the land-owners: the unskilled rural labourers, manual workers, lower income people in general – and shearers.

This unexpected depression, stunning in its rapid crash, took with it a whole society who had no security at all, never had had. Hardship, unemployment, illness, accident, orphaned children, abandoned oldsters, there were no social provisions to succour them. Until now, the new colonies had been forging ahead at least in a manner which allowed these victims to be absorbed into the system. Now there was chaos and deep resentment. Poor houses and charities had never been organised here on the vast scale of the British Isles. People starved, babies died, children were abandoned, families parted, and men headed from the cities to the bush expecting to find work. But there they found a terrible thing: the country was in the scourge of a drought the like of which has seldom been seen since white men began to graze and crop the virgin land. And it lasted for the decade. For ten years it sat, burning deep into the soil, rotting the carcasses of millions of sheep down into the fissures which opened across a greater distance than the scorched earth over which the French had retreated from Moscow to Paris. Small farmers were unable to go anywhere else, for who would take their bare, scabby little holdings? They lived on what they could grow; they had nothing to sell, and no money at all. The impoverished peasants of Europe and the peons of South America were their equals. The large properties were, in many cases, snapped up for a pittance by the large companies or by banks that had weathered the storm. Many a fine family whose forebears had trundled out into the unknown, making up the map as they went, was swept off its hard-won acres. Belts were tightened across the land and the walking workers who had been the mainstay of the pastoral workforce were now no more than pests on the boundaries of properties where once they had been expected and welcomed. The bewilderment, hurt and shame was compounded by the now certain knowledge that the society in which they lived was of such a nature. (The *Bulletin*, 'the Bushmans Bible', made this plain in 1893 after the bank crash, when the Church of England called for a day of prayer for the business community: 'We are shocked at the Church gathering together all the thieving businessmen and dishonest bankers in bell toppers and broadcloth to pray for a return of "the good old days".'

There were to be no good days for the shearers and their travelling mates for a long time. Unemployment stalked with the drought and the closing of the banks. Their star that had slowly begun to rise waned with the economic crash. They fought, raged, and attempted to hold the ground they had so stolidly won, but of course there was no hope.

Riding with the nomads, Julian Stuart had seen the earth-floored windowless, unlighted huts where shearers slept in three-tiered bunks and ate in the same dark squalor, where they were paid what the squatter decided and not paid at all if that autocrat decided a man's work had been incompetent. Although the shearer was the man whose skill was the key to the wool-grower's cheque, he was treated in a manner considered to be 'good enough for his type'.[1] Stuart had not been part of the events at Brookong but he knew of them. Squatterdom didn't like the introduction of unionism and sent five members of the Amalgamated Shearers' Union to jail for offences alleged to have been committed at Brookong station.

In 1889, at Burenda on the Warrego, the owner was shearing on the Free Labourers' Agreement (non-union) when Stuart rode in. A team of unionists arrived and attempted to enlist the men they considered to be 'scabs and black legs'. The manager of the station had three of the men arrested and charged at Roma Courthouse with 'rioting'. 'It was the year before the Jondaryan trouble,' said Stuart. 'Some shearers tried to get Queensland Shearers' Union rates and terms but the owner refused.'

Here was part of the crux of the matter. Some of the men at the roll call at Burenda had ASU tickets (Amalgamated Shearers' Union) and some had QSU (Queensland Shearers' Union). The owners wanted to shear under their own contracts or even ASU terms, which had lower rates and less advantageous conditions for the men than the QSU were asking. The lack of unity caused by the novelty of the birth of unionism and the colonial distances crippled the cause.

One can hardly blame the wool-growers – what employer would not do the same? The QSU were asking £1 a hundred sheep, the ASU 15s 0d. Stuart saw 'A QSU man draw his stirrup iron and said he was going to strike for £1 a hundred. They clashed. The stirrup iron landed first; the 15s 0d bob champion bit the dust. Then the police came and marched them off to Roma jail for rioting.' They were acquitted 'much to the disgust of the squatters'. Julian Stuart later wrote, 'Burenda shore "anyhow" that year, and we went back to the Barcoo with our minds made up to alter things next season – and we did, too.'

Yes, things were altered in 1890. The Queensland Shearers' Union had been organised in 1886 and by 1889 it had some thousands of members. A concerted action was flung across the Queensland colony, with unionists sweeping into every non-union shed they could reach. They encouraged the men to join the union and, when successful, if the squatter would not 'shear union', they moved the whole team out to take work on a station that would accept them on their own conditions. If no property would do this, they founded a Strike-Camp, a feature later to become a legendary part of the Queensland scene. It was here, in these camps, that it was claimed that 'the workman's patriotism was put to the test'.

At Jondaryan station on the Darling Downs, non-union men had been employed for the shearing season of 1890. It was a sign of the tide in the time of the working man that the Carriers' Union refused to cart the wool away from Jondaryan. They had thrown in their lot with other working men. Teamsters refused the freight and thus began the great struggle that continued throughout that drought-stricken decade and the next.

Jondaryan got the wool to the coast with its own transport, but the Brisbane water-side workers refused to load it. 'Any attempt to ship Kent and Weinholt's Jondaryan wool, shorn by non-union shearers, will cause us to take such steps as we deem necessary to stop such being shipped in your company's steamers.'

A message was wired to W. G. Spence, President of the ASU: 'Possibility of general strike in the back country to support action on the coast. Rely on you to guard New South Wales border and lower South Australian border. Will need

The military camped wherever shearers set up camp

5000 horsemen to patrol effectively. Will you arrange for them so that on Monday next, if then necessary, they will move to protect our flank? We must maintain our principle whatever the cost.'

Spence replied: 'Men will be found to protect the border.' The whole country appeared to be igniting.

'The fight was bitter, it could not be otherwise,' wrote Spence. 'The marvel of it was that it did not lead to civil war. At one time it was feared that it would come to that, and to prepare for such a contingency one of the leaders of the Australian Labour Federation and myself held a private conference at three o'clock in the morning and came to an understanding as to certain action to be taken by myself in the southern colonies. The men were becoming very exasperated at that time, and had they gone to extremes they would have taken possession of the country.'

This was no idle boast. Spence later admitted that 'the action of the Queensland Government and its grossly unfair administration of justice so provoked the men that at one time it was probable at least 800 rifles, and ammunition for same, were in possession of the unionists.

'The business people of one town subscribed £200 for the purchase of rifles and ammunition and one hundred rifles were purchased in Brisbane and taken from there by rail under the very eyes of the police and detectives and safely landed in the union camp.

'These were Mauser and Mannlicher rifles of up-to-date pattern.

'Men of the resource of the Australian bushmen would have taken possession with ease and without much bloodshed.'

Perhaps they would have done. The *Sydney Morning Herald* accepted a news item from Winton suggesting that the opposition was not strong. 'The mounted infantry men sent to the Winton district, with a few exceptions, are unfit for police duty. Several are undisciplined and untrained. Some are very bad riders; with a mixture of plain clothes and uniforms. The general impression is that a serious mistake has been made in sending such men to this district.'

The Maritime Strike of 1890 drew ships' officers, seamen, labourers, shearers and coalminers into the attempt to bargain collectively rather than the traditional method of the employer bargaining individually with the employee. This new, revolutionary tactic was causing stresses that employers had not known before.

For Julian Stuart, Jondaryan was the beginning of his march to the dock and imprisonment on St Helena. A consummate bushman who could turn his hand to anything, he was working his way across country as were the other shearers. He came to Jondaryan on the Darling Downs near Dalby. This was an 'early' shed. The men tried to get Queensland Shearers' Union rates, 2s 6d a hundred higher than the Amalgamated Shearers' Union rate, but the manager, a wily businessman, tried to play one union against the other. Eventually all the union men left and Jondaryan got men from Brisbane. 'Riff-raff who tomahawked the wool off somehow.'

Teamsters refused to cart the wool and when Jondaryan teams got it to Brisbane the city carters refused to handle it. It was thirty feet from S. S. *Jumna*, the ship waiting to sail to catch the London market, but now the water-side workers said it 'could stay there until the day of judgment if the Jondaryan owners did not agree to pay QSU rates – henceforth and forever more'.

Of course it didn't stay there. 'We went back to the Barcoo,' wrote Julian Stuart, and the news followed him: the Australian Labor Federation had paid £75 to the shipping line to help defray the cost of the hold-up. He was disgusted. But the wheelings and dealings had only just begun, and men such as Julian Stuart never did accept them.

Brave words have been written about the 'successes' of the 1890s, but there were no immediate successes for the shearers or their unions. They did develop in that time, however, the determination, the strengths and virtues, and the loyalty to a mate that for almost a century were to be admired as the greatest of Australian traits.

They were also men of their time. The White Australian sentiment was strong in them. At the Annual Conference of the ASU held at Bourke in February 1890, it was agreed to continue to bar Chinese and South Sea Islanders from joining the union either as shearers or shearers' cooks (By-law 62). The eight coloured men who had received union certificates prior to the passing of the by-law in 1887 could claim protection under the resolution.

On the face of it, it was discrimination against a man because of his skin, but they did have some other antipathy, perhaps the strongest of all: money. They were afraid the coloured workers would bring wages and conditions down to an even lower standard. Niel Black, one of the early Victorian squatters who had brought Chinese labourers out for his station, paid them the 4s 6d per month that was the regular going rate for Chinese on Australian properties, and he had been one of the early critics of shearers' aspirations. The following year, they had the example of Fernlees station using Chinese as strike-breakers on the shearing board in February 1891.

The union at no time attempted to bring coloured shearers into the union and, as the 1898 census was to show, there were 24 366 'coloured aliens' in Queensland alone, one for every five white males and almost every one of them labouring class. So along with a stand against a reduction of wages and the squatters' attempt to introduce 'Freedom of Contract' was a major complaint, the use of the low-paid 'black-birded' Kanakas and the 'coffin-ship' Chinese as well as the 'niggers', the Aborigines who had not only comprised much of the sheep station labour but had, in many cases, been shearers as well.

The 'anti-coolie' stand became one of the most adamant of the shearers' and bush workers' demands on the pastoralists. At Wellshot shed there was a white woman cook who was assisted by her Kanaka husband. 'The lads struck, and two of them rode 170 miles in 24 hours to get instructions from leaders,' Spence later wrote.

In their turn, the pastoralists, as always, said they would hire – and fire – whatever labour they wished at whatever wage they cared to pay. Also, they would ignore the ASU, QSU or any other 'bush workers' organisation, and would deal with each intending employee singly. This was 'Freedom of Contract'. The men who put their signature to the contracts were to be known to wool-growers as 'Free Labourers'. To their fellow workers, they were to be known as 'scabs' and 'blacklegs'.

The seeming arrogance and the disdain by wool-growers of workers' unions was tempered to a degree by their recognition of the strength and power of the ASU centred at Creswick (Vic.). William Spence (President) and David Temple (Secretary) were a formidable duo, idealistic and indefatigable. In their own compact, militant group, the pastoralists banded even closer into a federation, pledged to deal as an organised body with any situation that arose in their industry. With their wealth, interests and family relationships, they were as immovable as these up-and-coming adversaries.

It has been claimed that the wool-growers invested £100 000 to fight the unions that were uniting bush workers and shearers. The squatters would levy members in proportion to the numbers of sheep they owned, and members would be bound to this bond, subject to a penalty. The shearers did not have that kind of money, but in 1890 William Lane edited the first edition of the *Worker*, which was to be the unionists' spearhead and mouthpiece.

The lines were drawn.

Julian, not yet 25 years of age, was sent by the unionists to Clermont, the centre of pastoral lands. Two hundred and fifty mounted men with pack-horses rode across country while another group of shearers who had no horses left by train and met up with the young man at Clermont. By January 1891, he had set up a camp at Sandy Creek, three miles out of the town, with Julian as camp chairman flying the 'blue flag of the bushman's Australia over the camp'.

The men were armed – what Australian bushmen were not, are not, armed? When police were sent to order them to lay down their arms they replied, truthfully enough, that the guns were for kangaroo shooting, the occupation shearers traditionally followed between sheds. As for living in a camp being unlawful, wasn't that the way the nomadic shearers had always lived while awaiting the start of shearing in the various districts? That was what they were doing now, waiting for the season to start – on their terms. 'The sheds'll be shore Union or they won't be shore at all.'

This camp at Clermont was only one of many. Most railheads throughout western Queensland had their camp, some with up to 3000 men in residence. Run on sound economic lines, with bulk buying of commodities, their own butchery to handle the truckloads of cattle that were bought to feed the men, they were in the main aided by the townsfolk, even to the organising of concerts

and games to include the unionists. They were orderly, clean and well-controlled. The men knew they had the moral and financial support of inland and coastal unions, as well as the small farmers and many people of the small inland towns.

Shearers, small farmers and townspeople knew that ministers of government and their supporters were all directly interested in banks and stations and able to use all the power vested in government to defeat any attempt by working men to better their lot. These men controlled Cabinet and were able to place police and the military at the disposal of the Pastoralists' Union. As well, the public service could be coerced into acting against the unionists and this, of course, included railways, the means by which 'free' (or 'scab', according to who was referring to the non-unionists) labour was brought into the warring areas.

It was never even remotely suggested that all this idealism, mateship and comradely care for one another's well being was, or could be, played as a gentleman's game. Julian Stuart wrote and spoke of the 'chucker-out' at pre-strike meetings at Barcaldine, and was ruefully aware that 'trying to make unionists out of the wilful and deliberate offenders against the code that meant so much to us' needed tough knuckles.

He told in the *Worker* of young Harry Casey saying to him: 'Split me knuckle on jaw, I suppose I ought not to have done it. It'll cost me a quid or two if the police make a case out of it.'

'I reminded him that if the police "make a case out of it" it would cost him more than quids. It might mean two or three years gaol, for the lock-ups were full. Offences punishable by "two pounds or 48 hours" in normal times, were, under strike conditions, bringing the long sentences "that the judge gives away with a pound of tea".'

Apparently they had been out trying to convert musterers 'but a big surly sort of chap, the ring-leader, kept persuading them not to notice us. The young fellow, Casey, did very well. He pulled out his delegate's book and offered to give them tickets and told them about the camp, with tucker as long as the strike lasted. But the burly bloke chipped in once too often, Casey smote him on the jaw, gave him a hiding and then we rode on. We were heading for home when we sighted a couple of troopers talking to the musterers at the gate we would have to pass through to reach the road.

'Casey drew back into the scrub before they sighted us. He wasn't wanting three years if he could avoid it. "I'll have to do a break." He walked his horse up to the six-strand fence that separated us from the Western road. Leaning down, he knotted a red handkerchief around the top wire, went back a length or two, then rushed his horse at it and cleared it like a bird. As he bent down to untie his silk he said, "I'll get along to Muttaburra. Keep an eye on my things until I see you again." [Muttaburra was three hundred miles off.]

'The troopers were still waiting when I rode through the gate and they asked the musterer if I was the chap who hit him but he said no, it was a bloke with a silk handkerchief round his neck.'

Scotty McKenzie was one of the many Scots who organised or joined the strikes of the 1890s. He was a handy chap to have in a strike camp, for he always had a hack that was a racehorse in disguise, a pair of field glasses to see faces, notices and guns from a distance, and was in for anything.

Although innocent, he was arrested and held on a patently faked charge of having sawn through the piles of the Abor Creek railway bridge so that the train of free labourers bound for Peak Downs station would fall through. The case dragged on, and Scotty and other lively customers were held on a succession of charges that kept them away from the camps and thus made life easier for the military.

The newspapers had stated that the destruction of bridges and the derailment of trains were to be part of the strikers' campaign, and Tom Cooke, a big engine driver, came to the Sandy Creek camp to discuss it with the strike leaders.

'Of course, if you can't get the bastards any other way, don't worry about us,' Tom told the shearers. Then he quoted Adam Lindsay Gordon: 'What does it matter for one soul lost? Millions of souls have been lost before.' The strikers assured Big Tom that the wrecking of trains and the taking of life was not a part of any plan of theirs 'in spite of the words of the daily press'.

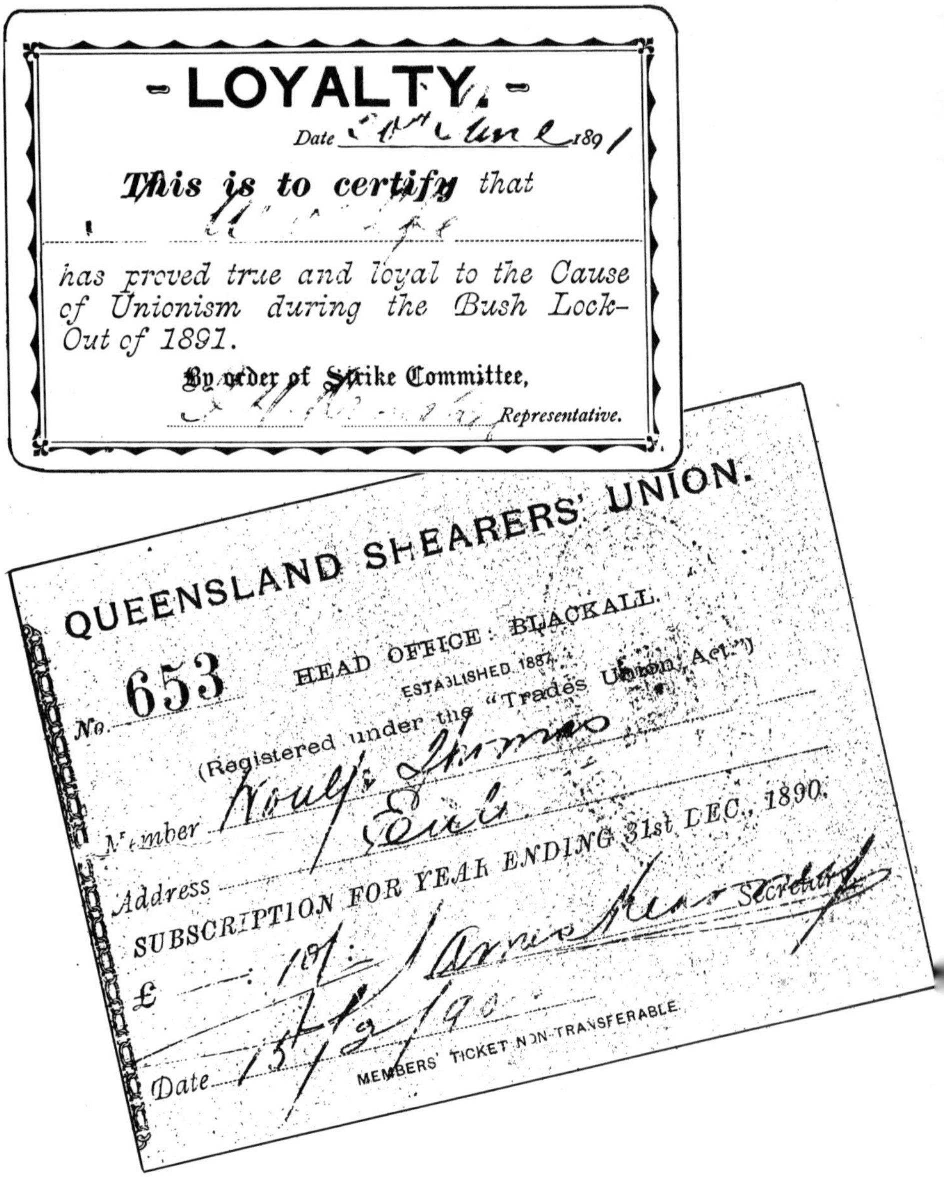
- LOYALTY. -

Date .... 189 ...

This is to certify that

....

has proved true and loyal to the Cause of Unionism during the Bush Lock-Out of 1891.

By order of Strike Committee,

.... Representative.

QUEENSLAND SHEARERS' UNION.

HEAD OFFICE: BLACKALL.

ESTABLISHED 1887.

(Registered under the "Trades Union Act.")

No. 653

Member ....

Address ....

SUBSCRIPTION FOR YEAR ENDING 31st DEC., 1890.

£ .... Secretary

Date ....

MEMBERS' TICKET NON-TRANSFERABLE.

The pride of the new unionists showed in numerous men keeping all their union tickets from first to last. Thomas Woulfe (Longreach, Qld) treasured among his cards that of 1891 when loyalty was a costly thing to the beleaguered men

SECOND YEAR'S LEVY—GENERAL SECTION—10/-

Labor Daily Levy

NORTH QUEENSLAND BRANCH.

7067

RECEIVED from Mr. Thos. Woulfe

of Longreach

the holder of Ticket No. 4872

Ten Shillings

on Northern Branch, A.W.U., Ten Shillings, being Labor Daily Levy for 1911-1912.

Issued at Longreach

Date 29/12/11 Issued by R. Bow.

Note.—It is important that each Member paying his Levy should SIGN THE BUTT OF THIS RECEIPT.

THIS TICKET MUST NOT BE ISSUED TO ANY MEMBER WHO HAS NOT PAID LAST YEAR'S LEVY.

SHEARING TI...

Amalgamated Workers' Unio...

OF QUEENSLAND.

1887-
1892.

CHARLEVILLE.
...ACH.

...DEN BRANCH.

SHEARING TICKET.

Amalgamated Workers' Union

...EENSLAND.

1887-
1895-

SHEARING TICKET.

Amalgamated Workers' Union

CHARLEVILLE.
LONGREACH.
HUGHENDEN.

OF QUEENSLAND.

1887.
1892.

LONGREACH BRANCH.

Ticket Number 1510

Member Thos Woulfe

Address Wellshot

AVAILABLE ONLY FOR THE YEAR 1895.—NOT TRANSFERABLE.

£1 : 0 : 0

Thos Woulfe

Date Aug 29/95

10

The Law

Because our romantic bush poets of the day didn't mention it we should not think the men were timid in the use of what the charge book calls 'obscene language'. In the Register of Police Charges for Adavale, from 1888 to 1902, not one man is charged for saying 'that blamed thing' or 'that ruddy man', which was as far as the fine upstanding sons of toil went in the books of Henry Lawson or Banjo Paterson. A good deal more than the reported 'groaning' and 'boo-hooing' went on. Just what did the words 'insulting' or 'obscene' language mean in the police charge book?

On 3 May 1889, one David Frank was given 48 hours in the lock-up at Aramac for 'using obscene language, to wit, the words, "I don't care about your fucking name".' On 27 December 1889, Constable Quinn had been called: 'Bloody dog. You fucking Quinn', by George Chester, who also paid 10s 0d for the luxury. But Edward Smith was fined 20s 0d for saying 'Fuck them. Bloody whores'. Profane language by John Stone, 'to wit "by Jesus Christ"' gained a caution and the order to leave the town before 12 noon.

And so they carry on. 'Fucking town'; 'Come across the fucking Creek. There is two gins there'; 'Fuck you'; 'Bloody bastard, bloody whore by Jesus Christ'; 'I don't give a fuck', (George Flynn, shearer, 18 March 1890, 5s 0d); 'You buggers'; 'Willfully exposing his person in a public place – "you fucking buggers"'. One man on the way to the lock-up shouted, 'You whore's bastard. Your mother was a cow. By the holy Jesus Christ.'

'Bugger the fucking bottle', cost the speaker 20s 0d, but 'Fuck the bloody police' brought an order to leave town or be brought up for sentence. 'The whole lot of you can kiss me arce' [sic] 10s 0d.' The language was standard. 'You pink-eyed bastard.' 'Go frig yourself.' 'I don't give a fuck for anybody.'

By 1891, 'obscene language' was the regular charge in the book but in particular, 'Fuck the bloody police'. Frank Taylor paid 10s 0d for the pleasure of 'riding furiously in Gordon Street and riding through two sitting-rooms from the front to the back of the Royal Hotel'. Mostly, however, it was 'language',

Acting as Commander-in-Chief for the Government, R.A. Ranking, government agent and police magistrate from Rockhampton, was directed by 'wires' (telegrams) from Colonial Secretary Tozer (Brisbane)

and knowing the speech of the bush at any time, why should 'obscene language' suddenly become a cause for arrest?

It is the period and the occupation of the offenders that is the key. From 1888 to 1902, when the great shearers' strikes were raging across Queensland, Adavale was in the midst of it, and any excuse was found to try and stem the rage that had burst the banks of the shearers' minds.

If one thing, more than any other, boiled the pot that had been simmering among the men it was the constant evidence that the law was not only seen not to mete out justice, but that the dispensers of the law felt no obligation to pretend it did. At Adavale, a tiny village surrounded by large sheep stations, the police records for the 1890s unintentionally reveal this. During the strikes cases were heard before Messrs A. H. and H. C. Pegler, both Justices of the Peace and joint owners of Milo sheep station.

Milo covered 4343 square miles, taking in other stations as part of the whole: Welford Downs, Tintinchilla, Sedan, Emudilla, Ambathala, Boondoon and Gooyea. In 1892, 566 217 sheep were shorn in the sheds on old Gooyea, Milo and Welford Downs – with blades.

This station was shearing non-union, and throughout the decade it determinedly refused work to union men unless they abandoned union rules. For much of that time the Peglers sat on the bench at Adavale convicting union shearers; during the time remaining they worked to defeat the cause of the men on whom they were bringing down judgement.

Shearers had formed no great part of the crimes and offences committed in the outback. At Adavale in 1889 a few were charged for drunkenness, disorderly conduct, obscene language, assaulting police and vagrancy, and one for aiding and abetting in a case of larceny. But from 1890 onwards the scene changes and the occupation 'shearer' is often mentioned in police records as the strike continued. So, too, did pastoral workers generally, as the men closed ranks. The names of the two Peglers regularly crop up in police correspondence beneath both the hats which each wore: the one to defeat the strikers on Milo, the other to dispense justice from the bench.

'*2 March, 1890:* Mr Pegler J.P. and manager of Milo described wanted man and promised if offender came to Milo he would give him work and inform constable' [in other words, keep him busy until the police arrived].

'*April, 1890:* Constable Manuell goes to Milo to arrest a man. Mr Pegler had ordered the man and his mates to leave the station and they were expected to go next day.'

Under pressure, on 18 July 1890, Constable Manuell 'respectfully reports that a system of gambling by means of tossing coins and betting on the results of heads or tails is very much practised here. And this bench of magistrates are of the opinion that this kind of gambling is not unlawful within the meaning of the vagrant act. The constable desires to know if there is any law prohibiting this kind of gambling, also, if publican is liable under Licencing Act for permitting such gambling.'

The reply from Inspector of Police at Charleville to Constable Manuell stated that, '*Two up* is not unlawful gambling'. (It had been argued before the Full Court.) So yet another means of convicting shearers was closed.

On 5 August 1890, Manuell writes about the police paddock being fenced: 'John Russell is at present shearing at Milo and it would not be convenient for him to erect the fence until after shearing which will be finished in about three months. Several other local persons would tender had they not been shearing at present.' (An indication of the number of non-union men shearing this year.)

On 21 October 1890, in reply to a telegram requesting information on any disturbances during the strike, and sent as a reprimand to the policeman for what was believed to be his reluctance to harass shearers, Constable Manuell reported that on the 30 September 'about 100 shearers and labourers were camped near town when it became known to them that they could resume work; they then formed themselves into a procession and marched through the town. The leaders made speeches about unionism in front of the stores and hotels. Afterwards the greater number of these men started to drink and skylark with one another, two of them were fighting and the constable separated them.

'On the 1st November about 100 more shearers and labourers came into town from Milo where they had been called out of work by the leaders of the strike. These men joined with the others and paraded the main street, carrying a banner and giving three cheers for Unionism. Afterwards the greater number of them commenced to drink and skylark with one another. Three of them were summoned for obscene language, two have been convicted, one yet to be heard.

'John Hood, shearer, aged 40, was fined £3 for obscene language and resisting police, as was Michael Quinn, aged 50. The other man, Queenslander Edward Dignum aged 27, was given six months hard labour in Roma gaol for larceny of a silk handkerchief and attempted larceny from the person and drunkenness, Magistrate A. H. Pegler J.P.

'The nature of the disorderly conduct was not such as could be made amenable to the law. It consisted principally of throwing pieces of old bags and bedding at one another, this was indulged in by the greatest number of them and they were all more or less under the influence of liquor. No case of property having been destroyed was reported to the constable and on receipt of the Inspector's telegram on the 12th Inst. he made careful enquiries at all business places and was informed that no property had been destroyed.

'The constable begs to explain that he did not report this matter to his inspector in the first instance because there was nothing unusual about the occurence: the same kind of disorderly conduct occurs here and in every other town in the back country whenever there are a great number of men in the country.'

For profane language, Michael Pender, labourer, was fined £3. 'Property found on prisoner: 6 union tickets, 1 purse – £3 11s 6d.' The same profanity cost Harry Harrison £3 and obscene language cost Henry Harrison £5. Their possessions: 5 union tickets each. 10 December 1891: Gerald Bachelor (assault, case dismissed). '5 union tickets found on prisoner'. The possession of union tickets would not have endeared these men to the magistrates, the Messrs Pegler, since it showed that any one of them could have been recruiting for the union on the Peglers' own land.

Sometimes Pegler gave out £2 fines for 'obscene language', other times £5 or £3. (The precise language is not always recorded, but the Messrs Pegler remained busy.) James Harris aged 29, shearer of NSW, 'a C. of E. man', was charged with 'Being an idle and disorderly person' (not for having a 'defaced coin' [double sided penny] and dice in his possession as recorded in property on prisoner). Discharged. But for obscene language, £5 or three months. A. H. Pegler, J.P.'.

On 30 April 1891, Constable Manuell asked Tozer for reinforcements, as '40 unemployed in town and as labourers at Milo have given notice they intend to strike; a greater number of unemployed will be in town next week. [Also] Troop Horse Rattler was staked about six inches below the off knee on the 16th inst. while the tracker Charlie was riding him through the bush tracking offenders who had set fire to Gumbardo woolshed.'

Special constables volunteered and were sworn in to serve in troubled areas

On 2 May 1891, he paid £2 4s 0d to constables Murphy and Fairbrothers 'for night allowances from Dec. to Feb. 1891. The constables are at present at Milo protecting property.' By 8 August 1891 he reports that, 'The new quarters were occupied by Capt Brown and the Mounted Infantry during the late strike and two panes of glass in the windows are broken whilst so occupied.' He also gives details of searching for, and gathering up bones 'scattered over large area of 17 Mile Creek' and 'burying these remains of a man'. Enquiries revealed that the man was a 'free labourer', information being given by Henry

Pegler, manager of Milo station, that he met the free labourers at Charleville (they having been engaged at Brisbane) and distributed them to the various stations, but George Wendle and another had failed to appear.

Two letters written to the Inspector at Charleville report that Constable Herbert Jackson had sent a police tracker out riding hell for leather 'in a foaming condition on return' looking for a horse belonging to Mr Pegler of Milo station because he had 'received the loan of a horse from the station previously and [had] always been treated with great courtesy'. This is precisely the favoured treatment of pastoralists that the Shearers' Union was complaining about.

The relationship between squatters, government and the judiciary was publicised when the following letters were tabled in the Queensland Parliament. Questions as to how friends of the shearers' cause came by the letters were bandied about in an attempt to divert attention from the disclosures. The shearers claimed that the baggage of Police Magistrate Morris's luggage had been mistakenly sent as that of another man of that name. The first letter was from F. Ranson, Secretary of the Queensland Pastoralists' Association. He was writing to C. A. M. Morris, the Police Magistrate in Barcaldine, in April 1891.

'The twenty specials were engaged yesterday, and will leave here to-morrow by steamer. They should arrive at Barcaldine on Wednesday next, where they have been told they will be met by you. They go up in civilians' clothes, and will require to be sworn in when they reach Muttaburra, or possibly at Barcaldine. See Ranking as to what would be best. Saddles, bridles, four pack-saddles, rifles, revolvers, and ammunition for each man go by same boat, packed up and addressed your care. The men carry their uniforms and blankets in their valises. They must not carry arms till they have been sworn in, and it might not be advisable then either. They are young gentlemen, that have been promised treatment as such. It has been clearly explained to them that no one holds position over another, except that they will have attached to them a sub-officer and constable of police. They could be divided into squads of five if necessary. Sub-Inspector White will be in full charge. Have twenty-four horses with you ready for them at Barcaldine waiting.

George Bunning, the manager of Darr River Downs, Muttaburra, wrote to C. A. M. Morris, on 2 May: 'I enclose a copy of wire received by me from Secretary, Pastoralists' Union . . . It is impossible for me to get down to see personally to these men. May I therefore ask you to take the matter in hand for me. I think it will be advisable to have them sworn in as specials immediately they arrive in Barcaldine; then, as there does not appear to be any boss man amongst them, kindly get police to put an official over them to conduct them up here with all speed consigned to care of P. M., Muttaburra. Upon arrival here I will hand them over to the authorities to guide their youthful minds in the way they should shoot. I am sending Messrs. H. Atkinson and Goss down with twenty horses for the men, and with instructions to report themselves to you and to receive your orders. I fear I am rather adding to your trials just now, but see no other way to get the men put through to Muttaburra without a hitch. . . .'

On 8 April 1891 the following advertisement, signed by Charles A. M. Morris, Police Magistrate and Assistant Government Agent, appeared in the press: 'The Government are prepared to provide free rations and protection in police or defence camps to all men, Union or otherwise, who are willing to return to their work at once. A list will be kept of the names of all such men, and those applying first will receive the first engagements. Plenty of men can find work on those stations where shearing is about to commence, and on all other stations about the Central District.'

This admission of an offer by the government to work as a kind of labour agency for the squatters was seized on by one of the Shearers' Union leaders, W. Mabbott, who publicly called attention to the advertisement. R. A. Ranking, the Government Agent, immediately saw the danger and wrote to Morris: 'Above is an extract from 'The Courier' of the 11th May. There is no doubt that Mabbott has intended to refer to your advertisement. Lane, of 'The Worker', says this is so. Of this advertisement Tozer [Colonial Secretary], of course, is ignorant. We must not allow Tozer to remain any longer in ignorance, or he will commit himself and the Government unwittingly. However painful it may be to us, we must, without a moment's loss of time, put Tozer in possession of the whole facts. Such an explanation will come better from you than from me. I rely on you doing it by return mail. You can easily say you feel that he ought to be placed in possession of the facts.... Let him understand that immediately you and I met you withdrew the advertisement. Don't lose a moment. It can make no difference to you, as you've got your promotion, and if you do not see your way to do it, I must, in self-defence. I could not let the Government go to Parliament to have such a bombshell burst on them unawares.'

While bush workers fought for their cause others answered the roll in the cities for a chance to get to the country 'for a scrap'

11

The Conspiracy Trial

The integrity of Julian Stuart was never in question, right until his dying day. The prisoners were sent to Rockhampton for trial. 'The town is crowded and all hotels are full,' wrote the *Argus* (major newspapers from most colonies had sent journalists and artists to cover the trials).

Squatters and unionists alike gathered daily to watch proceedings, with shearers 'gathered outside the court to cheer the prisoners as they were moved each day to and from their cells'.

When they were called 'prisoners' by the bench, Julian Stuart commented that he and his mates should, by law, be called 'defendants'. Police magistrate Morris promptly replied that he would 'clear the court, as obviously the prisoners were playing up to the public, and he would not have the court used for that purpose'. (Julian Stuart's descendants recall that 'for the rest of his life our father was proud to be referred to as a "union prisoner"'.)

The defendants at the conspiracy trial were George Taylor, William Fothergill, Hugh D. Blackwell, Thomas Joseph Ryan, William James Bennett, Alfred John Brown, Isaac Fry, Robert Prince, William Hamilton, Julian A. S. Stuart, P. J. Griffin, Henry C. Smith-Barry, Edward Hartnett Murphy and Alexander Forrester. After the indictment, which took 35 minutes to read, Mr Lilley, for the defendants, pointed out that the statute under which the defendants had been indicted had been repealed by the Trade Union Act of 1886. The judge overruled the objection.

Mr Virgil Power for the Crown read telegrams and letters seized from the various District Councils of the Unions and Labour Federations.

The letters proved, were proof needed, that Julian Stuart was one of the fourman committee managing the strike in the Clermont district as well as being in charge at Sandy Creek camp. 'This district must be looked after as the squatter must fight us here, once for all.' And, as for the shearers camped at Capella, twelve miles from Gordon Downs which was shearing with free labour: 'We have the best mob of our shearers here at the present time. We must keep them here to fight with them. We must fight down here and win at all hazards.'

The letters George Taylor had tried to throw to the crowd for safety when he was arrested now told the story Ranking had known they would do. The 'slight disturbance', as Taylor called the affair of the Fairbairns in the buggy (and the military called a 'riot'), took up much time. One of Taylor's letters claimed that when the military and police forces were on their way to the camp the next day to arrest men he asked the Government agent (Ranking) to send the pastoralists back – these men were following the party in order to identify the 'rioters'. 'Their presence in the camp would probably cause bloodshed.'

Another letter told of his being on his way to rendezvous with 100 horsemen who would meet the train carrying the free labourers for Peak Downs station. The following day he wrote, 'We had a great time yesterday with the blacklegs and military, but we were not successful. We must try something new . . . I feel as if I had given moral suasion a fair trial without the slightest success.

'The employers are not beating us by moral suasion either on their own or the Government's part. If getting guns and the militia are moral suasion, and they used plenty, the sooner we follow their footprints the better.'

One letter said: 'No stone must be left unturned to win over the first lot of these men – free labourers – the squatters are forcing the men to fight but before they win the squatters will get such a doing that they will remember for their lifetime.

'Bennet informs me that the Newcastle and Sydney wharf labourers will assist us physically as well as financially if we require it. In fact all along the coast the men are of the opinion that constitutional means are very little good and that other methods must be tried.'

There were letters to other unions. It was obvious that the Shearers' Union leaders had sought their aid in what they saw was the great battle. For two days Virgil Power read letter after letter, telegram after telegram, all, he claimed, establishing that some of the prisoners had recommended resorting to desperate means to secure their object. 'The first shot fired may cause the Australian revolution,' a telegram from Blackall claimed. 'The men are very determined and anxious to be at work. What work, I leave you to imagine.' He had in another telegram instructed the use of compulsion to bring out the men at Milo station.

The little town of Clermont was not only hundreds of miles from the Rockhampton court-room, but the crowded court-room could scarcely be expected to comprehend the distance in reality. Bridge where pastoralists 'stuck up' (left), town hall (centre) and unionist hotel (right)

The Railway Employees' Association had been asked to go out on strike, 'We have used our best efforts to constantly resist the intrusion of southern scabs.' (This was Thomas Ryan, Chairman for the Strike Committee.) 'Here everyone hopes the Queenslanders may smash things generally,' he had written when he attended the Adelaide union conference. After his return to Barcaldine he wrote, 'Nothing definite has been done besides staring down the soldiers' rifles which is an education in itself and may bear fruit one of these days.

'When horseman get numerous start an agitation drill. It will keep the men employed, be a weapon and divert Government attention from their antics. Keep rousing the men. Concerted movement and knowledge require time.' Ryan was an active letter writer and telegram sender. 'As far as I can grasp the situation I believe that desperate remedies will be resorted to that will gain recognition of our rights. In your meetings I would strongly advise that this view of the case be given earnest consideration . . . before your committee only: not an ordinary strike view but from the tone of determined resistance that is all over the colony.'

Two letters written by Henry Smith-Barry were both termed 'conspiratorial'. 'At the most, moral suasion is but a poor weapon in our hands. Is it ripe? Are the men's ideas strung up to concert pitch? For, then, in my opinion, the only certain method for working man to get justice is by his becoming expert in the use of the Winchester rifle.'

Some of the defendants' letters showed original thinking for the time. Murphy had found fault with keeping the unionists in big camps and suggested they should be scattered over the country in lots of fifteen or so, thus making it impossible for police to give protection. 'If you want to make a grand coup at any time you could concentrate a force and force a rising by putting a little more devil in the fight.

'What chance would police and the military have against bushmen in their own element?'

Alex Forrester's letters were explicit, if worded in rough code. 'Union mice very destructive in West country, several fires having been started through the agency of these destructive cusses.' Referring seemingly to the sheds that had been burnt down he wrote: 'Shearing in the open air would be a bugger of a game.'

They were great letter writers these early unionists. As Lyndall Hadow, Julian Stuart's daughter says, 'One can see them now, in the long hot evenings in their tents, by the light of a hurricane lamp, or even, if stores were running low, by the light of a tallow dip. Getting their protests on paper, or their requests for assistance, their details of movements of themselves and of the free labourers in their districts, one can see these leaders in their committee rooms in towns at the rail heads, conferring with those who came and went, those who brought news from other districts and those who sought it; and sitting down to write to their fellow-workers in other spheres.'

Charles Fairbairn took the witness stand against the thirteen bushmen 'in their chains' as they were described. He told how a party of about 200 unionists met the free labourers at Clermont railway station on 11 February, how the crowd frightened the horses in the waggons in which the free labourers were to be taken out to Peak Downs station, how they called out 'whoa' and 'boo hoo' when the drivers urged the horses forward. 'Fairbairn informed the court that

While the chained men stood in the dock, their comrades sat out the strike in tents and gunyahs erected in bush camps

he asked the police to interfere but the law men replied that they had no orders. "Do the police want to see a man's throat cut before they do anything?" the bench was reported as snapping. "How do they act in this country? They seem to have no idea of system. It's funny, very funny indeed. It's a nice, pleasant country where such a state of things exist," said Judge Harding.

Fairbairn told the judge that there were four policemen present at the time to which His Honour replied: 'Let me see. They all had six-shooters. Four sixes are twenty-four of them. There would not have been many who boo-hooed the second time if I had been one of them.

'It is monstrous to consider how any free nation can exist and such things are considered justifiable and allowable.'

The police, who had had a bad time of it throughout the whole of the uprising, now would have to bear with the majesty of the bench – and yet, of all men, the only group who were a match for the bushmen on trial were the bush police, those underpaid outcasts of the system who rode and lived as tough and as dangerously as any of the legendary men in the outback. But four police to control 200 boo-hooing, striking shearers?

Sub-Inspector Michael Dillon told how three of the free labourers who had arrived at Clermont the previous day (10 February) 'had to be escorted right through the township, a crowd of unionists following and behaving in a most disorderly manner . . . hooting, hissing and swearing.'

His Honour asked what the duty of police was when they found people behaving in such a manner, and Dillon replied: 'To arrest them.' 'Were any of these men arrested?' 'No, we were not strong enough. I had only five or six men under me. It took us all our time to protect the free labourers.'

'You might have arrested someone and you might have protected public decency.' Dillon made no comment, so Mr Justice Harding went on, 'From what I have heard there cannot be a decent person left in Clermont who has not heard foul language. The place must have been a perfect bedlam at times.'

'So it was,' said Michael Dillon.

'Then why did you not arrest them?'

'I believed there would have been a rescue.'

'And supposing there had been a rescue?'

'There would have been serious consequences.'

'And if there had been serious consequences would you have not been justified in doing your duty under the circumstances in the public streets?'

'Yes'.

'Then,' said Mr Justice Harding to the man who, of all men in the country at that time had delayed 'the firing of the shot that would start the Australian Revolution', 'How is it that these things are allowed to occur and responsible people have their houses assailed like this?'

'I acted as I thought best,' said Michael Dillon. (The reference to the 'houses being assailed' seems not to have been any other than rhetoric by the judge.)

The trial was now in its eighth day. The union prisoners in their chains must have sometimes thought how little meaning all these legal niceties had when they considered the bushmen living hard in the camps through the Queensland 'wet'. Perhaps they wondered if this coastal-town jury knew anything at all of the inland, or the sheds, the shearing, the mustering and droving, the yolk boils, the damper or johnny cakes? Did they know the industrial situation? The long dry ride from station to station with no certainty of a job at the end of it. The day your back broke and you drank your cheque to forget it . . . but then, they would know that last, for didn't the Crown Prosecutor denounce shearers in general each day and these striking shearers in particular?

William Lane wrote, 'The Bushmen have not two friends sitting together except in the public gallery and there, well, the squatters could not find two friends: only the public gallery does not count yet.'

The evidence against Julian Stuart was most damning in the shorthand notes taken by a Mr Bacon, 'a divisional board clerk'. (As this was done at a meeting of strikers at Clermont on 5 March, we must admire Mr Bacon's courage.) Stuart had moved that the unionists stand firm 'and maintain the fundamental principles of Unionism by constitutional methods. We are going to win or die. It amounts to this . . . we can die at any time. If we like to start the ball rolling some of us have considered right out to the final scene; and what deters us is that a great many innocent people will suffer, but we are not able to wait for time and eternity for the public sympathy to come.'

When, at length, the addresses to the jury were given, Lilley, for the union prisoners, congratulated the jurors on having reached the end of the case. 'If you have distinguished yourself by nothing else you have been concerned in one of the longest criminal trials [19 days] ever to have taken place in the colony.'

The judge directed the jury on the law applicable in this case, referring to the Trade Union Act, which provided for any number of men to combine for peaceable means and attempt to affect by argument or persuasion other men in relation to their trades.

'... yet, I can tell you there is abundant and ample evidence to convict the defendants on each count and, gentlemen, I think I may go further and tell you that if you believe that evidence it is your duty to find each and every one of the defendants guilty; and it is between you and your God how you perform your duty this day.'

Two hours later the jury foreman reported that the jury had not agreed; after being locked up for the night they again reported that there was no possible chance of their agreeing.

Justice Harding spoke: 'I shall make a few observations to you before you go out again. Perhaps there is some person on the jury who does not know his duty. Perhaps you have not understood me. I understand that you sent something across to me this morning but of course I was unable to look at it. I am willing to hear you now.'

The foreman said, 'Some of the jury have had nothing to eat since yesterday morning.'

HARDING: 'What do you want?'

FOREMAN: 'Something to eat.'

HARDING: 'Well, gentlemen, it is a reasonable request but you did not ask for it last night. If there are any matters upon which I can offer to instruct you, I shall be happy to do so. It may be, possibly, as I have experienced in Brisbane when I ordered refreshments for a jury after forty-eight hours ... that after you have had your breakfast you will agree.'

His Honour had much more to say to the still unfed jury, including that, if any of them were unionists and if they thought that for the sake of their union they might forswear themselves, they must not be so influenced. 'It does not matter that large crowds of men are outside the court shouting for these men when they leave. It is no part of the glory if that shout should be shared by you.'

Tozer received a telegram from Ranking: 'Judge has given them long exhortation on same lines as before and has allowed them breakfast. Evidently they will get no discharge until after conclusion of civil business ... Judge has clearly stated he will not discharge jury for some days.' That night he wired: 'Jury locked up for night. Still disagreeing.' The following day, Tozer was advised that the 'conspirators' had been given sentences of up to three years. A delighted, but pompous and unctuous Horace Tozer wired St George, Cunnamulla and Hughendon: 'I hope that better class of unionists will now return to work.'

It was a painful time for the shearers on that day in May when their leaders were sentenced to be cut off from them for so long a time. They tried as many tactics as were available to them. But the end was in sight when Ranking was able to wire Tozer: 'Free labourers landed in Barcaldine without any trouble and no secrecy.' And by 8 June, when George Pont, one of the last of the most wanted men was arrested, he said: 'Believe another month will see end of strike. Satisfied money failing and their only hope now is if New South Wales shearers go out.'

Everything was out of kilter in their world, and worse was to come.

Word arrived in the camps that the ASU had conceded 'Freedom of Contract', that the shearers' own W. G. Spence had worked it that way to force the Queensland Shearers' Union to amalgamate with the ASU. There had been 'a sell out of the workers' at a conference held in Sydney between the ASU and the Pastoralists' Union on 7 and 8 August 1891.

Later, a telegram arrived, 'Utterly untrue. We are firm and determined on this point.' It had been a fabrication; the rock on which the union was being built was still sound. Or so they said.

Little record of this moment has survived, but we need no memoirs to convey the smashed spirits of those men who had been 'cheered as part of the glory' and then, at the peak of their exhilaration, had learned that the one unyieldable tenet of their faith had been yielded. ('A ballot', wrote Julian Stuart, 'to test the feeling of the Queensland Shearers regarding this paramount feature of the struggle showed the following result... Against conceding freedom of contract to the squatters, 8462. For, 81... conclusive evidence that the men were still with their leaders.')

However one reads the newspaper reports, minutes of meetings (sketchy and partisan) and the few surviving oral and written memoirs of those days, one is still no closer to the truth. Perhaps the truth is simple: there was no one way out of the impasse. The pastoralists had said they would not attend a conference unless freedom of contract was first agreed to; the unionists needed the meeting but were in the hands of the wool-growers, for all their fine speeches about the wool remaining on the sheep without their labour. Perhaps after those first wild, exhilarating days of the pride in their manhood and skills of workmanship, some in that clutch of unions saw that they needed to be drawn into a manageable unit... but we will never know.

'The Capitalist Press' was blamed for the story that 'freedom of contract has been conceded', and they certainly did trumpet such a statement from the major city dailies to the small, but often professional, news-sheets that almost every bush town printed. But were they lying? Or was it true? The answer is all too plain.

It would take an 'Unconquerable Soul' to have endured, and prevailed, in the 1890s. William Spence, President of the Amalgamated Shearers' Union, was later called this and it did describe him well at the time. For six years he was made to look foolish, and since has been said to have been so; David Temple, the union secretary, appears naïve rather than foolish. For all those years they were in a most complicated bind.

The concept of unionism was very new in this country, which had begun as a convict settlement and remained so until 1852 in the eastern states and longer in the west. Now, less than forty years later, when there were still men living who had been transported here in chains, 'the working class' were attempting to organise on a scale comparable only to American unionism. And we lacked the roads, transport facilities and population of the United States.

One of the difficulties of moulding their disparate members into a solid phalanx was the assuredness of the squatter, that ease of manner worn by what had come to be the ruling class. The certainty that they were born to rule others was bolstered by their gentlemen's clubs and belief in the brotherhood of the privileged class. This made it simpler for squatters to organise. Indeed, landowners in all places have always known that to retain their acres they must be a club, even if, before 1891, there had been no need to organise in the industrial sense. This age-old freemasonry of land-owners had conditioned many men into believing that this was how God meant life to be, and an authoritarian 'gentleman' God was presented to them in their churches by the ruling class-brothers and actual blood-brothers of the squattocracy.

Now men were rising to deny much of what had been presented as Holy Writ, and they brought down upon themselves the righteous wrath of those who had lived comfortably within the established state of affairs. Even many of their own class, men and women who were comfortable only when there was no fury in their lives, were justifiably afraid of any upheaval in the settled scheme of things. Spence and Temple knew it – all labouring families know it – yet these two went forth to do battle for what they believed was right. In the colloquialism of a later century they were 'done like a dinner'.

The Pastoralists' Union report of their conference with the Shearers' Union shows this. Held in Sydney in August 1891, while some shearers were in gaol, some being sent there, and others were still battling to keep alive the spirit conjured up in the strike, the squatters humiliated Spence, the shearers' advocate.

It was inevitable. The ex-miner with no formal education was up against the fluent and polished speech and thought of men who had been well educated both in this country and in the finest colleges and universities in England. Temple, who was to die young, was all heart – this commodity was worthless in such a gathering.

Knowing that they had beaten the strikers, believing that the whole unionist movement was in rout, they calmly and methodically rendered Spence's arguments impotent. Messrs Elder, Wilson and Abbott repeatedly grilled Spence over Freedom of Contract. Unless the definition of this term was agreed upon, the pastoralists would not continue the conference, their definition being that men were free either to join a union or not, and to have equal rights to work on the shearing board, non-union beside union. It had been a basic tenet of unionism since its inception that union men do not work beside non-union men, but to get the squatters to come to a conference, and stay there, Spence was hedging.

SPENCE: Sentiment is very often a powerful factor in determining men's actions. I repeat that we are prepared to stand by the fact that the definition of Freedom of Contract has been agreed between us.
WILSON: You have not said that.
SPENCE: Mr Temple in his letter indicated that so far as it went we recognised it as being given, and fair, to the limit of its extent.
ELDER: He telegraphed to Queensland stating that he did not say so.
TEMPLE: As far as we understood it.
SPENCE: In history there is a distinct interpretation.
ELDER: We have nothing to do with history but we have a great deal to do with Freedom of Contract. We should agree to state the definition of contract which has been accepted.
SPENCE: Have we not done it?
ELDER: Then why not acknowledge it if you are a straightforward man?
SPENCE: We have acknowledged it but you may see we wish to put nothing in the provisional agreement which the men will not sign.

Spence was fighting his own battle for the strikers in a meeting with the squatters, by this time known as pastoralists. Back row: T. Williams, J. M. Toomey. Front: D. Temple, W. G. Spence, H. Langwell MP

So there it was. The truth was out. The hounds had the fox cornered and they went in for the kill.

ABBOTT: The Shearers' Union claimed over and over again that one of the objects of the conference was to define Freedom of Contract – are you prepared to accept it?

SPENCE: Your conditions in the past were that you would not meet us unless Freedom of Contract was admitted.

Far from being inept and blundering, Spence should be regarded as a consummate negotiator. He was hedging, yet being careful not to go so far that the squatters would walk out. And he had no option, for the matter had been decided before the two parties met.

At Young, the Union Secretary, J. M. Toomey, had met with his squatter friend G. H. Greene. Greene was a moderate man and both of them realised that the impasse must end, for the country could not continue in its present

Pastoralists' Union delegates to the conference, 8 August 1891. Back row: J. A. Campbell, A. Wilson. Front: D. Elder, W. E. ('Wingen') Abbott, A. Cottee

state. They agreed that Freedom of Contract as defined by Greene's organisation should be recognised. This agreement was sent to the Pastoralists' Union secretary, Whitely King, and in loyalty to two fine, conciliatory men, Spence conceded the most pertinent point of the Union and the strikers' tenet. The squatters were condescending; foremost in their minds was the fact that it was not they, but the shearers who had at last surrendered.

Even allowing that it is the Pastoralists' report of the meeting, it is obvious that Spence was trapped. He was reduced to objecting as to where Freedom of Contract would be printed on the agreement that individual shearers would sign at a shed.

ABBOTT: We want to put it on the front. You want to put it out of sight.

SPENCE: The men would not agree to it even if we did. That is the trouble we are in.[1]

12

'If a Little Blood were Shed'

As the wool-grower who wrote to the *Brisbane Courier* said: 'I honestly think that if a little blood were shed a great deal of good would result.' (He didn't state whose blood.) And Henry Lawson wrote:

> We'll make the tyrant feel the sting,
> O' those that they would throttle,
> They needn't say the fault is ours,
> If blood should stain the wattle.

The battle was joined, the first shots fired. Julian Stuart, in charge of the Sandy Creek camp outside Clermont was certainly not thinking of a boy scout romp through the countryside when he told his men, 'It's no use going out with johnny cakes in your pockets for ammunition'. But he must have had so little time to proselytise. With the hundreds of men camped there, the very collection and gathering of funds to pay for their food must have been a gargantuan task.

In the year 1891 the shearers stood alone. The other unions were not 'called out' in their support. With the weight of money and privilege brought against them they must manage to live without wages. There was of course no strike pay and with roughly 10 000 men on strike their upkeep in the camps was considerable. Copper miners of South Australia sent help, unionists of Wagga sent £1000, but with the cost per head of each striker being roughly 7s 0d per week for rations, money was needed. Even the means of communicating by telegraph was costly – their opposition had free access to the telegraph, as their long telegrams show. And all the while the numbers in the camps grew as the unionists met the trains and 'encouraged' the free labourers to leave the squatters and join the men in the strikers' camps.

The significance of the Clermont camp was this: several large sheep stations surrounded the little town as was the case in the west when most villages existed only to service the sheep stations. Of all properties within the wide radius of Clermont, Peak Downs Station, the property of George Fairbairn and Sons, was the one that most rankled the unionists. Fairbairn, a member

Rifle drill, shearers' camp. Preparing to receive cavalry (top) and skirmishing drill (below)

of the Pastoralists' Association executive and an absentee landlord who owned several stations, wielded both political and social power and now demanded special action against those who had tried to encourage his 'free' labourers to leave him and join the unionists.

These sixty-three 'free' men had arrived on a heavily guarded train on 19 March and were 'hooted' and 'groaned' at and 'fifty bells were rung', in the fashion of the day, before being whisked away. The unionists had caused a flurry on 7 March among the military and the squatters over the jostling of a buggy taking five pastoralists including Fairbairn from the Clermont railway

station – there were said to be 200 unionists at the station, more than half of them mounted, when Fairbairn clambered into the buggy on his way to the Commercial hotel escorted by Sergeant Dillon and seven foot constables.

Next morning ten mounted police, ten Moreton Bay Mounted Infantry and fifteen Permanent Artillery men under Major Jackson and Major Rickard rode to the camp with Police Magistrate Ranking and Inspector Aherne leading the way. They arrested two men out of the 200 that lined up offering themselves.

The law then rode to Wolfang Creek camp and from the 100 men there arrested a shearer named McCarthy. 'With the bushman's sense of the fitness of things, McCarthy declined to walk the 100 yards to the police buggy on the main road and it had to be brought to the camp for him.'

Julian Stuart was still free, still organising the Sandy Creek camp; but he knew his time was near. 'Seditious language' was cause for arrest, as was seditious libel. 'The squatters felt their self-assumed dignity as well as their profits were threatened by the actions of the unionists' and as they were also the law makers they warned that those who incited others to commit 'crimes' would be arrested for incitement to commit felony, itself a crime. Unlawful assemblies, people meeting and sallying forth to intimidate free labourers, would be charged with conspiracy to intimidate men, while others would be charged with the offence of intimidation by molesting workmen pursuing their lawful avocation.

But all the while the wool was growing on the sheeps' backs, the inexperienced men were not getting it off quickly enough, there were not enough bales going to the waiting ships. It was as if a fever, a flux was running through all concerned. Perhaps if more were gaoled for rioting, threatening, abusive language, sedition, conspiracy, anything – but gaol them and end this impasse, was the feeling of the Colonial Secretary, Horace Tozer.

Searching an incoming train for non-unionists

Strike camp, Wolfang. Julian Stuart, also in charge of Sandy Creek camp, Clermont, was responsible for the men from many stations including Wolfang

Like dragons' teeth, where one unionist was gaoled another sprang up to fill his place; and then Julian Stuart was taken. On 24 March, fifty-six mounted infantry under Ranking and Aherne rode out and arrested shearers concerned in what by now was being emblazoned across the press of Australia as the Peak Downs Riot – the 'Fairbairn in the Buggy incident', as Stuart was apt to call it. The arrested men were taken to Clermont railway station and there Stuart was charged, along with Taylor, with 'conspiracy'.

Taylor had in his pocket letters concerning strike matters and these he attempted to throw to the crowd that had poured into the streets and surrounded the unionists, but the police grabbed the papers and set off with the prisoners, Stuart and Taylor refusing to walk at a pace faster than one mile an hour although it was raining torrents.

'The importance of these arrests cannot be over-estimated,' the Melbourne and Sydney papers cheered. 'Some sensational revelations may be looked for.'

Stuart, aged 24, was on his way to three years hard labour. The arrests now came thick and fast and the small western towns had not had such spectacles since the days when the first bullock drawn waggons came in. At Barcaldine, 120 mounted infantry were drawn up in line across the street where the union office had been opened and infantry with fixed bayonets guarded the police station. Blackwell, Fothergill, Ryan, Murphy and Bennett, all office bearers or members of the strike committee, were arrested, followed by Martin.

'Shearblade' Martin as he was known, had once advocated the use of 'the queen of weapons', the pike, but in the event of it being used at the opposite end of the earth to its origin his colonial pikes were to be made of shear blades lashed to sticks. He was now arrested for seditious speech. His incantation had been somewhat as follows: 'There is only one way and that is to fight the squatters. Take up arms and fight. I have a petition in my swag for electoral reform but the only petition I believe in is 10 000 resolute bushmen with 10 000 shearblades. There are 10 000 bushmen on the plains for active service and I am one of them.' Someone, not Martin, then called for 'Three groans for the Queen' and Martin was reported as saying 'I suppose you are groaning for Old Mother Brown, John Brown's widow?'

Back at Clermont, the magistrates used the lock-up as a temporary court-house, fearing a demonstration if the prisoners were brought out into the street of the little town. There were the eleven 'Peak Downs Rioters' and the three 'conspiracy' prisoners of whom Julian Stuart was one, all in chains, all remanded without bail.

From all points of the compass the union organisers were organising as never before. Roma, Blackall, Winton, Hughendon, Charleville, Normanton, Clermont, Barcaldine, Cloncurry, Thargomindah, all the centres were in action. Free labourers were being 'persuaded', 700 unionists paraded up the street of Barcaldine in a torchlight procession, others galloped off widdershins into the scrub, leading the military on a fools' chase they could never hope to win against men raised in the bush.

'The arrest of the strike leaders,' wrote the *Argus* on 26 March 1891, 'instead of intimidating the men, seems to have aroused to renewed action the committees of the joint unions and in a few hours they elected a strike committee to replace those who were arrested.'

The conspiracy charges read: 'That you and each of you did at Goondiwindi, Hughenden, Roma, Blackall, Clermont, Charleville, Muttaburra, Barcaldine and divers other places conspire, combine, confederate and agree together to force and endeavour to force certain of her Majesty's well beloved subjects to alter the mode or method of conducting their business, thus imperilling the peace of our sovereign lady the Queen, her crown and dignity.'

The arrests continued until there were twelve 'Peak Downs' Rioters', eleven 'conspiracy prisoners', one (Isaac Fry) for riot, and one for sedition. All were treated as state prisoners and allowed meat, tobacco, newspapers and writing materials. As well as these men there were 'rank and file' members arrested in groups around the country side for 'boo-Hooing', 'groaning', 'insulting language, obstruction, intimidation'. In the towns circling Clermont, hundreds of military and police had been rushed to the centre by 6 April: at Barcaldine there were 30 officers and 768 men; Muttaburra, 1 officer, 14 men; Delta, 1 officer, 10 men; Clermont, 12 officers, 218 men; Winton, 3 officers, 51 men; Hughendon, 2 officers, 44 men. They were all kept busy, the shearers saw to that. There was no option by way of a fine; the men were to be gaoled and they made sure they would be available.

The case against the Peak Downs' rioters opened in April with the prisoners charged with 'intimidating, molesting, obstructing certain persons hired etc.' Their penalty of three months hard labour surprised the public.

Then Julian Stuart and the others charged with conspiracy were brought in, clanking along, hand-cuffed together to a heavy chain, and remanded to Barcaldine by special guarded railway train.

Their arrival at Barcaldine must have been a grand spectacle. The whole available military strength of the district was out to meet them, staff, horse artillery, field artillery, mounted rifles, infantry, volunteer and ambulance corps. Three hundred and nine officers and men marched through the town, the Mount Morgan Band and the Scottish Pipe and Drum Band played in the middle of the town. The prisoners climbing down from the train were delighted – but the display had not been mounted for them but for 'free

Shearers at a strike camp collected round the cooking place

labourers', to show the sort of protection Government forces would afford them should they work for the wool-growers. It was all very wonderful.

Julian Stuart recorded that the crowded streets remained silent throughout this display and until the arrival of the prison train when 'a crowd of 1000 were soon on the spot outside the town where the train had been stopped hoping to avoid any demonstration – cheering frantically as the shearers climbed down handcuffed to their long chain and escorted by police armed with rifles and sword bayonets. Two companies of mounted infantry, a company of Brisbane Volunteers and the Mackay Infantry took some time to clear a passage through the crowd and the escort was reformed in the boggy ground of the railway line beside the wool-scouring works.

'It was late before the procession headed by Mr Ranking started for the town at the head of infantry and police with the Brisbane volunteers surrounding the

Mounted police were sorely tried in isolated areas, being ordered on long marches day and night

Shearers drilling

prisoners. The pace was slow, the crowd harrying the procession on all sides, climbing trees and buildings along the route, and at one point becoming so demonstrative that the mounted infantry were needed to clear the road for the prisoners and their escort.

'After the men were safely lodged in the gaol, the excitement subsided with mounted infantry and police patrolling the streets. In the Union Camp that night there were camp fire songs to the beating of drums, the men not in the least daunted that some of their numbers faced arrest in the morning because of the demonstration.' Few men in Australia's history had been accorded such warmth of welcome by their peers as these 'horny handed sons of toil'.

The Queensland 'wet' had begun, the steamy, sweating, fever-ridden months of the year when life was a misery in the makeshift water-logged camps. The arrests went on across the face of Queensland, twenty-one prisoners in a batch being brought in to Charleville with eleven of them mounted in hand-cuffs and the other ten in special mail coaches being leg-ironed. The military and police were running out of handcuffs and chains and necklets, and also running out of lock-ups in a countryside when often a log outside the pub had sufficed to chain a man to for a day or two if no other accommodation was available.

'No unionist must hesitate to step into the offices rendered vacant by the arrest of leaders, no matter what persecution might be attempted,' ordered a manifesto issued by the Strike Committee at Barcaldine. 'It is our bounden duty to protest against injustice and assist in every way we feel to be constitutional our fellow unionists in this struggle for our rights.' It all sounds so hackneyed now, but then it was so very new.

The same manifesto deplored the 'brutal methods of ironing, chaining and gaoling' their leaders by a supposedly impartial government, and pointed to 'the Government's pre-arranged military measures showing their adherance to the squatters cause'.

'No matter how often members might be arrested,' the union declared, more would step forward to take their place, even to 'filling the gaols as full as they will go or as full as the government can pack them: but let each man who goes to gaol go confident in his own integrity.'

13

Queensland Me Mother – Are Ye Sick?

Horace Tozer, the Colonial Secretary, must have been working a twenty-four hour shift seven days a week, as was Government Agent R. A. Ranking. During the year hundreds of military were 'called out' including men from the Wide Bay Mounted Infantry, Moreton Regiment, Brisbane Field Battery, Townsville Mounted Infantry, Darling Downs Mounted Infantry, Queensland Volunteer Rifles, Charters Towers Mounted Infantry, Moreton Mounted Infantry, McKay Mounted Infantry, Volunteer and Argyle and Sutherland Highlanders Rifle Clubs, Brisbane Ambulance Corps and the Royal Warwickshire Regiment. (A 'call-out' is the means whereby Government can call in military forces to assist in times of national emergency. There was no coercion about this in 1891, it being the first time troops had had an opportunity to see 'action' since the Eureka Stockade emergency in 1855.)

A shearer

Hundreds of armed police were sent to the troubled districts, 'specials' were signed on and all were needed, as the striking shearers rampaged across the sheep lands of Queensland. Partly in frustration, partly in rage, mostly in an organised attempt to disrupt totally shearing for that year and bring the squatters to conference, the shearers set off on a campaign of arson and destruction. They burned many wool sheds, including Milo, the domain of the Messrs Pegler, Justices of the Peace, and attempted to burn many more; they burned wool – 576 bales at Gumbardo station alone, after they burned the shed – wool stores, outstations, huts, wool trucks, railway sidings, hay-stacks and waggons. They fired mile upon mile of grasslands, cut fences, smashed gates, killed bullock teams and cut horses' throats. They wrecked railway bridges and tore up railway lines in an effort to derail trains bringing the 'scab' labour in from the coast and across the border.

Monster meetings of up to 1000 shearers were held; they marched in procession 2000 strong through towns. Bodies of anything from fifty to two and three hundred were constantly reported as riding through the countryside, leading their pack-horses, and usually armed. Amalgamated Shearers' Union President W. G. Spence knew them to have at least 800 rifles and ammunition in the camps set up in each shearing area and run like military encampments.

When shearers were arrested, whole townships came out into the streets to show their sympathy, and when men were sentenced to hard labour for terms ranging from three to seven years 'penal servitude', the crowds round the court-house doors cheered them. Magistrates showing sympathy to their cause were deposed, as was C. Carkeek who had shown interest in the veteran unionist William Kewley (a friend of the world champion shearer Jack Howe); Blackall citizens were so incensed by the treatment meted out to Carkeek that they held a protest meeting.

When, on 19 June, the Brisbane *Courier* published a purported proposal for a 'union republic *coup d'état*', no one was surprised. As the poet George Essex Evans had once written, the cry could be heard again: 'Queensland me Mother – are ye Sick?'

Reading summaries of telegrams received and sent by the Colonial Secretary's office in 1891, one is unsure whether the reaction was justified or if panic was spreading at the speed of a crown fire in eucalypts. One thing is certain: the threat of near-total anarchy, combined with the vindictive reaction to it, brought Australia to the edge of the precipice. For four months the country teetered on the brink of the abyss: we can never be closer without toppling over into total anarchy or civil war.[1]

Telegrams sent by Horace Tozer, Ranking, police, army officers and squatters give an indication of the uprising's extent.

'*16 February, Sharwood, Clermont*: 100 armed horsemen threaten to burn sheds and destroy property. Suggest strong police action immediately.'

'*19 February, Police Magistrate Moray, Clermont*: Request, if riots commence to withdraw or spill all liquor in public houses.'

Telegrams from squatters asking for assistance poured in: '*24 March, Augathella*: Large number of unionists gathering at Oakwood today. Parties of men going towards Oakwood fired the grass all along the roads, broke down and burnt the gates and posts and cut down netting fences.... Wires from Adavale urgently ask for protection as large numbers of unionists gathering there.' '*Barcaldine*: Aspects here most serious. Expect procession 2000 strong Sunday. Government must be prepared to assert her rights promptly and protect country. If not sufficient forces in Queensland call upon other colonies immediately for assistance. Mob rule might break out any moment.'

G. A. Peddle, owner and manager of Biddenham station at Augathella, constantly sent telegrams: '70 mounted unionists under Fanning tried to force interview with Oakwood shearers yesterday, police prevented. [This was the unionists attempting to encourage 'free' shearers – 'scabs' in their terminology – to leave the shed and join the strikers in their camp.] Mob went on after smashing gate to setting fire to fence. All gates on Charleville road destroyed. Mob expects to join 70 others from Tombo. How much longer is the government going to allow this sort of thing?' (Peddle was eventually appointed a J. P. so that he could deal with the 'mobs' – and no unionist expected unbiased judgements from a magistrate whose sheds shore non-union.)

The Fanning referred to became well known to the squatters; they constantly referred to him. When Fanning came onto a property, the squatter immediately went on fire guard. They declared that bottles of kerosene were found beneath sheds after his departure, that sheds were fired the day after he had been seen

Shearers coming in

there, that he was an agent who provoked the men to outrageous acts but mostly arson.

Men who burnt woolsheds or destroyed property were called 'rioters'; those who attempted to coax 'free' shearers to leave the sheds and join the unionists were called 'intimidators'. '*Blackall*: Hold 22 warrants issued for apprehension Lorne rioters and Northampton intimidators. Require sufficient force to execute same. No police except myself. W. Britton, Police Inspector.' (The shearing shed at Lorne had been burnt and destroyed; Northampton had suffered an incursion of unionists who forced their 'free' labour to leave.) '*Winton*: Unionists getting very strong here. If matters go to extremes at Clermont or Barcaldine they will go in for revenge here. Principal talk . . . anarchy and defiance of Government. Special constables are being sworn in.'

By 10.10 p.m. that night it was raining at Clermont: 'I have police out all night patrolling. Raining here. Winton wants protection badly. Unionists there threaten revenge for arrests here.' By daylight things seemed brighter: 'Quiet here now. Swagmen leaving union camp. Isis Downs wool shed reports all quiet along the line.'

Military men were anxious to try their mettle; it was the first chance since the

Sir Horace Tozer directed the war against the shearers

Defence Act of 1884 which had provided for the formation of Volunteer Corps. Colonel Chauvel, later to be known in the Boer War and World War I as leader of the Australian Light Horse, offered his services, and from Cooktown, William Armit wired: 'Beg to volunteer for a command wherever required in disturbed districts should you require tried officers.'

There were frequent telegrams from property owners and their friends: '*Hughenden*: 19 mounted unionists came here yesterday, intimidated free labourers and union hands who have all left. No safety without protection. Develyn, Cochran and Chisholm, Lansbro Downs.' To this the Colonial Secretary replied: 'Government are aware of necessity of protection of persons and property and had already despatched a larger force of military to Hughendon.'

From Mt Morgan, Winton, Cloncurry, Townsville and Bundaberg, the wires were humming across thousands of miles of Queensland, and the shearers were forming their camps and picketing roads and railway lines along which free labourers would pass. At Roma, a worried police officer telegraphed Tozer that 'Unionists at Noondoo and Mungindi determined and have decided to bring shearers at Noondoo out by force if necessary. Men coming across border from Bourke for that purpose'. And Colonel French, in charge of the military at Bundaberg, reported that 'thirty men had heavy march in mud and water, steady rain falling all the time. Mobs of roughs on Gympie platform hooting men and preventing their falling in. Dispersed them with fixed bayonets'. Cunnamulla wired: 'Bands of armed unionists are patrolling boundary fence between this colony and New South Wales and intimidating teamsters from loading wool.'

Day after day, night after cold night, messages were sent from a score of little villages. Rain made everything miserable. Clermont police station (Julian Stuart was in charge of the unionists' camp there) called for bales of blankets and 100 yards of oil cloth for the constables camping under canvas. J. Ahern, Travelling Inspector, rode hundreds of miles during the emergency, arresting shearers and protecting woolsheds from incendiarists.

The pastoralists and their union were in constant touch. G. A. Peddle wired Ransom of the Pastoralists' Union from Augathella: 'Seventy unionists rode past Biddenham this morning towards Augathella. Passing our huts, before the women and children, they gave utterance to abominable obscenities. Fanning and Forrester's arrest seems to have scattered them. They should be followed up and made to disperse.'

The piles of the Abor Creek railway bridge were cut through before the train carrying 'free' shearers and labourers to Peak Downs station passed over. It was for this that Scotty McKenzie was arrested [although he was innocent of the charge) and it served to keep him away from the camps and out of the sight of the military and the police.

Evesham, Croona, Amby, Aramac and Rodney Downs stations all called for help. The chaos placed tiny, understaffed police stations under great pressure in the little wool towns.

From Tambo the police ask Brisbane for instructions: '15 chinamen driven from work by intimidation report themselves destitute and apply for rations. What shall I do?'

'If without money, relieve,' the hapless policeman is told.

W. W. Hood of Burenda station gives a picture of the type of men volunteer-

Torchlight processions of seven and eight hundred men from the shearers' strike camps paraded streets of western towns

ing as 'Specials' (or 'scabs'): 'Wish to have men sworn in as Specials in same uniforms as mounted infantry for sake of moral effect, as patrol being in uniform prevents them being made marked men by unionists. Men require no pay.' (Colonial Secretary Tozer replies: 'Regret existing law not possible.')

Ships shuttled up the coast from Brisbane with 'free' labourers to work as shearers. These had to be met and the men accompanied to stations. Trains had to be accompanied and met. Camps were raided, shearers arrested, and brought to court; evidence was tendered and, when men were remanded to a larger town, they had to be taken there, sometimes by train, sometimes by coach over the rough, gravelled roads. 'Free' men who had been taken out of sheds by unionists were working a scheme whereby they came into the police station for protection, 'penniless', and the police had to ask permission to ration them, then go to the union camp to retrieve their swags. 'Many more would desert if arrangements could be made for feeding them till sheds open,' said a police telegram. Emerald wired for 'twelve rifles and ammunition for Special Constables watching Bridges,' and more military offer their services: '*Sydney*: Tenterfield and Inverell companies Mounted Infantry have volunteered to Sydney authorities for service Queensland if required. [signed] Chauvel, Captain commanding Tenterfield M.I.' Tozer replied: 'Gratefully acknowledge offer but strike matters so far eased off that, at present, services not likely to be called into requisition.'

In fact, 'strike matters' had not eased off, and one wonders why Tozer refused this force when he so readily accepted the Augathella 'specials'. Could it have been, as the shearers were claiming, that the military were more trouble than use? The unionists certainly led them a merry dance, setting false trails and soberly riding off to one station with the military who tagged on behind later finding that half the shearers had ridden off in another direction to 'encourage' non-union men to leave a shed. At one time, a Major Ricardo was led up a back road 'by a black boy' in the opposite direction to the one he should have taken. His troupe of uniformed soldiery trotted along behind him.

(When the 'emergency' ended, several officers were commended for notable work by the Defence Force: 'Arduous marches performed. Long, weary marches undertaken and well carried out.... Detachment marched 109 miles in 32 hours partially through flooded country, having to swim rivers.'

Of course, it could be said that the union men were doing even greater 'marches', using their own horses and helping provide tucker for the camp, on no income at all, while the army rates of pay were quite good for the times, a private earning 6s 0d per day. This was the same rate of pay as men enlisted in the AIF earned during World War II.)

Townsville Mounted Infantry about to leave on patrol

Across the land shearers were causing havoc that disrupted the shearing of vast flocks of sheep by non-unionists. They had some staunch friends among the bush workers as Constable Manuell from Adavale reported: 'Proceeded ten miles on the Kyabra Road to interview John Dalton who is in charge of the "Kyabra Team" [bullocks]. Asked if he required protection, Dalton replied that he did not require no such thing and if "any of police or infantry" were to come and guard said teams he, Dalton, would at once leave and have nothing more to do with the teams. There are three teams loaded with wool (42 bales on each wagon). Dalton is a staunch union man and so is both his drivers, Mathew Dalton and John Seeres.

'Dalton has been lying at the camp for the past six weeks being unable to travel owing to the flooded state of the country. Dalton further states if weather was permitting would not yoke up his bullocks as same would be against union rules. Dalton considered it would be more advisable for him to remain in camp during the present trouble until such time as it was settled one way or another. Asked if he would push on with the teams if police accompanied him, Dalton said no, he would not drive a yard while escorted.

'Dalton is a married man having his wife and family at the teams.'

Unionists at Winton started their own postal service because, apart from the expense of sending telegrams, they dared not trust the post office. Telegrams between Taylor and two other delegates, Hamilton and Griffon, had been seized and given to Tozer, as a result of which the two delegates were now in custody. 'Documents seized at Barcaldine', wired Ranking, 'disclose absolutely revolutionary schemes.'

Vigilant police continued to report on the movements of numbers of unionists camped in various places, many arrests were made and Tozer and Ranking duly told of them. Riot, arson and conspiracy were now the common charges – the charges of obscene language were no longer needed – and bodies of police travelled great distances between camps and lock-ups with their parties of prisoners. The Commissioner of Police was advised by Inspector Stuart on 31 March: 'After arrest of Fanning . . . mob of Unionists divided. About forty went Tambo direction, one hundred and sixty towards Augathella camping on Buckey's Creek, twelve miles from Oakwood, and burned down the out station there on Sunday night-Monday morning. Accompanied by Captain King, seventeen infantry, five constables and Mr. Barclay started after them, found they had gone. Followed them up and found them camped close to Augathella, surrounded them and arrested fourteen. Visited camp again this morning arrested seven more. They burned and broke up nearly all gates between here and Oakwood. Charged them with inciting to riot, conspiracy and arson. Their names are Mathers three, Morleys two, Gavins two, McMahon, Jermyn, Hastie, Creagh, Lee, Wilkinson, Nolan, Trainer, Richards, Brown, Latreille, McNamara, McGrath, Brennan. Two of them identified as having actually fired hut. Captain King and Infantry worked splendidly, also Police.'

Detachments of soldiery enthusiastically camped out – their first opportunity to shoulder arms since the Eureka Stockade battle

The military and police were spread thinly over such a large area in an attempt to contain the uprising. In reply to a telegram from Ranking, Inspector Stuart reported that their 'distribution strength' was as follows: King and seventeen infantry were at Augathella; Lieutenant Thorne with ten infantry was escorting prisoners to Charleville; Lieutenant Chauvel and fifteen infantry were guarding Oakwood shed; five infantry were guarding Biddenham; Sergeant Geraghty and one constable were escorting prisoners to Charleville. One constable was at Oakwood, 'myself and seven constables here'. 'Here' was Adavale, where about two hundred unionists were camped.

Unionists refused to disperse and suggested that they, too, be arrested. Police

Newspapers gave pages to illustrated stories during the troubled period. 'Disaffected shearers on the Barcaldine Downs try to hinder the passage of free labourers under military escort'

and military strength was quite inadequate to cope with the number of men and afford protection to surrounding stations, as well as shuttle back and forth across the state with accused men, witnesses and escorts. Some men were sent from St George to Rockhampton, others to Roma, while Roma was sending men to Barcaldine. Police were meeting men sent to Brisbane from Charleville and sending others to Rockhampton. The immense planning took many telegrams, as did the reports to and from the Travelling Police Inspectors, who raced from one troublesome spot to another during the whole of the emergency in order to investigate, act and report.

From his War Office in Brisbane, Horace Tozer, that indefatigable Colonial Secretary, sent the steamer *Warrego* up the coast to Keppel Bay near inland Rockhampton with the captain holding copies of telegrams that had been seized from unionists. These were to be handed over to the court trying the 'conspiracy' prisoners, among them Taylor and Julian Stuart.

While the hundred or so prisoners were going through the courts, the brillant Tozer directed the committal proceedings by telegram. He received a telegram from Ranking, now at Rockhampton for the trials, for a copy of 'Statute 6, George IV' to be sent to him. 'Had better be sent by steamer tomorrow without fail,' said Ranking. 'Only copy here in law journal which might not satisfy court. His Honour complained yesterday of absence of Act 3, George IV.'

Tozer had the copy away by steamer that night.

It was this Statute 6, George IV that was to cause so much anger and distrust of the law in the next few weeks of May, 1891.

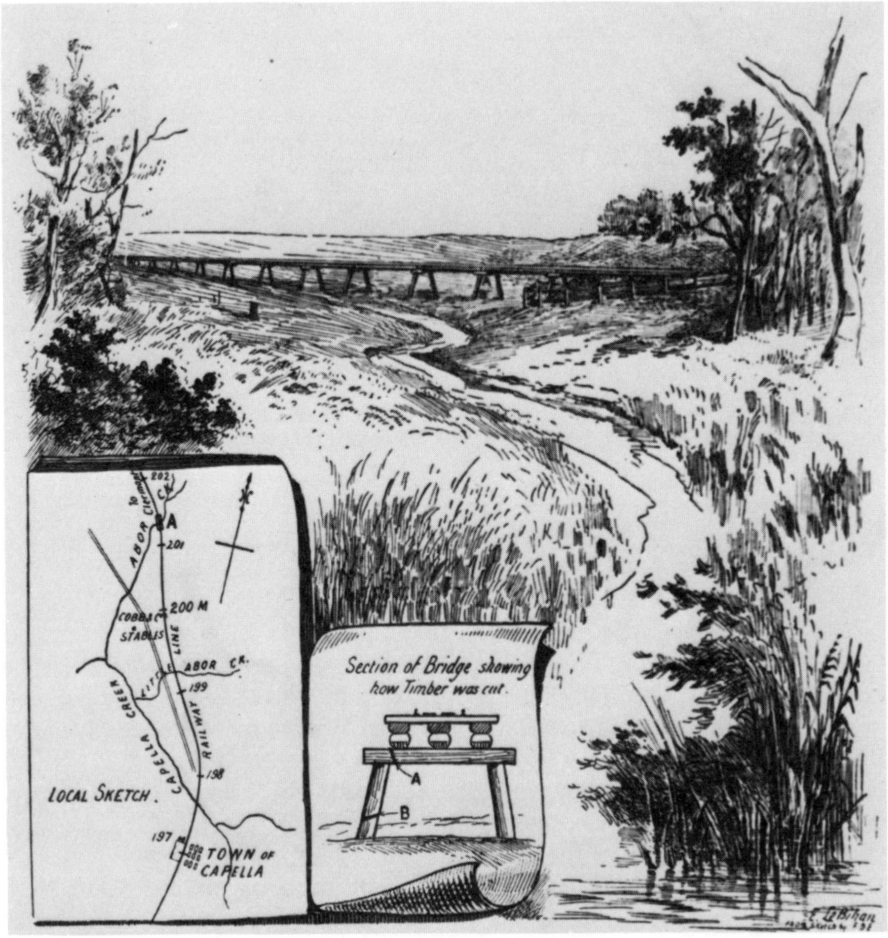

The Bridge at Abor Creek

Converting a free labourer

A rouseabout

14

As Time Rolls On

St Helena, 'the rock' as the union prisoners knew it, dated back to convict days. The chain and irons the new prisoners wore were not new to this place. (The 'dog chain' to which the men were hitched on their way to serve their three years hard labour was on exhibition in the Police Museum, Brisbane, for many years.)

'It was the brutal handcuffs . . . which were bad,' wrote factual Julian Stuart. 'They were of the old convict pattern, non-expanding, and in some cases were too small for the wrists of the men who had to wear them.

'Harry Smith-Barry was my chain-gang mate and I noticed his wrists were swollen and that the iron was biting into the flesh. We spoke to the Sergeant in charge of the escort but he could do nothing about it because Smith-Barry was wearing the largest pair in stock.

'When we were taken from Barcaldine to Rockhampton T. J. Ryan, "Tommy the Rambler" [who had once lifted 30 "scabs" from a coach], and I were yoked together and we had to take turn-about for a sleep, the other staying awake to take the weight of the chains.

'It was cold and grey and miserable arriving at The Rock. At daylight we were called on deck to have a cup of coffee after a rough passage from Rockhampton on the *Otter*, a little government boat that had tossed about like a cork in the bad weather. [That was the last coffee for three years.]

'It was rough when we left Keppel Bay and grew worse until we had to run for shelter and anchor off Burnett Head.

'We had the chains on and there was no accommodation for passengers, so we slept on the floor with a straw pillow or two among nineteen of us. Everytime she nose-dived into a wave the sea came down the companionway and the saloon was soon awash. We asked to have the chains taken off so we should all start fair if it came to a swim for it.

'Our police escort hesitated and seemed to think we might cause trouble if we were unyoked but the Captain exerted his authority and ordered them off.

While the press pictures his comrades riding free through the bush, cooking johnny cakes in the bend of the river, and meeting the trains to 'boo-hoo' and 'groan' at 'scabs', the young prisoner sat in his cell on St Helena Island for three years

'I had been on a trip or two to the South Seas and Billy Fothergill ('Bosun') to whom I was chained and I got yarning...'

They yarned on, the storm blew down, the anchors were got up and the ship got under weigh. They had thought they would go to Brisbane, but instead were 'shanghaied' direct to the penal island 'because the authorities feared that mysterious danger, a demonstration, as they had when they panicked and shanghaied us away from Rockhampton.

'Now we were placed on the... dog-chain again and put ashore. A tramline connected the jetty with the stockade, the tram being drawn by a horse, the driver was a convict, and a warder with a rifle sat at each end. There was no shortage of either men or guns about the island.

'One long-sentence man hanged himself while we were there and two were flogged, one for taking a knife to a fellow prisoner and the other, a young fellow, for heaving blue metal at a warden.

'Sunday morning was the favourite time for yarning and exchanging ideas. Bill Hamilton, the biggest of our crowd (they were almost all over six foot in height, twelve stone in weight) and Big Bill was over six foot and fifteen stone, was a splendid single-handed pitcher and when he had the flute was always sure of an appreciative audience. Bill was one of the first men in Queensland to shear 200.

'When we all met at St Helena from Hughenden, Townsville, Clermont, Rockhampton, St George, Roma and Charleville, 21 of us all told, we took stock, for some of us were strangers to one another. This was 1891. We were all determined that in serving the various periods allotted to us we should not waste the time altogether and the resolution was well carried out.'

Some coached others, some learned. 'We discovered too that there was talent among us. Paddy Griffin was an artist at step-dancing. Hughie Blackwell excelled at chess. "Happy Jack" McNamara beat all-comers at draughts. "Smiling Dan" Murphy knocked out the stockade bully one morning before breakfast – wardens and prisoners alike had stood in awe of this man; and Bill Bennett generally had the intellectual section grouped near him for he was profound and sound on logic, philosophy and mathematics.'

A matter of pride among the strikers at St Helena was that they were all first-timers: none had been in trouble before. 'It was said to be a rare thing in that place.' Within his first year on the prison island Stuart was twice approached by members of Parliament beseeching him to 'sign anything to get out boy. Tell them you were misled, only get out. There are thousands waiting outside who will think none the worse of you for doing it.' But as Stuart tried to explain, they hardly knew the complexities of the situation.

While they were in gaol William Lane, editor of the *Worker*, wrote a book about the shearers, and Stuart says that 'Lane was read with an avidity that I have never seen equalled in the Queensland bush. Everything that came from his pen was read.' (Stuart was not afraid to argue against or criticise Lane publicly and at all times was against Lane's desire to upset the social order while having nothing to replace it with. Most of all, Stuart at no time had the smallest interest in leaving the land of his birth to found a New Australia in Paraguay.)

The book, *The Workingman's Paradise*, was written around the events of 1891 but the prisoners were not permitted to have it sent to them. In time,

The men were released in November 1893 and posed for this picture ashore in Brisbane. Left to right: H. C. Smith-Barry, W. Fothergill, A. Forrester, J. A. Stuart; G. Taylor, P. F. Griffin, F. H. Murphy, H. E. Blackwell; A. J. Brown, R. Prince, W. J. Tinnett, D. Murphy. Reclining, William Hamilton (later elected to parliament, became Minister for Mines and President of the then Legislative Council of Queensland). George Taylor (second row) was later elected to the Western Australian Parliament, becoming Colonial Secretary and Speaker of the House of Assembly

however, they grew to know every word of it. 'Forrester Lewis was suffering from rheumatism and was sent to Boggo Road for special treatment and while in the infirmary got hold of a copy. He set to work and learnt it by heart. By the time the rheumatism was cured he had memorised it. When he came back to St Helena he eclipsed the exploits of Scheherazade, the lady with the record for tale-telling in The Arabian Nights. The hours in the wards, cells, work room and exercise yards that were as a rule tedious and wearisome to a degree almost unspeakable, were, for the time being, brightened by Forrester's repetition of "Billy's Book".'

Stuart wrote little about those three years lost from his young life, but four lines from Lane's book were to sweep across him 'eternally and always' as he heard them repeated by the returned convict.

> The harmony of Heaven and the music of the Spheres
> The ceaseless throb of nature and the flux and flow of years
> Are rudely punctuated by the drip of human tears
> As time rolls on.

In 1893 he left St Helena with tuberculosis. He had been accused of malingering when unable to walk, now he was to have his pick of a country gaol: he must not die for it would look bad. He remembered all his life the day he left St Helena: 'It was the day the first big batch of Labor men was returned to Parliament!' When his condition improved he was returned to St Helena to complete his sentence.

'Shearblade Martin had now finished his time and been appointed State Organiser by the Australian Labor Federation. He came to visit me with two ladies. They remarked on my peculiar diet – raw meat juice to drink and raw meat shredded very fine, to eat. "That's nothing," Shearblade told them. "He's one of those wild shearers from the Barcoo of whom you read so much. Those shearers all drink blood and eat their meat raw".'

In hospital he had been able to read. Now, back on the Rock time dragged. 'But the months were not without incident. One of the women warders used to sit on watch near the window of my cell and read aloud to the empty yard all the strike news and the labor items. I never saw her face, but often blessed her kindly voice.'

His three years hard were up in 1894 – just in time for the strike. 'I went through the Riverina and noted that the squatters were getting ready for it.' He noted something else: 'Labor militant is not the same as Labor triumphant.' When the Labor parliamentarians were asked to hold meetings in Sydney and so discourage free labourers from enrolling for the sheds 'the parliamentarians were lukewarm. I reminded them that these men the squatters are trying to crush are the men who sent you from the sheds and the coal pits to parliament. You fought with them then, why can't you fight in the Assembly the same as you could on the creeks and billabongs? Must a man lose his punch because he's got a cushioned seat?'

But now Julian Stuart must tread warily. Like the other conspiracy prisoners, his sentence of three years hard labour was followed by a bond of £100 and two sureties of like amount each that he would be of good behaviour or do an additional twelve months.

Shearblade Martin was arrested again at the start of the new strike on a charge of being a man 'of evil fame', a charge from some antediluvian statute. He was later given fourteen years on yet another charge arising from the 1894 strike.

Stuart's lungs were reluctant as he pushed through the unemployed of Sydney and yarned with Mary Cameron (later Gilmore) about going to Paraguay. 'But I had been reared on a farm and the practical side of it didn't appeal to a farmer's son.' Even though William Lane quoted poetry at him to persuade him: 'He was always crushing me with poetry because I was young.'

Eventually, he took his poor lungs to the ripe rosy air of Kalgoorlie and worked and wrote for Labor for the rest of his life, while time rolled on.

15

The Unconquerable Soul

He had the perfect curriculum vitae for an Australian labor leader. Although born on the island of Eday in the Orkneys (7 August 1846), there were few native-born Australians who could boast his background. On their arrival in Victoria in 1852, his parents went by bullock cart up into the Mallee and there, 'two sleeps from the Murray River and forty miles from the nearest settlement', the six-year-old William and his older brother, aged thirteen, began tending sheep. The boy shepherds worked for Messrs Bell & McGuinness, living in a shepherds' hut and cooking their own food after folding the sheep at night to protect them from the dingoes. One night, the hut was gutted by fire and from then on the boys 'slept out' covered by a Tarpaulin. They lived on 'post and rail tea', damper and molasses sugar supplied by dray from the homestead. Apart from the weekly arrival of the men on the dray, the lads saw only Aborigines.[1]

Later the family set off, once again on the bullock waggon, this time for 'the diggins'. They settled at Creswick, on the famous tent-dotted field with, at times, as many as 40 000 men. The goldfields had 'made such a deep impression on my youthful mind that nothing but the grave will efface it,' he later said.[2]

William was 'shovelling pipe-clay underground' on 3 December 1854 and saw the men mustering to head off to Eureka. By the time he was fourteen he held a Miners Right and with his brother followed the 'rushes in a horse-drawn dray. He had already been receiving a full share in a co-operative party of miners. He was said to be 'supple of limb, that more than accounted for the deficiency in age and strength'. He left mining for butchering for a year, then, at eighteen, went to the big mines around Ballarat 'filling all sorts of positions from tool-sharpener, trucker, to manager and director'. (There is a photograph of him from this period in the dashing uniform of the Ballarat Rifle Rangers.)

The Amalgamated Miners' Association began in 1874 with some success but fell away to 250 members by 1878. In 1874 Spence had initiated a trade union in Clunes district, and by 1878, as secretary of the Creswick Miners' Union, he led 600 members into the Amalgamated Miners' Association of Victoria. For this, both he and the President, John Sampson, were black-balled by the mine-owners. On 11 July 1882, a meeting of miners was called to discuss the threat of reduction in wages in the mines. Spence was in the chair. Aged 32, he had been organising around Creswick, building up the Association, and now was appointed general secretary. He was to remain in the post for fifteen years.

He was successful. Every miner in the district joined the Association, causing Creswick to be called 'the holy ground of unionism'. He was resourceful and energetic; he travelled to every area in the state wherever mining was carried out and he organised. After he became general secretary, he increased the sphere of activities to Broken Hill, Tasmania, Queensland and New Zealand, working up a loose federation of coal, gold, silver and copper miners with an 'intercolonial council' controlling it. (When he left the mining organisation there were 23 000 members in the federal body.) Though it was said that the body was 'moderate and conciliatory but firm on fundamentals' it had called twenty-nine strikes by 1890. At that time he was a superb negotiator. The Association became the AMA of Australasia from 1884.

He was approaching the period of the greatest challenge he, or any Australian for that matter, had attempted: the organising of the bush workers.

He might have been thought an unlikely saviour for such men. He was a leading temperance advocate, a member of the militia, a Justice of the Peace, prominent in debating societies, a borough councillor and, as always, a teetotaller. He was a Sunday School Superintendant for the Presbyterian Church of Creswick and had often preached with the Primitive Methodists and the Bible Christians. He was that man who is usually overlooked in Australian histories: the devout but liberal protestant with a total belief in the brotherhood of man.

In 1886 he was appointed Foundation President of the Amalgamated Shearers' Union of Australia. By this time his reputation as an industrial organiser of working men was well known. But it had been one thing to organise the miners; it would be another entirely to get that concept of political organisation from the goldfields to the outback. It would become one of the astounding stories not only of Australia, but of the world.

The men he was to get the message to were strung out in isolated shearing sheds across a land so huge that some of it was still being settled – over two thousand sheds, most of them in 'enemy' territory. In 1982 it is difficult and very tiring to visit the outback sheds, but we must always remember that these men travelled by cart and horse, camel or bicycle.

The men, too, were different. They were the most independent, the most rakishly audacious fellows, daring, quixotic, impetuous, with the prejudice of experience that was not equalled even by their adversaries the squatters. Any man other than that Presbyterian Scot may well have felt daunted by the prospect, but he knew that it took courage to live what seemed to others to be adventure, hardship and the life of chance of the shearer. He knew they were resolute men – they had to be if they were to survive – and his idealism admired the staunchness of their mateship.

His introduction to this new world came about through the zeal of Mrs Ellen Temple, of Creswick. Later, Spence was often to tell of how she called him 'to the work of organising the shearers'. Moved by stories of the conditions for shearers as told her by her shearer sons, she urged them to 'see Mr Spence', and in 1886 the Amalgamated Shearers' Union was born.

W. G. Spence, first president of the Amalgamated Shearers' Union. Spence, Sunday-School teacher for much of his life, confirmed teetotaller and member in 1863 of the Ballarat Rifle Rangers, believed the union was a weapon for moral change. He wrote: 'Membership of the Union makes a man a better husband, a better father, a better and more active citizen.'

This call was to the field that suited Spence best. The organisation of the bush workers was like a far-flung mission field to him, as opposed to the conservative miners' officials whom he had found to be a severe handicap to effective organising.

The bush workers were of quite a different type. Of them he wrote: 'Unionism came to the Australian bushman as a religion. It came bringing salvation from years of tyranny. It had in it that feeling of mateship which he understood already.' It was hard work organising workers scattered over half a continent and working less than half the year, but it was done thoroughly.

Forty members had been enrolled by the inaugural meeting in 1886 with prospects of an immediate enrolment of 100. The call to brotherhood was strong and magnetic. It was the most remarkable movement in Australian history. It was passionate, idealistic, exciting – and necessary. In 1909, with the AWU embracing the ASU and other groups, the *Worker* said of Spence: 'He now presides over an organisation approaching 40 000 members. In 21 years it had put an additional £4 000 000 into the pockets of bush workers. That was not gained without some sacrifice but it eloquently preaches the gospel of co-operation.' A strong believer in Labor journalism, he fought for the *Worker* which started as a small, unpretentious sheet in Wagga in 1891 (for the first two years called the *Hummer*). By 1908 it had 32 pages and some famous contributors.

'A labor paper', said Spence, 'is always organising'. Unionism for him was in the nature of a religion. 'Discontent properly directed is one of the essentials of progress. A feeling of injustice will bring about discontent in a man's mind. The man who, in the press or pulpit, denounces injustice is noble, but nobler far is the man who goes out and fights it.'

He saw unionism as something that would assist the law-makers, could change the atrophied mind of society, and he worked for a state based on justice and sound moral principle. 'To defend oneself and family is good, but to suffer in unselfish defence of others is, to my mind, the highest pinnacle man can attain.'

He was well read, as many a man from the loneliness of the bush has been, and it was said of him that: 'He who has fought the good fight of so many bitter industrial struggles for the benefit of his fellows has a most likeable face, which suggests at once the good heart he owns. There is no more courteous nor kindly man.'

The bush workers held him in great regard, but pastoralists called him the Dictator of Australia. Others said he was a Benevolent Despot, 'The keystone of the democratic arch, and the boss of all Australian states rolled into one'.

He was 'genial and quite unperturbable, standing out as the most remarkable man.' As an historian stated on 17 May 1890, 'he dominated a conference in Brisbane where the owners of Jondaryan station agreed to recognise the Queensland Shearers' Union', and this at a time when the maritime unionists had refused to handle Jondaryan wool.

It was a great victory for Spence. Up until this time few, if any, of the large properties had recognised the unions. But Spence, weary because of his heavy work load and travel, then made a series of errors. Intensifying his efforts for maximum unionism, he helped precipitate the maritime strike that began on 15 August and ended on 2 November 1890 with the workers defeated. The Riot Act had been read, police had escorted waggons of wool down Sydney streets to the wharves, and now the unionists were in rout as the wool was not only loaded, but its freight paid for by the union. Some say that Spence's 'unique mixture of inspirational socialism and hard-headed unionism evoked a confused vision of all employees in one big union but produced little understanding of the problems of city workers. Essentially, his great success with the bush workers had limited his industrial understanding'.

He was a fighter for Federation of the states, for Australia-wide associations of workers and the civilisation of capitalism. He defended the union from the International Workers of the World (the 'Wobblies') who were organising 'the

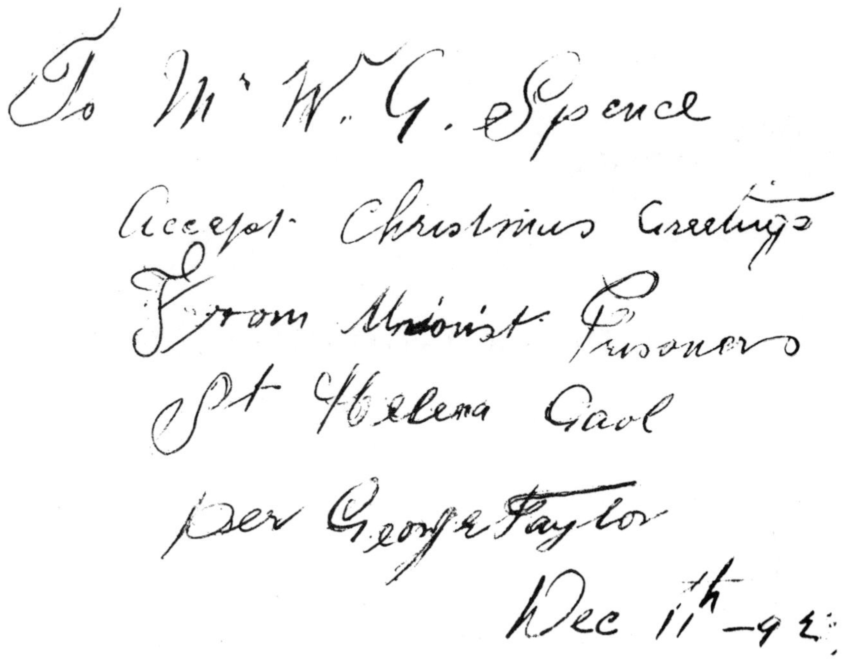

To Mr W. G. Spence
Accept Christmas Greetings
From Unionist Prisoners
St Helena Gaol
per George Taylor
Dec 11th -92

Union prisoners on St Helena in 1892 sent their leader a card: 'Here courage, strength and love you see, Symbols of what I feel for thee' was printed on the front. George Taylor delivered it

one big union'. He saw the union as the industrial wing of the Labor Party, stating as he did, that the organisation of the bush workers and the strike of 1890 marked the foundation of industrial trade unionism in this country.

Later, in 1898, he took the seat of Labor in the NSW Legislative Assembly, pressing for Federation, using the example of the disparate unions forming into the AWU as an illustration of strength from uniting. In 1901, after Federation, he was elected as member for Darling and was Commonwealth Postmaster-General in 1914–15. In 1916–17, when Vice-President of the Executive Council, he was 'tricked by Hughes' (according to Labor men) into voting for conscription. In deference to his past achievements he was permitted to resign, the only member of the union not to be expelled for his action.

Rejected by the Labor Party, he lost his seat but was re-elected as a Nationalist candidate. Prime Minister W. M. Hughes, writing to him after the 1919 elections when he failed to be re-elected, said: 'I know your indomitable spirit too well to think you are despondent. Yours is the most serene, the most optimistic yet the most resolute and unconquerable soul I have ever known.'[3]

As much as any man he had prepared the way for Labor in politics. He had seen the first representatives of the working class elected to the Australian Parliament, and the first Labor Party in the English speaking world to take office.

The man who 'attempted and accomplished the apparently impossible' was aged 73 when he left Parliament, and he died seven years later in 1926.

16

The Band of Unhappy Amateurs

They were days that come only once to the men of a nation. There were pure and clean-cut goals to strive for. Manhood Suffrage was one. And any who think that this was no momentous thing have never seen the framed certificate on the wall of an old bush worker: '. . . [So-and-so] is entitled to vote by reason of Residence and Manhood.' These certificates are still to be found today, ninety years after the battle to vote had been won by itinerant workers such as shearers and seamen and others whose work caused them to be known as having 'no fixed abode'.

'Rise ye! Rise ye! Our Glorious Dawn is Breaking!' sang Henry Lawson to the bush workers. Plural voting had been ousted. 'One Man, One Vote, We're waking from our Slumbers.' There was now one general roll on the basis of Manhood Suffrage. And they asked for more: free, compulsory and technical education to be extended to all; eight hours to be a legal maximum working day in all occupations; repeal of the Masters and Servants Act; elective magistrates.

It was like the Holy Grail and the Golden Fleece all wrapped up together. 'Toilers will not bend or bow!' They had managed to serve God and Mammon in one wild endeavour.

For the first six months of the year they had kept the newspapers across the country in fits of excitement with strike stories of burning, rioting, kidnapping, train wrecking 'and every form of intimidation'. The country had rung with cries of 'Freedom of Contract' and 'Fight the Union! Crush them!' Queensland was overrun with a far greater ratio of troops and police to shearers than England was using to suppress the mutineers in India. And by June, the great bush battle had been lost, the union put to rout, Freedom of Contract was rampant, and unionists were unable to get a pen in any shed where their faces were recognised, no matter what name they used. They had lost in all ways, it seemed, because they had earned no money during the strike, there were no jobs for them and the best of their mates and leaders were in prison.

Yet they seemed not to recognise these signs of total defeat, for they rejoiced at what young Julian Stuart remembered of the day he was carried by stretcher back to St Helena Island prison: 'I'll never forget it. It was the day the Labor candidates got into parliament!' To everyone's surprise, including the Labor men (it was not yet a party), 36 Labor men were voted in to the NSW Legislative Assembly on 11 July 1891.

They had no firm policy, no structure, no financial security (although payment of politicians had now been introduced) and no organised backing among the scattered bush community. It was sudden, dramatic, and unexpected. It transformed the political outlook of the whole continent. The working man had seen the power of unionism to protect himself at the source of law and authority.

In May 1891, South Australia elected three Labor men to the Legislative Council and eight were elected to the Assembly in April 1892; Western Australia had eight men pledged as Labor at the elections in April 1901, followed by twenty-four members in June 1904. Tasmania elected four Labor men in 1903 and seven in 1906. Queensland elected sixteen men in 1893 and Victoria, ten in 1892. There was a scent of Parliamentary democracy to encourage them.

'We were a band of unhappy amateurs,' one of them later wrote in the *Bulletin*. Only two of the men were shearers: others were bricklayers, miners, saddlers, seamen, a stock and station agent and a medico who is listed as a 'free trader'. Ten only were native born, the great majority (18) being born in England, 2 from Scotland as well as the USA, 1 each from New Zealand, France and Ireland. There were 14 Anglicans, 6 Methodists, 5 Presbyterians, 5 Roman Catholics, 3 Congregationalists; 1 Church of Christ and 1, Denahey, had the courage to put a dash in the column for religion. C. D. Clark was a Temperance journalist and the New Zealander, A. Rae, had arrived in Australia as recently as 1888.

If these men were 'a band of unhappy amateurs', one at least of them did something about it. Tommy Ryan was gaoled in 1891 with Julian Stuart and after his sentence expired in 1892 was elected MLA for Barcoo. But when the time came for him to seek re-election in 1893 to the Queensland Parliament, he refused and is reported as saying: 'The friends are too warm, the whisky too strong and the seats too soft for Tommy Ryan. His place is out among the shearers on the billabongs.'[1]

The election of those first men in 1891 was not only a beacon for the bruised and beaten shearers that year, it was a constellation that kept their spirits soaring and their confidence rising in readiness for the next battle which they knew must come: for they had not only begun a crusade they wholly believed in, but they now had nothing to lose.

The country was reeling from the effects of strikes and lockouts of a severity scarcely surpassed anywhere else in the world, but they had headed towards what democratic government is meant to be: at least two parties, one in government, the other as strong opposition to represent all manner of people.

Strangely, it is always overlooked that the origins of a third party, the Country Party, may be traced back to the same struggle that saw the formation of the Australian Labor Party. In March, 1891, pastoralists claimed that the growers needed to form 'a powerful country party' because of the need to resist 'the wild theories not unmixed with miserable envy' of Labor unionists.[2]

17

The Scab

May the Lord above – send down a dove
With wings as sharp as razors –
To cut the throats of bloody scabs –
Who cut down poor men's wages.

The 1890s left a legacy of bitterness that this country may well have avoided. But once having crossed into such territory there was no turning back. The gall spilled on friend and foe alike.

One of the legacies is the anthology of verses about scabs, blacklegs, one-time-mates-who-now-are-more-loathed-than-the-employer-himself.

A scabby form lay rotting
Upon a filthy bed,
It smelt a danmed sight nicer
For the weeks it had been dead.

There's no maggots on his carcase,
There's no worms to crawl and feast –
The blow-flies turn their noses up,
They will not blow this beast.

So what's the use of scabbing,
He cannot go to hell –
The maggots and the blowflies
Have blackened him as well.

Merv Lilley believes the poem was used during the 1890s when the shearers refused to sign the pastoralists' contract at the Fairbairn family's Logan Downs station. The Fairbairns then imported 'free' men (or 'scabs', whichever way one views it) and the trouble began.

Spence believed that the first organised effort to introduce these men into the west of New South Wales had been in 1891. Mr Nutting, of Fort Bourke

The *Bulletin* portrayed strikers as living on rabbit meat while non-union men shore under police protection

station was in charge of them. About seventy members of the Amalgamated Shearers' Union met them and formed two lines between which the scabs were to pass. It was agreed that moral suasion should be used as they passed along to where they were to cross the river; it was also understood that they were not to reach the station, about four miles down the river on the opposite bank.

Bourke then had twelve horse cabs, and eleven owners out of the twelve had volunteered to help cart away the swags of those who consented to join the unionists. Before the crowd had got half-way to the bank, the unionists had been largely added to and very soon scabs and unionists were all mixed up, the cabs busily carting away swags and scabs as well.

When the leader and his remaining scabs reached the river crossing at the entrance to the punt, Mr Nutting suddenly drew a pick-handle from under his coat and raised it to strike a unionist. But a blow from someone's fist went straight to his jaw and he fell, unconscious. A loaded revolver dropped out of his pocket. The scabs were all marched to camp. The six constables on the ground had enough sense not to interfere, and perhaps saw no occasion to do so.

'Mostly the alleged capture and forcible abduction of "freemen"... was in reality carried out in such a way that moral suasion was not exceeded.'

There was one reported shooting. Unionists wanting to get scabs to leave Ti-Tree station had had such a long ride to get there that their horses were knocked up. Mr Head of Waterloo station was sympathetic and lent them horses, although he knew their game. On their way to the hut at Ti-Tree to interview the men, a shot was fired. No-one knew by whom, but it is believed to have been by a constable. One of the unionists who had previously been noticed all of a tremble, fell down out of nervous fright when the shot rang out. The press next day reported the man as having been shot.

'It was often the way things were done that made for success without risk of being arrested for breaking the law. In 1889 there was a big camp of unionists at Walcha, NSW. One day it was learned that a large number of scabs were to be brought through by coaches. There was a considerable body of police in the town under an Inspector, and when the coaches were coming these were ranged up each side of the road, the unionists crowding closely behind their backs.

'Gentlemen' soldiers were led a merry dance through the bush by experienced shearers and added little but colour to the 'Shearers War'

'Tommy Ryan, a well-known character among the unionists, squeezed his way inside the ranks of the police without being interfered with. Ryan was a big, powerful man, very strong in the arms, and so simple and good-natured that he was a general favourite. The coach containing the scabs was cunningly driven close behind the mail coach, with the view of deceiving the Unionists. The squatters in charge of the scabs depended on the police for getting safely through, but the crowd became so great that the coaches had to stop in the centre of the two lines. No sooner did the coach stop than Ryan calmly lifted the scabs out one by one, just as you would lift a child, saying at the same time to the police, "We won't hurt 'em; but they must come out you know."

'No sooner had each surprised scab landed on the ground than he was hustled out amongst the unionists and lost. He was so bewildered at what had happened to him that he gave no trouble. In less time than it takes to tell, the coaches were empty.

'That day, before the coaches arrived, a young constable, fresh from Sydney, drew his revolver and pointed it at Ryan, but Tommy coolly put his clenched first alongside the muzzle of the revolver and said, "That is my weapon." The inspector at once ordered the constable to put away the revolver and threatened to send him back to the city should he dare do such a thing again without orders.

'In all, about thirty scabs were lifted that day, and the man who had engaged them lost his money and his men.'

These non-union men were being brought north from distant places at considerable expense and trouble but the unionists appear to have been their match. A team of 202 men were sent by ship from Melbourne to Queensland in 1891 and three members of the Wagga Wagga branch of the Shearers' Union smuggled themselves down to Melbourne and onto the ship. It wasn't the best of voyages. There was not enough food to feed the men on board and Burgess, Barron and Belmont suffered with the rest, but began to organise almost as soon as the lines were cast off. None of the men were allowed ashore at Brisbane, and when the ship arrived at Rockhampton the train was run up alongside the ship and surrounded by military. Burgess managed to escape. After their train moved out he took the next train to Barcaldine where he gathered together a group of shearers and rode to Rodney Downs. Others went to other stations where the non-union men had been drafted.

Those 'three staunch men' on the boat, persuaded 150 of the 202 on board to come away to the camp. 'It must have cost the Pastoralists' Union at least £500 to take the crowd to Queensland... Similar dodges were resorted to through every strike and whenever possible a good organiser travelled with the scab crowd [and] – generally demoralised them from a pastoralist's point of view.

'Amongst record rides none can beat that of one of the leaders in the Queensland trouble, who rode one horse 86 miles in six and a half hours. It saved him from gaol, as his proven presence so far away from where the disturbance took place six hours before was accepted as an alibi by the justices.'

At the Bourke camp a leader known as Number One had a picked party to assist in 'rousing out scabs'. (These men were known by number only to avoid names being quoted after a raid.) They are said to have had at least one narrow escape when they opened a door of a hut and a constable stationed inside fired – and missed.

Another group of unionists were sleeping out in the bush when, just at daybreak, a body of police who were pursuing 'free men' kidnappers levelled their rifles at them and shouted to surrender. 'One of the sleepers, who was nearest to the muzzles of the rifles, when awakened, raised himself on his elbow and calmly opened his shirt front at the breast, and told the police, "Shoot away, you bloody bastards."'

The shearers' strikes were conducted very much in military style and capturing the enemy was a part of the strategy. When the Queensland men were hard pressed a message came to Bourke (NSW) from St George (Qld) asking for some volunteers to 'lend a hand'. Donald McDonell, a youth six foot 2 inches tall, was one of the first to volunteer, and with a dozen others rode off. The Bogan was in flood and a mile wide. They had to swim. They found a boat up the river but no paddles. One horse could not be induced to face the water so he had to be hauled across behind the boat [which was] drawn by a cable of fencing wire. It took from two in the afternoon until eight at night to get all hands over. He rode on, over the border, picking up volunteers, until they reached St George twenty strong. One of their first tasks was to interview the non-union men at Noondoo station. Police were patrolling so the men could not go by road. Instead they rode across country, cutting those fences which could not be jumped. They waited until midnight to go to the hut, leaving their horses tethered behind them. Some of the unionists had shorn at this station and were familiar with the interior plan of the shearers' hut. They also knew that one of the non-union men was a large, powerful bully who slept with a loaded revolver under his pillow. It was decided that Donald, and another tall, hefty shearer named Driscoll who knew the bunk the bully slept in, should rush the hut and disarm the big fellow.

At midnight they raided – and found the bunk empty. They knew immediately what they were up against. Word of the unionists' raiding party had been sent to Noondoo and the manager had sworn in seventeen special constables . On the unionists' arrival he read the Riot Act.

'No notice was taken of... this however as a general argument [resulted] in the scabs being morally persuaded to leave with the unionists. This made the party a very strong one, almost sixty men.

'They rode away towards Doondi, another station belonging to the same owners.... as the first of the cavalcade rode along the road they heard that yes, shearing was under way – with scabs.

'They at once made for the shed and invaded the hut, some of the scabs from Noondoo being the roughest in dealing with the Doondi scabs.

'One man was sitting at the table quietly cutting an apple with his pocket knife... big Jack Driscoll put his hand on his arm to ask him what he was going to do when the man suddenly made a dash at Driscoll with the knife and drew blood. In a few seconds the man was thrown outside. One of the converts from Noondoo had lifted a nail-can and bashed the apple-eater on the head with it, cutting him severely.

'Amongst those in the hut was a constable in plain clothes... to identify men... when the raiders had returned to St George word came that they in their turn were to be raided by police.

'A number of the more active quickly went to look for their horses – before daylight. Donald McDonell did not, although he had been one of the most

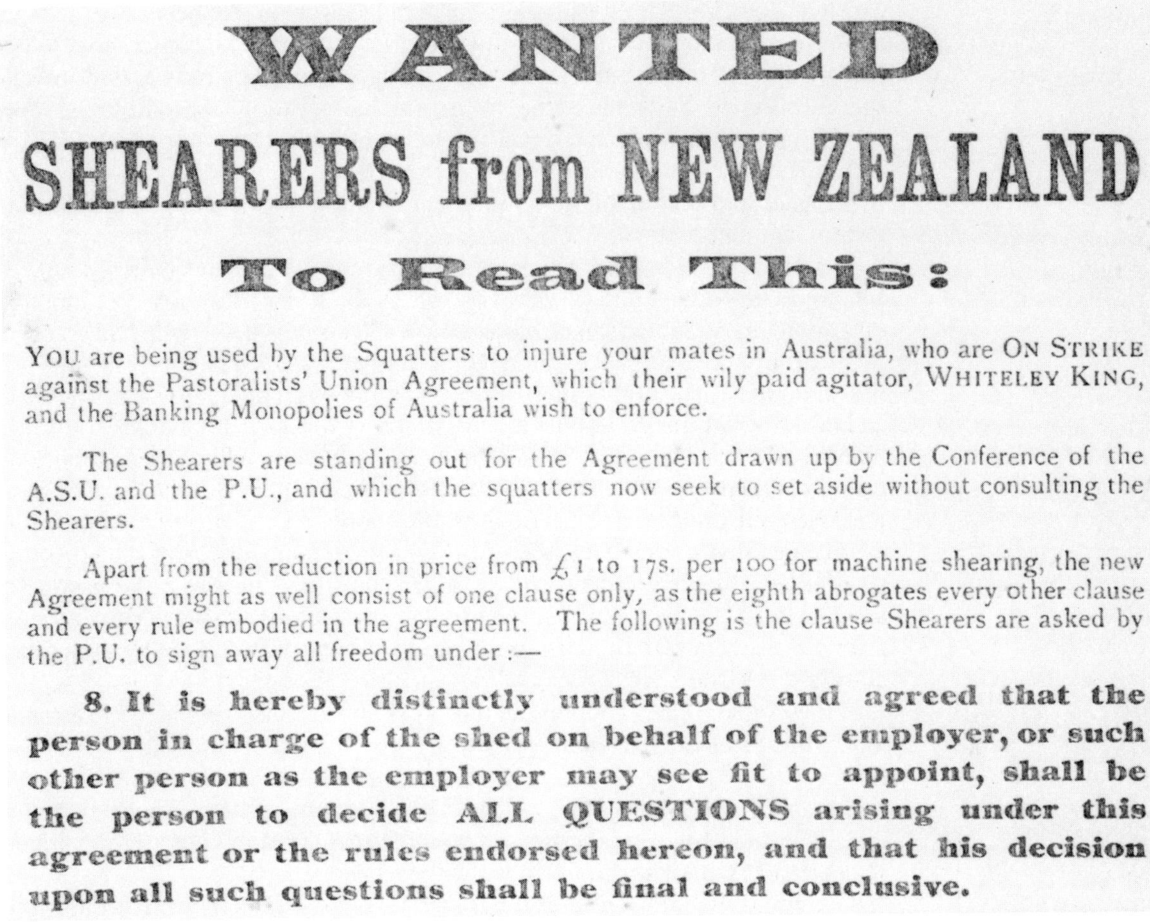

WANTED

SHEARERS from NEW ZEALAND

To Read This:

YOU are being used by the Squatters to injure your mates in Australia, who are ON STRIKE against the Pastoralists' Union Agreement, which their wily paid agitator, WHITELEY KING, and the Banking Monopolies of Australia wish to enforce.

The Shearers are standing out for the Agreement drawn up by the Conference of the A.S.U. and the P.U., and which the squatters now seek to set aside without consulting the Shearers.

Apart from the reduction in price from £1 to 17s. per 100 for machine shearing, the new Agreement might as well consist of one clause only, as the eighth abrogates every other clause and every rule embodied in the agreement. The following is the clause Shearers are asked by the P.U. to sign away all freedom under :—

8. It is hereby distinctly understood and agreed that the person in charge of the shed on behalf of the employer, or such other person as the employer may see fit to appoint, shall be the person to decide ALL QUESTIONS arising under this agreement or the rules endorsed hereon, and that his decision upon all such questions shall be final and conclusive.

Part of a broadsheet printed by the *Worker*

obvious at the hut by his speech and actions. He lay in his tent asleep while the raid was made. When the noise awakened him, he sat up, and was about to pull on his trousers when the flap of the tent was thrown open and the inspector with the constable who had been on duty at Doondi looked in. The constable was asked by the inspector if he knew this man. The constable looked at Donald and said, no, he did not. The flap was dropped and Donald escaped at least three months imprisonment.

'It was always said that had he stood up, his height would have identified him as he stood a head taller than other men. Donald McDonell later became secretary of the Bourke branch of the AWU, and later still General Secretary, a member of State parliament and the Labor Cabinet of New South Wales.

L. T. Maher was a well-known converter to the cause. 'One of the quietest and best behaved of men. Once he was chased with a waddy by the Hon. George Lee, MLC, and the general opinion was that of all men, Maher never provoked such treatment as he met with at Merriwee. At Combelong station the boss had set a boar-hound on him and he had to leave there hastily also. But he returned the following evening – disguised as a swagman. He was sitting on his swag outside the hut when the boss rode up and asked him if he wanted a job. He said yes, and was engaged as a sheep brander.

'That night, in the men's hut, he whispered to the rep., the bell was rung, and the boys called together to listen to an address from the "Sheep brander". He had gone by morning, but after cut-out five days later met all hands in Bombala and enrolled the lot in the union. When the boss of Combelong heard of it he would not give any of them a pen the following year.'

And so the stories go on. Perhaps their retelling made worthwhile the long, tiring rides, the beating of the heart when the hut door was flung open, the surprise when met by a constable's fire-arm.

Certainly the raids were still going on when the new century began. In 1902 at Coree and Widgiewa there were two camps about five miles apart containing, in all, roughly 500 men. 'Some were returned "contingenters" who had scabbed it at Carney,' wrote Spence, referring to the Boer War. 'They went in and the boss, Horsfall, had another mob of scabs coming by train. There were 400 men at Narrandera railway station to give them a welcome [as Spence puts it]. There had been two police at Widgiewa but Horsfall had got twelve more at his own expense to shepherd the thirty-two men coming on the train. These men were all in one carriage. Leaving the train standing at the platform, the engine ran this carriage out along the line to a point opposite the shearing shed. The Unionists had two miles to run, but they cut across the paddocks as quickly as possible and managed to capture six of the scabs. Horsfall had a vehicle ready to take the scabs on to the shed but it was delayed by somebody cutting the harness etc.

'Cameron, the men's representative, was summoned but got off. Horsfall took steps to claim £5000 damages from the union but was glad to settle for £250.'

It may all sound like a *Boys Own Paper* from this distance, but when one multiplies the happenings across and down the eastern states it is as near a scene of chaos and gun-less warfare as we will ever see.

Sometimes the raids failed, but mostly they bore fruit, and even if it appears to us now as being in some ways a clumsy manoeuvre, it was one of the few exciting things to happen to young men in those drear years, whether they were union men or scabs.

In his memoirs Spence spoke of blacklegs: 'We put up big posters in coloured ink all over the country appealing to the traveller by the wayside to stand by the Union. Members who blacklegged were warned that there was provision of a fine up to £10 which would be imposed.

'The names of men who had blacklegged were published in the *Shearers' Record*, our then official organ, and a good many fines of £10 were collected, as well as entrance fees of 50/-. We made the union a terror to evil-doers, whether they were squatters or shearers.

'Fernlees started with Chinese. Mount Abundance started with scabs under police protection. Barcaldine, Bimerah, Beaconsfield, Wolfang Downs, Bowen Downs, Northampton and Minnie Downs all started with scab labour.'

18

The White Wings

Whitely King, secretary of the Pastoralists' Union, unwittingly lent his name to a whole army of men when the 'free' labourers became known as 'White Wings.'

By 1894 there were many of them. The bank crashes of that decade threw unemployed into the maelstrom and the great outback seemed attractive to them. The vast numbers of young men looking for work, at any wage, was partly why the squatters in 1894 repudiated the agreement they had so solemnly made in 1891.

'I am directed', Whitely King wrote to the shearers of the AWU, on 14 April 1894, 'to inform you that the Pastoralists' Federal Council on behalf of the Federated Association of New South Wales Victoria and South Australia, having found the shearing agreement made at the conference held in August 1891 unsatisfactory, this Union hereby gives notice from this date its members will discontinue working under the said agreement.'*

The most objectionable alteration to the shearing agreement lay in clause eight: 'It is hereby distinctly understood and agreed that the person in charge of the shed, on behalf of the employer, shall be the person to decide all questions arising under this agreement and that his decision upon all questions shall be final and conclusive.'

'They would be at the absolute mercy of the boss!' Spence cried. As well, there would be a reduction of 3s 0d per hundred sheep. But what most rankled was the power the squatters had assumed to thrust men back to past days when, whatever their complaints, the owner was the sole judge. And now, they not only had distant memory to remind them how this power would be used, but knew how readily it would be used after the events of 1891.

The union tried to prevent war; they appealed to the pastoralists, to the Conciliation Court, to the Premiers, to banking institutions. But all in vain.

'It was a battle royal. It was a straight out contest in which there was no place

*Only 3 states were party to this decision.

for compromise and no quarter would be given,' wrote Spence. It was of course a straight-out bid by the pastoralists to kill the union once and for all.

The fight started in July, and by the middle of August thousands of men were on strike and had to be maintained in the various camps. There was a camp at The Springs, eight miles from Llandillo; there was a big camp at Hillston by 15 August and another at Ivanhoe. Gingie men refused to sign PU and so did 150 men at Willandra. There was a camp with 800 men at Wilcannia; Weilmoringle men struck on 3 August and stood out courageously for ten weeks without cost to the union. There was a big camp on the 70 Mile near Marfield station and another at Forbes. New South Wales was now finding – as Queensland camps had found in 1891 – that all sorts added to the legitimate numbers they had to feed. In the camp at Mossgiel there were over 100 'scabs'.

'All sorts of bipeds were raked up by the P.U. agents. We watched wharves for arrival of boats, and daily shepherded labour and P.U. officers so as to intercept the unfortunate creatures they were alluring into their net.'

It sounds a very lively and hectic undertaking, to say nothing of the colossal difficulties of running their camps with this motley mixture of men thrust together. There were good and bad, old and young, unionists and 'free' all camped together, and any outbreak of violence was watched for by police to enable them to close the camps down. Some camps were closed, but there was never a case of such a closure being made on legitimate grounds.

'All peace and calm,' was how A. A. Dangar described 1893 when, as president of the Pastoralists' Union, he addressed the 1894 meeting before the beginning of the shearing season. 'But now labour is endeavouring to bring about a suicidal strike.' He does not state that the Pastoralists' Union had reneged on the 1891 agreement by lowering the shearing rate for 1894, but he was aware that many wool-growers were unhappy about breaking a gentlemen's agreement.

He put forward an excuse for the decrease in the 1894 shearing rate. 'Many of the members have incurred expenses putting in machines, and as shearers who use the machines earn more money than those on hand shears . . . the rate has been reduced to 17s 0d for machines, and kept at 20s 0d for hand shears.'

Dangar spoke heatedly 'of the vile charge . . . that pastoralists place secret signs on discharge certificates issued to Shearers' Union members to prevent their being taken on elsewhere.' (Julian Stuart had had to change his name three times to get pens because his name had gone on ahead of him or was in some way marked on his discharge.)

'We are not supposed to have anything to do with politics, but I . . . suggest to shearers that they . . . not try to kill the goose that lays the golden egg for, depend upon it, while that goose is being killed, and the violent portion of the labor party exalted, so will capital become scarcer, and the number of unemployed will increase . . .'

The shearers already believed the squatters had been vindictive after 1891. The Wagga Branch of the General Labourers' Union of Australasia, claimed in May 1892 that hundreds of men were so destitute they were happy to work for a miserable wage just to get tucker. They believed that 'this wholesale discharging of men all winter is part of the great conspiracy to wipe out Australian unionism. Be that as it may, there are three men anxious to obtain the work offering for one and consequently a reduction in wages is the outcome'. They now decided to amalgamate all the Bush unions, shearers and labourers, under the name of the Australian Workers' Union.

Each group was as truculent as the other, each refused to give any ground, each viciously denigrated the other, scoring points as if it were a great game. Inside at conference and outside on the battle ground of the stations, each side thrilled to each petty victory. To this day some gleefully tell of those chases and rides and hold-ups as though they were extensions of school boy pranks instead of events that changed the social and industrial way of a nation.

While the bush workers embraced unionism with the fervour of religious clinging to the cross, the pastoralists surveyed the gloomy economic prospects. Seventy-two of those who joined the Pastoralists' Union in its first year were in severe financial straits. It is believed that half – or more – of the sheep farmers of New South Wales were mortgaged to banks or finance companies. For instance, William Alison, who with G. N. Griffiths called the inaugural meeting of the union had, in 1889, applied to Goldsborough Mort & Company for an advance of £175 000 on the security of his two stations, Canobar and Meryule. By 30 January 1891, that firm was preparing to auction Canobar.

The shearers had had one hundred years to distrust and fear the squatters and smart under their arrogance. Now the squatters were afraid. 'The question is,' George Mair wrote to A. A. Dangar, 'whether we shall preserve to ourselves the control of our business, or put ourselves under the control of our workmen.' Then he added a surprising thing: 'A shilling or two per 100 sheep would make no great difference in the annual balance sheet.'

But they didn't pay the shilling or two. Instead they lowered the rate, and the shearers launched such a furious attack in return that the air thrummed with the anarchy and destruction of it.

Strike camps began again; food was provided for hundreds of men

Whitely King had been most successful in New Zealand and a batch of 'free' men was recruited; 110 being put on the *Haurato* for Australia. But Mrs Garmson, the New Zealand union secretary, slipped on board and hid until the ship left Lyttleton. On the Sunday, she emerged, and though many of the men were sea-sick she mustered them and 'addressed them on the foolishness of their fighting against their best interests'. 'A little boy of six, kept calling them "blacklegs" and they could not stand it.' Of the 75 who left at Wellington, two long wanted by the police were arrested and Mrs Garmson travelled with the rest to Sydney, 'sowing the seed of union ideas and principles.' On arrival thirteen tried to escape but Whitely King had his organisation ready and all were shuttled out to country sheds.

Once again, as in 1891, 'free' labourers were 'encouraged' to leave the sheds and camps were filled to overflowing. In some ways it was more bitter than the earlier strike. There were the same long, hard, midnight rides, the escapes and the challenges, the monster meetings and the clandestine gatherings. But overall, the lines that had been drawn only four years before were now seen to be finite; the old days would never come again.

The rivers were used in a way not done in 1891 to transport 'free labourers' by paddle steamers. Many of the stations were on the river, their sheds right on the bank for loading wool bales into the fleets of steamers. For half a century they were the transport for the wool-growers, and it was simple to bring workmen down this way, too.

Gundabooka received 20 men by river, 'huddled like sheep in the hold of the steamer'. Later, in an attempt to see if another steamer had brought more men, George Carter, Camp Commandant, led his men along the river bank in darkness. They were blocked by a pumping plant and, on swarming the bank above it, came on twenty-five armed police, one of whom asked the officer-in-charge whether he should shoot.

Police were plentiful on the river that year. With a view to blocking the steamer bringing free men to Killara, Carter and his men used a waggon wheel to spin a rope of eight strands of number eight fencing wire which they then stretched across the river by the aid of a Spanish windlass on two trees. After all their work and ingenuity, the rope failed to hold the steamer as it swept past with a load of police and their charges. Shearers from Killara camp later crossed the river and 'yarded' the 'free men' into a hut where the police were camped.

Two men – believed not to have been involved – were sentenced to seven and four years gaol respectively for that 'riot'; Carter was forced to change his name to 'Ned Williams' and head south while police expected him to be riding northwards. Before dawn broke 'Ned Williams' had left Killara camp for Mossgiel. Here he called for thirty volunteers and started for Mulurulu, ninety miles away and shearing 'scab'. Two attempts to get the men out had each failed.

'Ned' and his party reached Manfred the second night and, as they rode up, saw a posse of twenty-five police ride to the homestead. He let it be known that he and his men were off to Till Till in the morning and they turned their horses out. But not too far out.

By midnight they were in the saddle again and before daybreak they let their horses go in the house-paddock at Mulurulu. The horse boy saw them and attempted to make a dash for the huts to sound the alarm, but they captured him

As a speedy messenger service the Bicycle Corps was formed

and made him guide them to the shed by a road hidden from the homestead.

As day broke they rushed the shearers' hut, rounded up the men and set off for Mossgiel camp with the men's gear on a waggonette. But there had been a slight flurry: the manager had fired a revolver and the shot went through Ned's coat. Ned (or Carter) grabbed the firearm and stunned the manager with its butt.

They rode hard then. A posse of police pulled them up but Ned persuaded the sergeant they were from another camp. They spent the night at Manfred and police surrounded the camp. The men piled sacks of potatoes over the wanted man and he was saved again.

He rode on to Ivanhoe camp with two of his mates, then towards Hillston, but hearing that the police were on his tracks turned towards Willandra. On the way he met two constables who had lit a fire and were making tea, so he had a mug of tea with them.

It is an ineffectual enough story, but multiplied by the scores of similar adventures going on at the same time it gives a small idea of the chaos spread over the huge area affected by the strikes and the excitement and dashing adventures which fed them.

Perhaps the most famous event that took place in 1894 was the burning of the *Rodney*. 'Forty-five scabs were being taken to Tolarno by the steamer *Rodney*,' William Spence wrote. 'It was moored to a gum tree on an island near Morara, when a dozen men waded through mud during the night of 6 August 1894 and captured the lot. The boat was burned, the barge being cut adrift. S. C. Robinson, James Fox, James Pender, W. Duncan, W. Goadie, J. Evans, R. Studholme and T. McDonald were arrested and tried for burning the *Rodney*. Thirty witnesses came forward and proved an alibi and the accused were let off.

'A second lot were run in and found not guilty, and eventually they arrested Thomas Bonner for the offence and gave him seven years. He was not within 100 miles of the steamer when it was fired, but that did not matter to the authorities.'

Captain Pickhills told C. E. W. Bean 'I came along a few days after the *Rodney* sank,' he said. 'At that time I was master of the *Jandra*. When we saw the *Rodney* there was just her boiler and paddle-wheels sticking up – that was all that was left. The shearers burnt her; but they never interfered with me – they never would with anyone that was not meddling with them, and I had a good name along the river . . . that was what happened to the *Rodney*.

'It seems impossible today to find out exactly how it happened . . . The man most likely to know told me he was entirely in the dark. "The *Rodney* was coming up-river during the strike with two hundred shearers, free labourers, on a barge," he said. "They were waited for where the river runs through Tolarno run, at an awkward corner where the steamer has to go close to the bank. About two hundred men were on the bank waiting for her. How they got on board is a bit of a mystery; anyway they turned her nose to the bank and drove her ashore and set fire to her. Three men were arrested, but there was not a bit of evidence, and they were acquitted".'

Hundreds of men were arrested as the strike went on. Seventy-four were arrested and marched into Wilcannia Court on the one day on a charge of damaging property at Mt Napier Well. Eventually 22 were tried at Broken Hill, after 110 jurymen were called (69 had been challenged by the accused, 49 by the Crown). Four were found guilty of riot and sentenced to seven years imprisonment.

Inducing free labourers to leave their work usually earned a charge of 'riot' and there were many cases of this. Some differed a little from other such inducements. 'Twelve scabs were stolen from Salisbury Downs and taken into the hills of Coonamberry [and] . . . lodged in a cave . . . they got away. The police tracked ten of the unionists down and they were fined £10 each. Times were so lively that it is recorded that thirteen jaws were broken in one day at Bourke.

Police were still being ordered to escort non-union men and to ride with non-union wool teams. These mounted police were undoubtedly the men most hard-pressed by the uprisings

'Grassmere Shed had started with a bad lot of scabs and a number of unionists went out to interview them,' wrote Spence. 'On nearing the shed they were fired upon and William McLean and John Murphy were shot, the former in the chest, the shot piercing the lung; the latter in the shoulder. They were shot by a crawler named Baker. His action was highly approved by Judge Stephen, and the President of the Pastoralists' Union sent him a token of approval while it is said that he was paid £70 to help him get out of the way later on. He was reported as having been seen in Victoria and was last heard of in Tasmania.

'McLean was as fine a young man as ever lived and he died from the effects of the wound, together with the coldness of his prison cell. Nothing in the nature of self defence can be set up as an excuse for Baker, who to all intents and purposes murdered McLean. W. McLean, J. Vergin, J. Murphy, J. Jones, H. Cramak, Hector Osborne, Albert Herr and Alfred Montgomery were charged with riot and assault at the Broken Hill Session on 17 October before Justice Stephen and all sent to prison for three years.

'It nearly came to civil war at Dagworth Shed. It was held by police and scabs and to protect themselves they had the sheep mustered and allowed to run around the shed during the night. It was thus impossible to get into the shed without the sheep giving the alarm. When the unionists attempted it, the police fired on them, and about forty shots were exchanged. Whilst the firing was going on some daring spirits got around the back and presently the shed was ablaze, whether by accident or design is left a matter of opinion. It stopped the fight and the shearing too.'[1]

A coach load of free labourers was travelling to Calga and on the way had put

The *Rodney* at Echuca wharf, prior to being burnt up-river by an unidentified group of shearers

up at the Telegraph hotel, Gilgandra. 'A very fine young woman who was working there gave them such a dose of jalap in their tucker that they were unable to work for a week after. Her name was Martha Davis.'

On both sides the edges were blurred. Some pastoralists disliked the way things had been done, and some managers for the big companies were solidly on the side of the shearers.

'Marfield station, owned by the Australian Mortgage Land and Finance Company, was managed by John Cameron. He wanted to shear Union but the company would not permit it. The men went into camp on the stock route near the Gypsum Palace hotel but the station sent supplies to them until the directors heard and stopped it. After nine weeks, the boys were delighted when Cameron came galloping into the camp, waving his hat and shouting, "Marfield has given in!".'

All through the second half of the year the ferocity continued. The Messrs Pegler at Milo saw their scab labour stoned by unionists. Dagworth shed was burned, as was Ayrshire Downs and a number of others. Unionists were once again using the desperate measures of 1891.

Some of the woolsheds burned were old, some new and costly. The value of sheds burnt between June and November 1894 was estimated at:

Redcliffe	£1000
Manuka	£2000
Cambridge Downs	£5000
Ayrshire Downs	£5000
Casillis	£1500
Dagworth	£ 500
Pullies, horses, sundries	£12.10.0
	£15 012 10s 0d

'Value of telephone wires cut at Oondooroo, fencing and gates at Manuka and value of Warrego shed not known.'

19

'The Whole of the Country is on Fire'

The social realities of the 1890s were mass unemployment, starving and deserted families (country hospitals reported having admitted children collapsing with hunger and malnutrition), a shattered trade union movement, a ragged army of workers from whom hope had been bled, and vindictive employers.

As for the squatters, their moment of truth was upon them. Before the drought of the 1890s, much of western New South Wales carried almost one sheep to every three acres. When the effect of overstocking on the fragile grasslands system was revealed, the carrying capacity was seen to have been far too high. By 1907 most of the properties affected had dropped to one sheep per fifteen acres. Momba (1 923 000 acres) which had carried 420 000 sheep before the drought was now reduced to 74 016; Weinteriga (523 220 acres) by 1907 was down from 150 000 sheep to 49 500; Teryawinya (917 337 acres), with 100 000 sheep before the drought, had dropped to 46 108 when it was realised the land could no longer accept such numbers. Even salt-bush died out in some areas, and no one had thought of that small, grey-green shrub dying. It covers hundreds of square miles in a patch, and grows throughout most of central Australia. Sheep produce good wool on it, but when the sheep were let loose on a land that nature had previously husbanded, fragile balance was broken. The rich western lands of New South Wales were destroyed within twenty years of the arrival of sheep. In 1884, one and a half million sheep grazed on 6 million acres in the Cobar (NSW) district. By 1891, when 25 inches of rain fell there, no grass grew: only a green moss covered the land. For nearly twenty years grass refused to grow around the Cobar area where it had been killed off.

By 1895 the west of New South Wales was a desert: with no grass, the sheep were too weak to travel and a loose sand blew until even fences were covered. When rain eventually came it did not draw forth the green grasses that had come through the soil since the beginning of time. A Western Land Commission investigating doubted that the desert grass would ever grow again.

Growers had to try to get their sheep a thousand miles down the eaten-out stock route to the next pasture. Some far-out stations tried to drive their flocks down to Wilcannia and over to Bourke. One mob of 10 000 died on the way, at Louth on the Darling River. Only a few hundred lived 'and these were sold for a few pounds'.

The sheep growers were desperate, vindictive. Banks foreclosed on mortgages, property owners harassed their managers, managers bullied shearers, and shearers were out of work, dangerous, and biding their time. All this, and now the shearers were sounding the tocsin!

This final blow enraged the squatters. Frightened by the prospect of losing their land and life style, they thrashed out at the only tangible opposition: the shearers and their union. They banded even closer together and went into bat-

The Prodigal
OLD AUSTRALIA (*to New Australia*): 'Well, young 'un, you have had enough of true "mateship", and have come back to the old camp. Come in, you'll find some cold veal in the pantry, and when you've pulled yourself together a bit you had better take a pick and shovel and dish and make up for lost time.'

tle in the way they knew best, summoning up all the forces their social, legal and political position owed them. Convinced that they were in the right, they set out to humiliate the enemy.

They made no attempt to conceal their control of Government, police and the judiciary. One comes across small but pertinent pointers to this in police records: Sergeant John Nagle, ordered to recite the qualifications of a man to take his place upon the bench as a Justice of the Peace, reports 'that Mr Herbert Austin Smith is Manager of Gumbardo station and a fit person for appointment as a Justice of the Peace'. There is no need for him to write any more.

LAW AND ORDER (*to Queensland shearer*): 'Throw off those villainous togs; they're simply making a fool of you. Leave that sorry thing to cowardly foreigners and go to work like a man!'

There might have been nothing remarkable about this if one could forget that the shearers had burnt Gumbardo woolshed in 1891. But under the circumstances, one cannot believe Mr Smith would perform on the bench in the unbiased manner the post proclaims itself to be. Remember that the Irish Coercion Act was still in force in Queensland, making the holding of meetings by shearers illegal. The frustrated, angry, out-of-work men marked the sheds that had been the scenes of their humiliations and were owned by men who had ridden rough-shod over them.

Again men rioted and conspired and went to gaol for it. E. Cowling, D. Bowes, J. Loyola and J. Martin were convicted of firing Ayrshire Downs shed

Milo station was continually mentioned in the report books and from the following entry we learn of conditions there and of the Peglers' use of state police. 'Mounted patrol to Milo station. Between 2–300 men camped about the shed waiting for shearing to start. Patrolled vicinity of woolshed . . . Shearing commenced Milo. There were a great many more men than were required but no disturbance of any kind.' The unwanted shearers would have had to head off 'on the tramp' as they had done all those long years before the great movement that was to mean security for them had begun. Frustration jostled with bitterness.

The unfairness of the social system did not go unnoticed by the men. They knew that the Police Magistrate at Barcaldine was a fair man and that he had been an influence for good amongst them because he had truthfully reported that the unionists in camp were law-abiding. For this, he was removed to Muttaburra – which was known in the service as 'Siberia'. When Magistrate Morris appeared to favour the shearers by fairness, he was brought back into line with promotion. The men knew that the secretary of the Pastoralists' Union stayed at Morris's house in Barcaldine and had on at least one occasion held a meeting of pastoralists there. They knew that both Ranking and Morris had been placed at the service of the wool men. The Coercion Act had been rushed through Parliament in a single night's sitting, and anyone having, or suspected of having, fire-arms could be arrested without warrant and tried

The much despised Milo station woolshed was burnt in spite of the diligence of the Messrs Pegler as frustrated shearers rampaged across the countryside

privately. No person other than the magistrate would ask questions. No-one could see such a prisoner tried unless the magistrate permitted it. A witness could be compelled to give evidence, even if he incriminated himself. He could be kept locked up incommunicado – there was no end to it. No end to the indignities, and the retribution, and no end to the outbreaks in retaliation.

'Two bottles of kerosene found hidden close to shed today. Men demoralised since visit of Union delegate Fanning last week, I have put on extra watchmen. [Signed] W. Hood, Burenda Station.'

Whitely King added a note to this: 'We have no doubt of this diabolical intention to destroy the property [of] our members. Fanning, you may remember, was convicted of intimidation during the last strike and completed the term of imprisonment prior to his visit to Burenda.'

Fanning's visits to the men's huts on any station was likely to be followed by a firing of the woolshed, or the owner finding kerosene ready to start such a fire. His skill was such that in one night's discussion he could encourage and train men to act in his absence.

King ran his organisation as an adjutant at army headquarters might do, signalling a widely dispersed army and receiving their reports.

The pastoralists were, like the shearers, learning the haphazard rules of this guerilla-type war as they went along. At Roma, they had served a notice warning the union not to go on the Mt Abundance run to interview the shearers and labourers as they had done before. The Roma group therefore sent in a delegate from another district, 'who procured the interview with the men at midnight, leaving the station at daybreak. The result of that visit is that the free men are being subjected to abuse and intimidation to join the union'.

King wrote: '... Very imminent danger overshadows us. With men going about the country fanning the flames of discontent and hatred and exciting a desire for revenge amongst the men against their employers, there is no knowing how soon a serious outbreak may take place.' Mr Ebb Smith of Warribona had written to say, 'The whole of the country is on fire and has lost 300 square miles of grass'. At Whynebah, the property of the Pastoralists' Union President, W. Allen MLA, his manager counted 'ten separate fires in sight of homestead'. Fanning was still riding across the land.

At the end of 1891, when Whitely King visited New Zealand (his home country) the *New Zealand Times* published a lengthy interview with him headed 'N.Z. shearers in Australia earn £30 000'. (King insists in the article that £40 000 is nearer the mark.) He told reporters that the Pastoralists' Union had brought 500 New Zealand shearers over that year and many more came independently.

He refuted stories in the New Zealand press that as soon as the PU and the ASU came to an agreement 'the outsiders ... were the first to be sacked and their expenses were not paid as promised'. 'Absolutely untrue!' said King. He believed such rumours were the work of Laborites and added that a Labor member of Parliament was introducing a bill into the House 'to prevent "aliens" from performing labour or service in NSW'.

The editorial in the *New Zealand Times* said that 'The proof that the ASU never had a grievance worth complaining about lies in the £30 000 the New Zealanders have brought back with them'. He adds, without elaborating: 'Australian unionists' violent opposition to New Zealanders [coming over as free labour] was threefold: lynching, lying and law.'

On his return to Australia on 1 March 1892, King wrote to owners of sheds on the Paroo, Darling and Cobar, asking if they were willing to leave all arrangements for filling the shearing board in the hands of the PU Council. He stated that the PU was receiving large numbers of applications for pens from New Zealand, 'men who satisfactorily completed their agreement with the PU last year [and] were given the understanding that if they performed their duties to the satisfaction of employers they would be given the first sheds next season ... The council is unable to carry out this undertaking unless owners assist by keeping their sheds open for these men.

'Already I have on hand 200 applications and every mail from NZ brings applications from men who say they are willing to engage for the next season only if it can be effected through the PU. Our agent in NZ also advises that men are writing to him daily and he could send 200 men at any time.'

More than any other group involved, the police must have been sick to death of it all. They were asked and ordered to ride miles over drought stricken country to 'pick up a madman', 'look for a body in a water-hole', 'bring in badly decomposed remains'. They were still performing prodigious feats of horsemanship. After leaving Milo woolshed a constable was sent off to pick up William Marshall. Taking two horses he rode 244 miles in four days.

When 1894 dawned they were still on call whenever squatters were repelling shearers; men were still being held in gaols without charges being laid.

Delegates from the Amalgamated Shearers' Union of Australasia met in Sydney in February, 1894. Front row: J. Slattery, C. Poynton (Vice-Pres.), W. G. Spence JP (Pres.), A. E. Rae MP (Vice-Pres.), A. Poynton MP (Treas.), H. Langwell MP, T. H. Williams MP. Titles now gained by the men, such as Justice of the Peace Spence and four MPs, indicate successful toil of men who only four years ago were considered outcasts

Of the 26 names listed in the Register of Strike Offenders and Suspects for 1894, only one man has any offence noted beside his name: 'no. 5 John Amor, Intimidation at Oondooroo station. Aged 30. Height 6 foot.' For Jack Robinson, no. 22, the entry reads: 'Was cooking with Steeve Marshall at the Winton strike camp.' It was still being noted against arrested men that they carried union tickets in their pockets.

20

The Dark Days

The union was under seige, it could do nothing to help the men. The method used by American employers for fighting labour by court injunction was imported and the Pastoralists' Union attempted to seize the property and funds of the ASU, including the *Worker*, the new paper with which the union had hoped to educate the men.

An injunction was served on William Spence and other office bearers to force them into the insolvency court, on the charges reported by Banjo Paterson: 'A press correspondent then visiting Warrana station who was known as "Banjo" Paterson, wired some sensational tales to the city press, although he knew from personal visits to the camp that there was no coercion or disturbance. Some of the scabs went into Coonamble and signed affidavits in which they swore they were detained against their will at the camp.'

Spence was at the time in Coonamble and knew no complaint had been made at the camp. 'The men were orderly and well behaved and it was publicly admitted in Coonamble town that there had never been less disturbance or so few cases of police offences. They had an excellent chairman in the camp, Harry Murray.

'The false affidavits were good enough to secure an injunction and force us to break up camp, as well as make it necessary to call the fight off.'

They called the years that followed, 1895 to 1897, the 'dark days'.

The pastoralists reduced rates in Victoria to 13s 6d for merinos (and found) or 15s 6d and men find themselves. The shearing rates in New South Wales had also been cut by 3s 0d. Things were so desperate in the country for labourers that men flocked to the Pastoralists' Union offices in each city anxious to accept work at any rate at all.

Loss of heart had spread like a devastating bush fire, and the one idea of hundreds was simply to get work. The Bourke secretary found that every shed in his district was filled with names of men who had accepted PU terms, and this pattern was repeated elsewhere. Eventually, the ASU appealed to members to fight for union rates of pay, but to accept the best agreement they could get.

The troubles dragged on through the dark days of the end of the century. Work went on at Vindex station, Winton district (Qld)

The shearers knew that the lower rate offered – supposedly because of the bad season the growers were having – was unfair; even at 20s 0d per hundred the rate did not mean as much as it had in earlier years for then it had been merino sheep, now it was Vermonts, sheep bred with vast wrinkles so more wool could be cut from the jowl-like flaps of skin. But the pastoralists were adamant.

Despondent, unionists failed to pay union dues. With the union weakened and union spirit so bereft of vigour, the year turned out badly with a universal reduction of shearing rates. Few organisers could be sent out. Membership continued to fall. The Victorian *Worker* ceased publication in 1894, owing the printer £1000. Branches closed in all states. Spence himself went out organising in 1897. 'The mixture of loyal shearers with those who had black-legged on them and those who were "real scabs" made things bad in the sheds.

'It meant asking staunch men to forgive the men who had acted a traitor's part and allowing all hands to start with a clean slate again if they would promise to stand by union principles.'

This dark fight centred in New South Wales. By 9 August there were 110 men in camp at Bullarora and 150 at Coonamble. Warrana was out, as were Nebea, Calga and Bumba. Coree in the Riverina, Bundure, Yanko, Wunnamurra, New Coree, Cobram, Moulamein, Noorong, Murray Downs, Mungadel were all out, with men in camps.

Urangeline, Poonboon, and Yamma had agreed to shear union as had Tarella and Windourin, but Nyang, Warramee, Windourin, Melool, Yanabie, Nowranie, South Yathong, Widgiewa, Willerah and Wandock were out. There were

Big Burrawang, Coonamble district (NSW) had planked drafting yards

camps at Warren and Eurabalong and sixty men camped at Nyngan. There were 100 at Bugaldi and 50 near Trangie and others at Moulamein.

Some observers believe the struggle was aggravated by the return of soldiers from the Boer War, 'Friends of the PU'. Be that as it may, it was an entirely different thing that caused the collapse of the strike.

William Monaghan Keogh, the manager of Warrana station, took out an injunction against the union keeping any of his employees in camp. Keogh had brought 40 free labourers from Melbourne up to Gulargambone railway station, 25 miles from Coonamble camp. He had coaches waiting for them and six hand-picked men, pugilists, to whom (it is said) he paid £5 each. Their task was to prevent unionists from stealing his labour away. The union camp had a corps of fifty cyclists acting as scouts and these sent word to the camp to await the arrival of the coaches across the river below Gulargambone. The professionals were no match for the bushmen and the free labourers were taken off to the camp.

Keogh served an injunction on General Secretary McDonnell, Branch Secretary Ike Smith and President W. G. Spence – an *ex parte* statement. As a result the men were enjoined not to 'cause, induce or persuade any employees of the plaintiff by words or by any means whatsoever to violate or attempt to violate any contract entered into by him with the plaintiff'.

The order was accompanied by a direction for sequestration of the personal estates, the rents, issues and profits of the men as well as those of the AWU (including the *Worker*). It was game, set and match. The costs claimed on the injunction motion and the sequestration motion totalled £1568 12s 3d, the lawyers cost £1878 14s 2d. The camps were disbanded, the men rode away, and a strange yet thoroughly Australian period of our history ended.

It took until 1904 for union membership to recover, and by then Queensland had amalgamated with the other states, making the AWU a united body over a land that in 1901 had become the Commonwealth of Australia. They set themselves immediately to prepare a case to secure an award from the new Federal Arbitration Court.

They were unsure how the pastoralists would respond to this, 'but we resolved that if they decided to fight our claim on legal technicalities we would withdraw the case [because the Union did not have the funds to afford a protracted appeal]... and fight it out industrially on the old warfare plan.'

But the squatters had also had enough, and they agreed to meet the shearers in May 1907, when they came to terms on the wording of a shearing agreement which was later made part of the award. This historic award was handed down by Mr Justice O'Connor on 20 July 1907, the first case in which evidence on oath – 37 witnesses – concerning the pastoral industry as affecting the workmen was given.

Too Old to Rat

I don't care if the cause be wrong
  Or if the cause be right –
I've had my day and sung my song
  And fought the bitter fight.
In truth at times I can't tell what
  The men are driving at.
But I've been Union thirty years,
  And I'm too old to rat...

– *Henry Lawson*

21

Nigger Meat

There was an outbreak of communal-living fever in Australia in the 1890s. During the great strikes, men camped together in their thousands and shared toil, companionship and whatever money they had. When the strikes failed the greatest communal adventure of all was put forward: free men would sail from this land of fetters and create their own Elysian fields.

The shearers at Barcaldine had made a mighty stand,
But the squatters brought the troopers in to dominate the land;
And Lane, the labour leader, had raised the clarion call,
'We'll build a brand new colony where workers will share all!'

Young Mary Cameron (later, as Dame Mary Gilmore, to become one of Australia's best-loved poets) was going with them.

Oh! The men of the New Australia are ready to cross the sea.
  To form a band in a far-off land, and show what men they be;
For the men of New Australia have turned from a hopeless strife –
  Taken the Right as a guiding light to lead to a nobler life.
They are gath'ring in from out of the West, and in from the Central
  Plain,
  From where the pelican builds her nest and Drought a king doth
  reign.

William Lane, the charismatic 'leader', was at the helm.

The workers of the 90's, they heard his call and cheered;
But the squatters and the bosses, they hampered him and jeered.
Those gallant people, wives and all, assembled at Balmain;

Lane told them of their destination: far-off Paraguay in South America. The newspapers of the day ridiculed the scheme, but the emigrants were determined.

On board the *Royal Tar*, en route to hoped-for Utopia, 6 July 1893

They say there's injuns whoopin' round for New Australyin blood,
Muskeeters big as kangaroos, 'n' fevers in the mud,
'N krockydiles, 'n' horrid wars – at least the papers say –
But there's no Australyin squatters over there in Paraguay.

The aftermath of the strike years when brave deeds, braver words, and all hopes were crushed had to be assuaged in some way and how better than by building a workers' paradise? But it wasn't only disappointment that drove the emigrants away from Australia.

But work was scarce and times were bad for folks in Bogantungan,
And Queensland families were faced with poverty and hunger.
Chorus: So is it any wonder mates,
When faced with strife and hunger,
Those workers went to Paraguay,
Like this man from Bogantungan.

The *Bulletin* called it 'one of the most feather-headed expeditions ever conceived since ... Sir Galahad pursued the Holy Grail'. But then, the *Bulletin* never did care much for idealism, never did believe that democracy is better served for having a good leaven of blind faith and vision to counter the venal cupidity of others.

Lane was an impractical theoretician who confused socialism with sentiment, and he led his people into hardship, loneliness, death and poverty such as they had never known. One of the poems for children written at Cosme in Paraguay (where descendants of these Australian emigrants still live), unintentionally tells us that they lived no better than slaves had done on cotton plantations in Virginia a century earlier.

When we all lie abed at the tooting of the horn,
  That we may all be fed you go to hoe the corn
We clap our little hands, shake our little feet,
  Oh! my, but Cosme corn is sweet;
Oh! folks they must be saucy when they call it nigger meat.
  Oh! my, but the corn is good to eat.

And Lane's utopian chimera fell away.

Some couldn't stand the hardships, and ere a year was spent –
Some had forgotten their workers' vows, and forgot what
  mateship meant.

Most of those who set out with such high hopes scraped together, borrowed, or begged enough money to bring them back to Australia.

Shearers hardly flocked to Lane's call to form a new Australia in Paraguay, but a few sailed off on this maddest of all romantic ventures. Several union organisers are listed among those who sailed, including Alec Forrester who had been gaoled on St Helena for three years for conspiracy. Forrester had been a barman for a time in a Wilcannia hotel (NSW) and later an organiser for the Shearers' Union in Queensland. Like many shearers he was a cavalier regarding names. In Australia he was Alec Forrester; in Paraguay he was Alec F. Lewis. Another man who went was Jim Mooney, who had been a shearing organiser in the sheds along the Darling and around the Riverina. One of the women who joined those who chartered the *Royal Tar* to sail to Utopia was Clara Jones, who had been working as a nurse at Clermont hospital during the strike. It is believed that it was she who signed herself 'Union Girl, Clermont' beneath the rallying lines printed by the *Worker* on 1 May 1891.

Then forward my comrades and rush on the foe,
  And deal them a death wound in every blow.
Our swords are unsheathed, we have taken our vow,
  That no half-way measures will do for us now.

The following year Nurse Jones hoisted a red flag over Muttaburra hospital to celebrate the election of Thomas Ryan, the first Labor candidate to the Queensland Legislative Assembly. 'When a member of the hospital committee pulled the flag down, Nurse Jones simply hauled it up again.'

Clara was a girl with her heart in the right place and so probably were the rest of the people who fled to what they believed would be personal freedom, but the majority returned to the old Australia, leaving only a handful in Paraguay to cope with the fever, insects, heat, disease, discomforts and poverty.

This strange, stillborn babe of the strikes of the 1890s epitomises the canker (or maybe it is the saving grace) of the Australian working class: no man starves to death in this country even though he may not eat as well as some; no man lacks schooling, although he may not get the education of others; no man goes unclad nor quite freezes to death on a winter's night, although he may wear hand-me-downs and pull thin blankets over himself in bed. In short, things are never so bad here as to push him over into revolution against the state, but because man expected an egalitarian society when he came down to the other end of the earth, his folk-nerve continually twitches in revolt at the reality.

22

Lambing Down

I'm a broken hearted shearer
I'm afraid to show my face
For the way that I've been lambed down is a caution to the snakes

sang an anonymous shearer who had been lambed down. He speaks of getting 'a nobbler' at a certain place in town, where the barmaid 'was a caution to lamb a fellow down'.

Oh, she tossed me up at Yankee Grab: to keep me on the booze
And somehow or other I always seemed to lose.

Oh I've sold my good old horse and I'll get some work I hope
I've some tea and some tobacco and a half a bar of soap,
And that's all my five years brought me since last I left town
But it's nothing when you're used to it to do a lambing down.

But when the money is all gone he finds the credit has gone with it.

Then when the shearing's over
They gets their bits of papers
They to the nearest shanty go and there they cuts some capers
They call for one and then for two in a style that's very funny
Till the land lord says, 'Hey! This won't do. You bastards ain't got no money.'

Lambing Down was almost a part of the system. In retrospect it appears a tragic waste of the effort of those men who carried out an immense amount of brave, monotonous, hard work year after year for nothing more than a few days' feverish drinking.

When sheep were lambing, station hands used to be sent into the paddocks to see that everything went smoothly. The process is still called 'lambing down'. By a gentle metaphor the words were transferred to the assistance which in the old days publicans gave men who came in to get rid of a cheque. The man handed the cheque over the bar. Sometimes he asked to work it out, sometimes he expected change. It mattered little what his original intention was. The end result was he lost his cheque, the money he had worked for all those hard and often lonely months (a shepherd might go to town for only one week a year).

The details were the only things that differed. Some men woke up a few days later in a room empty except for champagne bottles. 'Oh yes,' he was told as he regained consciousness, 'You certainly hung one on! Shouted for the town! With champagne!' The patient was graciously permitted to stay a day or two to gather his digestion and his wits together and then given rations and a swag 'for which he was pathetically grateful' and sent off to tramp to look for work for another year.

The publican's rationale was, 'I'm licensed to sell grog and by God that's what I'm going to do.' The philosophy of some was, 'If we don't get their cheques here the next shanty will.'

Drugged liquor was sold in both large and small hotels, and some claim they were drugged or 'tempted' to spend their money.

Tally Town

A shearer tramped from Borderland
  When Sunset shed cut out;
He scarcely felt the miles of sand,
  The hot plains baked with drought.
His cheque was in his ragged clothes,
  The future wore no frown,
When blazing white before him rose
  The roofs of Tally Town

From Tally Town the shearer fled
  When Christmas time was past;
His cheeks were drawn, his eyes were red,
  His breath came rough and fast;
Along the track he ran and yelled
  Till devils struck him down.
The landlord banked the cheques he held
  That day in Tally Town.[1]

One effect of the strikes of the 1890s is that they appear to have not only consolidated shearers into a firm, proud group, but to have either weeded out, or smartened up those whose larrikinism sometimes tumbled over into hooliganism. C.E.W. Bean (later to become Australian Official Historian for World War I and a great admirer of all things Australian) said in 1909: 'The shearer is, taken all round, the highest type of worker.' He noticed that 'the shearer got away with his cheque intact far more often than the rest [of the pastoral workers]... the class of shearers has been changing this last twenty years, and the change is pretty well complete. Between the shearer and the station hand there is all the difference between a country man and a city one.... only the other day when a western shed finished shearing, the nearest town was pandemonium for a fortnight. Today if the shed cut out at 4 p.m., most of the shearers are... on their bicycles... on their way to Melbourne by 6 p.m. Some shearers will clear the town without a drink. Some may set aside a couple of sovereigns for a wet night; a few still get side-tracked and lose their cheques in a

Lambing down – a shearer knocking down his cheque

country town. And that is one sign of a great difference which has grown between the shearers and station hands in the outside country.' He pointed out that the shearer was a travelled man, he could look after himself in a town. 'The station hand too often can not. He knows it. He has been there before and within ten minutes of his getting there someone has touched him for his money.'

But as old Ray Watley, a shearer of the 1920s, says, 'By the time a Broomie reaches town with a few bob in his pocket he's a gun shearer!'

For a century there have been those who have cried, 'The shearer cannot hold his money! It's useless to pay him big money! He loses the lot in the first pub he comes to.' That may have been the case where men had to stay in a town for long periods awaiting transport after cut-out, but if it was no longer true in 1909 it is less so today. Many a small town in pastoral areas has faded away because of this very fact: the wages of the men in the pastoral industry were the little towns' main industry.

Wool-growers encouraged, even engineered the building of such small towns near their properties. Then their workers did not disappear to distant cities, they always came back after their cheques had gone. Most little towns grew up around a hotel, the main building. Some towns were only a hotel. James Bonwick wrote in the 1880s: 'Payment was an order on the nearest publican, who of course never handed over a cent. A man was compelled to stay there and knock his cheque down. If the cheque was drawn on somebody in Sydney, how could a poor devil get away to Sydney – perhaps a four or five hundred mile trip, without a penny?'

Shearers could not often have got the better of a publican in a shanty but Mick Ryan tells how it did happen once at the Lake Eliza pub in New South Wales: 'Some shearers came in with cheques and the publican put these off with the story that he didn't have enough money to cash them but they could order what they liked and he would keep an account – the old trick.

'The shearers drank for a day and a night, when one of the more sober of the men suggested a race to be run on their push-bikes over a course of ten miles. They were to compete with waterbags and swags up and the publican was to act as starter and judge. They lined up and wobbled into the heat haze, and the publican waited for them to return. He is still waiting, and he found when he gave up and went back into the bar that their cheques had gone with them.'

Of all the 'Christian' religions that have been set up in Australia, none have adhered more closely to the spirit of Christ than the Bush Brotherhood of the Anglican Church. Mostly very young men who had taken a vow of poverty, they set off for the great outback to help the working men spread over its thousands of kilometres. There is a nice story of them playing cricket with shearers on Sunday afternoons. 'I make a deal with them,' a Brother recorded. 'They come to my show in the morning, I come to theirs in the afternoon. Many of them come in from a hundred miles away.'

The Christ of the Never-Never

He works where the hearts of all nations
  Are withered in flame from the sky,
Where the sinners work out their salvations
  In a hell-upon-earth ere they die,
In the camp or the lonely hut lying
  In a waste that seems out of God's sight
He's the doctor – the mate of the dying
  Through the smothering heat of the night.

By his work in the hells of the shearers,
  Where the drinking is ghastly and grim,
Where the roughest and worst of his hearers
  Have listened bareheaded to him.
By his paths through the parched desolation,
  Hot rides and the terrible tramps;
By the hunger, the thirst, the privation
  Of his work in the furthermost camps;

By his worth in the light that shall search men
  And prove-ay! and justify each –
I place him in front of all churchmen
  Who feel not, who know not – but preach!

– *Henry Lawson*

23

The End of Silence

William Russell of Dennistoun, Van Diemen's Land, said something very surprising in a letter to his brother George in Victoria for 9 January 1846: 'If you or any of your acquaintances stand in need of a steam engine, the Russells of Kirkcaldy have shipped one on board the 'George' and commissioned me to dispose of it. It is of six horse-power, similar to those used for driving threshing machines in Fife, but might be applied to many other purposes; perhaps you might think of shearing your sheep by steam power, when labour is so very high. A plan for the erection can be sent along as a guide to those that may put it up.'

Was he joking? No, the Russells were not known for their humour. William had thought that it was possible for sheep to be shorn by mechanical means.

When machine shearing came, wool-growers welcomed it, even if unsure of its mode of operation. Edward Henty of Merino Downs (Vic.) wrote to his manager in 1884: 'I expect it will take three of our horses to work it six hours. I have been present at two of our Society's trials and the machine makers all had exceptionally large horses and only a pair each and we have none like them.'

Horse-powered machines were tried, even human-powered contraptions were demonstrated, but as early as 1860 inventors were working on mechanical means of driving shears. In 1868 James Higham, a Melbourne compositor, had taken out the first patent for a 'sheep shearing machine'.

At the same time, Frederick York Wolseley was carrying out his experiments and it was the machine that he and John Howard evolved that revolutionised the sheds.

It has often been written that shearers showed considerable resistance to machines, but many squatters also were conservative and for years refused to use the new invention.

By 1887 the great Toganmain (NSW) had machine shearing, but the first full machine shearing in a shed was at Dunlop, on the Darling River, near Louth (NSW). The strike of 1886 had affected this shed and doubtless the owner was hoping that the machine would solve all his problems.

In 1866, T. Millear, Edgarley station, Willaura (Vic.) designed a machine and other apparatus. Plodding horse provides power, shearer sits sheep in iron chair, opens up neck and shears points and belly. Rest of body was shorn on table in front of him; sheep were turned over by means of rollers. This latest labour-saving device was said to be sure to replace shearers

Dunlop station was (and is) a big shed. The great gum pillars and beams for its frame were brought up the Darling from the Murray by barge and unloaded on the river bank. There they built the shed and roofed it with shingles. The property covered almost two million acres and the old homestead, a quarter of a mile from the woolshed, was the heart of a busy little township with its store, blacksmith's shop, offices, men's quarters and saddlery.

For many years the station washed its own wool, the scour standing beside a big lawn on which the fleeces were spread to dry, looking like snow with the thermometer sometimes at 120°F.

Dunlop, owned by Sir Samuel McCaughey, was considered a good shed to work in, and when the call went out in 1888 for shearers for its forty stands, twice the number needed applied as usual. But when the selected men saw the machines, they refused to sign the agreement: the Amalgamated Shearers' Union had not known that the 'new-fangled fleece removers' were to be used.

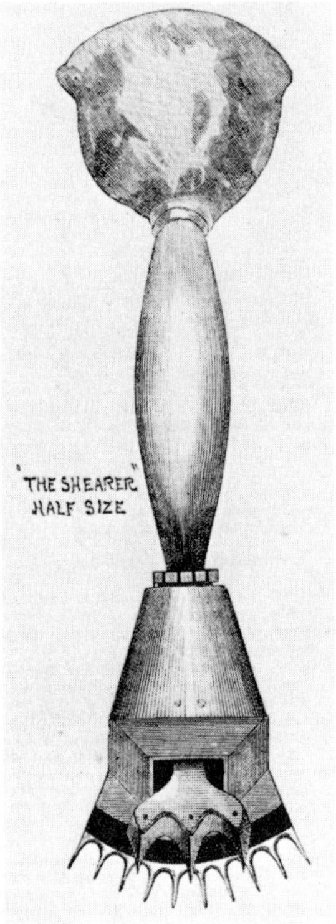

These particular machines that made history at Dunlop were devised by two English migrants, Frederick York Wolseley and John Howard. Wolseley was the brother of General Lord Wolseley and son of a clergyman in Ireland. He had come to Australia in the 1860s and learned the sheep business on his brother-in-law's station on the Riverina before buying Euroka station, near Walgett.

Wolseley had patented two unsuccessful shearing devices based on horse clippers before meeting the young English mechanic, Howard, who had learnt his trade in the Birmingham works of Clark and Co. Howard had also been trying to adapt the horse clipper to sheep-shearing, and the two men became a team, Wolseley providing the money and facilities on his Walgett station to enable Howard to continue his experiments.

By 1885 Howard produced the first satisfactory machine shearer, a hand-piece equipped with a fork and a springing prong. To all intents and purposes it

Wolseley sheep shearing demonstration at Goldsborough Mort & Co. wool stores, Circular Quay, Sydney, 1887. Frederick Yorke Wolseley (left) leans on walking stick

was basically the same hand-piece as in use today; only the method of driving has changed radically. The Howard-Wolseley patent was driven by an endless rope which ran through pulleys on each shearing bracket, a steam 'donkey' engine providing power.

The new machine was first demonstrated in 1887 at Goldsborough and Company's wool store in Melbourne. The shearer performing on this historic occasion was not an Australian but a man from Khartoum, Hassam Ali, said to be the most skilful of Howard's assistants. The demonstration took the form of a straight-out contest between the machine and blades with a gun shearer, Dave Brown, first shearing three sheep with the blades. Hassam then went to work on the sheep and, with the machine, took another 2¼ pounds of wool from the three already-shorn sheep.

It was this 2¼ pounds of wool that convinced pastoralists that the blades were on their way out. As a result of the Melbourne trial, it was decided to equip a few sheds with the machines for shearing the following year and, of these sheds, Dunlop was ready first and called for shearers.

When the men refused to sign the contract they maintained it was because of a clause in the agreement requiring shearers to supply their own combs and cutters. But the truth went much deeper, down to the basic fear of working men, that the machine had the power to rob men of their livelihood.

The Sydney *Morning Mail* had warned, 30 April 1887: 'The introduction of the machinery may render the employment of expert shearers unnecessary as it would not require more than a few days to enable any man of ordinary intelligence to learn to shear the sheep properly with machines.'

There was something else: shearers were already discontented over wages and conditions, and a trial of strength between the pastoralists and the union needed only such an issue to bring matters to a head. Shearing rates were then 12s 6d per hundred sheep with shearers providing their own shears and working from 6 a.m. to 6 p.m. (Saturdays to 4 p.m.) in the toughest of living conditions.

They picked up their swags, crossed the Darling River and camped on the other side, leaving a flock of 184 000 sheep unshorn and the steam blowing in the new boilers.

John Howard, who was there in the capacity of an expert (machine supervisor, mechanic and cutter sharpener), talked with the shearers, but it took him three weeks to persuade them to start work with the machines, and by that time some were sullen and in a dangerous mood. Consequently, the new experiment started off with not only the usual teething troubles of any new venture but with a lack of co-operation by the operators.

There were upside-down combs, and shears were released while in gear so that they wriggled about on the floor like snakes; an expensive cotton rope, brought from Sydney to use as a driving belt, was hung up to stretch and during the night was severed in several places.

By 1911 the firm could boast having erected over 50 000 machines and shorn 80 000 000 sheep a year

'Big' Burrawang (NSW), as the men called it, one of the first to change to machines

At first the tallies were poor. On some days 43 sheep was the highest any shearer recorded (in comparison with a record 275 in one day with the blades). The shed was in the doldrums and for a time it seemed the experiment would fail. The change in attitude began after a month when several men, including some New Zealanders, began to feel that the future lay in these machines and that they might be worth studying. At the end of five weeks, one of the New Zealanders shore 100 sheep in one day. Howard quickly offered a bonus of 2s 6d (50 cents) for each sheep shorn over 150 a day and tallies began to rise. It was not long before the shed was averaging 120 sheep per man each day.

It was that tally board that heralded the end of blade shearing in Australia, even though the blades died hard. (Mulwala in the Riverina was shearing with blades in 1914. Many studs still use blade shearers and as late as 1960 Jim Carew ran the Denistoun shed (Tas.) with a full board of blade-shearers. The following year, 1961 was the last time that that shed shore all blades.)

The partnership of Wolseley and Howard broke up. Wolseley became a successful manufacturer of shearing machines in England, where he died in 1899, and Howard worked with a firm of machinery merchants, Dangar, Gedye and Malloch Ltd, until his death in 1920.

One of the real difficulties experienced by the bladesmen in changing to machines was the positioning of the sheep. With blades, a sheep can be shorn in practically any position on the board, or even on the ground, provided the shearer can comfortably reach the part of the fleece being removed at the time. The sheep can kick and work itself out of position or even turn round without seriously inconveniencing the shearer. On the other hand, with machine shearing, there is a constant muscular and nervous strain to hold the sheep in the correct position, to prevent the sheep kicking and to enable the shearer to get the maximum amount of wool off with as few 'blows' as possible. A standard method of holding the sheep and removing the fleece must be used, and a correct position in relation to the shearing machinery must be maintained.

The successful establishment of shearing machinery took place in spite of an innate suspicion of anything new. By the end of the century, shearing machinery had been installed on numerous properties in all the Australian colonies. Some of the largest in number of stands were:

Big Burrawang	NSW	88	Northampton Downs	Qld	58
Mahonga	NSW	88	Wingadee	NSW	54
Brookong	NSW	80	Kallara	NSW	48
Canonbar	NSW	80	Toorale	NSW	46
Tuppal	NSW	72	Grong Grong	NSW	44
Killarney	NSW	66	Narromine	NSW	42
Coree	NSW	60	Goonoo Goonoo	NSW	42

In 1894 Dunlop shore 284 000 sheep and the adjoining station, Toorale, put through 220 000. The two sheds, under the one ownership, shore a total of more than 500 000 sheep in the one season. (At that time, flocks of 100 000 were as common as those of 10 000 today.)

With the establishment of machine shearing in all states and improved methods of transport the way was clear for shearers to be organised into 'teams', and the contract system of shearing began. Shearers were engaged by the contractor

Within a few years shearers posed at the machines for postcards to send home. 'Shearing in full swing in West Monaro Parish.'

and the sheds under contract divided into 'runs'. When shearing was completed at one shed, the whole team, including shearers, rouseabouts, pressers, cooks, expert, engine-driver, penner-up, woolclasser, boss-of-the-board and book-keeper, moved onto the next shed in the run. In some of the larger sheds as many as 150 men would be engaged in shearing operations in or around the shed.

Electric power-driven shearing machinery was introduced in many states during World War I, although its development was retarded for a time due to the lack of trained electricians.

From 1918 onwards, big changes took place. Soldier settlement and other factors resulted in the subdivision of large holdings into small ones. This subdivision, together with increased tallies by shearers, meant that large installations of shearing machinery were not necessary. Internal combustion engines replaced steam-driven units; overhead machinery progressed from plain bearing to semi- and full ball-bearing drives; the hand-pieces were changing to the self-levelling, chicken-foot, fork-type in general use today. All these modifications and changes in design have contributed to a smoother, more efficient drive and faster, cleaner shearing of the sheep.

'In the old days you only needed a roof high enough for a man to crouch under,' one old shearer said to me, but when machines came in, many sheds had to have the roof lifted to accommodate them.

From 1940 until the present day, the transition from large holdings to smaller holdings has proceeded at a fast rate. More shearing sheds have been erected, flocks are still being reduced in size, and where previously sheds of twenty stands or more were common, they have gradually been reduced until today, sheds of four, five and six stands are in the majority.

For small flocks and for crutching sheep on properties where it is inadvisable or impractical to bring the sheep to the main shearing shed, portable shearing plants are used. These are sometimes used in the paddocks or in a small shed, some single-stand units driven by a small engine being fixed on to any suitable post.

The coming of the machine to the shearing shed was all very wonderful, but it ended forever the silence and calm of the bladesmen.

24

The Bradman of the Shearing Board

They call him 'the Don Bradman of the Shearers', and so long as sheep are shorn in Australia the name of Jack Howe will be remembered. On 10 October 1892, at Alice Downs station (Qld), he shore the world's record of 321 sheep in 7 hours 40 minutes with blades. 'And all he used was a pair of glorified scissors,' later machine shearers wailed as they failed to better, or even equal, his world record until 30 September 1950 when W. E. Reick shore 326 merinos at Brinard (Qld).

In 1916, Jack Howe gave an interview to the Catholic *Advocate* over drinks at the Queensland Irish Association. He was reported as saying, 'In 1892 Messrs Coleman and Sons Eucalyptus Company, Cootamundra, had a standing offer appearing in several newspapers of a gold medal valued at £10 10s 0d for the highest tally of sheep shorn by hand in any of the colonies; also a £5 5s 0d gold medal for the biggest tally shorn by machine. I won both, as it happened.'

He had shorn 237 merinos at Barcaldine Downs in the same year, 1892. This latter effort was remarkable in that Howe was not accustomed to the use of the new tool. He was in fact only a novice with the machines which were installed in very few sheds at that time.

One week before his record blade tally, Howe had put up a tally of 1437 sheep shorn in 44½ hours, an endurance record that still stands.

The 'ringer of ringers' was a 14-stone Adonis. 'A quietly smiling man', is the most commonly used phrase in writing by those who knew him. Photographs show him to be the handsome, well muscled, moustached, curly-headed man one imagines the son of a circus acrobat might be. His mother, a Miss Stokes, came to Canning Downs in 1840 as companion to Mrs Patrick Leslie, the two being among the first white women on the Downs. In 1854 she married Jack Howe Senior, an acrobat and circus performer who had come to Warwick with La Rosia's Circus and had also performed with Ashtons. 'His most famous act was to spring-vault over the backs of fourteen horses,' his grandson Henry records. The circus performer took a position as stockman in charge at Killarnay Estate.

Jack Howe, the legendary ringer of ringers

Howe wearing the gold medals won for both blade and machine shearing

John Howe Senior, father of the champion, had himself a small claim to shearing fame: he had once shorn nine South American llamas. An Edward Daveney had overlanded the Peruvians from Sydney to the Darling Downs. The date has not been recorded but is said to have been in a 'depressed wool-sales period' and the llamas were expected to provide an alternative crop of expensive mohair. Henry Howe says: 'When my grandfather attempted to shear them they expectorated a vile effluvium. They spat on him.' But he did manage to shear them.

When his son Jack grew, he helped his father shepherding and assisted at shearing time. He quickly became an expert with the shears (this was before the advent of machine shearing), not with speed but by care and skill. At that time it was a popular custom at the cut-out of the shearing for the cleanest and most careful shearer to be awarded a bonus of a £1 or £5 note by the owner, and Jack Howe was a consistent winner. Some say he had been taught the finer points of shearing by a Chinese shearer.

In the early 1880s he left the Darling Downs and, after trying his luck at gold prospecting, came to Tambo, a little town in Central Queensland near the head of the Barcoo River.

His sons describe their father at this time as 'a fine, well-made fellow – one of the most perfectly developed and best set-up young men to be seen. He was a good athlete, a very fast runner, and a good jumper. He measured 27½ inches round the thigh, he had a 14½ inch forearm, a chest girth of 47½–50 inches, and he weighed 14 stone. His right calf was 18¼ inches and his left calf 19¼ inches, because of the pressure taken by his left leg in shearing. He had a hand the size of a small tennis racquet.'

The first shed he shore in Central Queensland was at Lansdowne near Tambo. 'He used only to shear about a hundred at that time: he'd got in the style of only doing small tallies on account of shearing on the Downs, where he used to shear them "clean and pretty" in order to win the bonuses.

'After Lansdowne he went to Langlo, and there achieved some fame by sacking the cook. The old hands at Blackall [said that] at Langlo there was a big gun shearer, who was ringing the shed. He was "a flash cove" who "liked himself" and who also thought a lot of his stomach. Well, Jack Howe thought the cook was over the odds and told him to pack up and get out. When the gun shearer came in there was no cook and therefore no tucker. "What the hell has happened to the cook?" he roared. "Oh," volunteered a shearer, "Howe sacked him." "What bloody right has that snapper got to sack a cook?" yelled the gun shearer. "Why doesn't he let a shearer do it?"

'Howe was annoyed and decided to take the "flash cove" down a peg. "I'll show him who's the snapper from now on," Howe is reputed to have said.

'Up till that incident Howe was still shearing his 100 a day against the 150 of the gun. But next day when the tally went up the board stared in amazement. The "snapper" had shorn 211 sheep. It was the first time he had ever shorn more than 200.

'The gun shearer tried his best to catch up with him, but the more he worked the further Howe got away from him. At last the fallen gun threw in the towel and disappeared from the shed.'

That was the first of the big tallies of 200 and over which Howe cut at the head of the Barcoo Waters, and from that time he became known all over the Barcoo. Gun shearers from all parts made it their business to come in contact with him and try him out. He invariably 'rung' every shed he shore in, and put up some wonderful tallies.

In 1892 Jack Howe was shearing at Alice Downs. One of his sons later recorded in the *Courier Mail* (19 April 1930) that his father's pen-mate, Harry Dunne told him the following details.

'In 1892 I was shearing at Alice Downs, and Jack Howe was there also... There was a very loud sort of chap there, tall and fairly good looking. He said he had come all the way from where the dog died in order to ring Alice Downs. I asked him, "Where did the dog die?" and he replied, "Three miles the other side of sundown." The first day... this chap led off with 197. Jack Howe and I did about sixty and seventy each. We always had to start off steady for a week to get our wrists strong. The Blackall girls used to ride out on a Saturday morning – it was only twelve miles – to have a look at the shearing and also for the outing.

'In those days the tally board was hung up in the shed, with the name of each shearer and his tally opposite. When the girls saw the tally board and the name of Mr X with all the big tallies for every day of the week and the small tallies of Jack Howe and myself they thought he was wonderful as they thought how poor we were. So Mr X got all the lime-light...

'After we had been there a week I said to Jack Howe "Tomorrow we will have to find out where that dog died." Jack said, "Righto! I'll be there for the funeral."

'Next day Howe turned out 2 sheep short of 250. I followed with 211 which was a fair jump from being under 100 a day. The chap who came in from "where the dog died" had only seventy. When the girls came out the following Saturday and saw where Jack Howe had shore an average 260 a day and I was just on 200 a day they came round our stands and never went near Mr X at all.'

For some years after introduction of machines the blades were used in many sheds – in some, blades were used one side of the board while machines whirred on the other. Kahmoo, Cunnamulla district (Qld) were still using blades the year Jack Howe shore world records using both methods

On that Saturday morning Howe shore 145 sheep, and on the following Monday shore the world tally of 321 sheep. He shore 1758 during the period from Monday to Monday night the following week. In the Central west it used to be said that every second shearer you met claimed to have been alongside Jackie Howe that day at Alice Downs.

The name of the great shearer is perpetuated in the shearers' garment known as the Jackie Howe and is said to have come about this way. A big shearer found the sleeves of his grey flannel 'bluey' restricting and ripped them off at the shoulder saying, 'I'll make a Jackie Howe out of it.' The name has stuck to the shearers' singlet ever since. A strange memorial to the handsome big shearer who smashed all shearing records in the days when wool was king.

'Mum used to go . . . off her bloomin' block when she'd read about these Jackie Howe singlets,' said Henry Howe. 'The big stores in Brisbane would send out catalogues with Jackie Howe singlets in them, so, one time in Brisbane she went round to see a solicitor to see if he could stop them advertising. You see, she used to make my father's singlets – she was a dress-maker, and why dad wanted them tight around his biceps was the sheep when they'd kick get their hoofs in, especially when they're doing the long blow, he liked them tight around his arm so they couldn't be torn by the sheep's hooves. And mum used to make them with sleeves tight like he wanted. So she went crook when she saw the advertisement and said "He never wore those sleeveless things!"

'The lawyer pointed out that it wasn't defaming Dad's name, it wasn't even put on the tag on the back, his name wasn't on it. But of course it can still be called a Jackie Howe singlet whether he wore it or not.'

The photograph taken in 1895 of Howe in a singlet shows the garment to be of a woven mesh-type material, not at all the blue or grey flannel worn by men of the period. Harry Howe says it definitely was not flannel, but he can't recall the material except that the ones his mother made were cream and neatly crocheted around the neck and sleeves.

According to his son Henry, Jack Howe was a union man through and through. 'There used to be a 400 gallon ship's tank beside the Universal hotel for rain-water – the shearers didn't like bore water in their rum or whisky. You could see this tank right beside the footpath so he painted Vote for Ryan on it and beside this he painted Kerr and a big white rat: Kerr had ratted on the Labor party you see.

'My father followed T. J. Ryan's career as a barrister in Rockhampton and he thought he'd make a good man in parliament and talked to him about it.

'When he came out west on his electioneering tours Mum would do up, tight stays and all, crimp her hair . . . did she do up! Then she'd get the Sociable out – in those days every hotel had a Sociable, a big buggy with one seat in the front and two behind – and she'd tie a big calico banner along the side of the Sociable with *Vote T. J. Ryan* painted on it and then set off to go round and pick up all the pensioners.

'It was a long walk to the police station where voting took place and any old man there was always brought back to the pub for a drink.

'Mum used to pick up old Mrs Kewley too, Bill Kewley's wife. She weighed about 20 stone, lived in a little hut on Cemetery Road. Mum got there this day and poor old Mrs Kewley was getting done up to vote and she'd got out the bodice she'd got married in and had tried to put it on and got her arms pinioned behind her and couldn't escape. "Oh, the good lord sent you to save me," said Mrs Kewley to Mum. Oh, they got dressed up all right! Put their best on to vote. Old fellows would put their black suits on and a red rosette for Labor in their buttonhole . . . Old Bill Kewley used to dress up something grand. Bill Kewley was the man arrested in Blackall as they thought he had the strikers' funds during the big strike, but Mum and Dad were sleeping on the funds, they had it under the mattress on their bed and no-one but they and Kewley knew, so old Bill Kewley let himself get arrested.

'When T. J. Ryan came west Dad would first pick him up at the railway station in the Sociable and later in his T-model Ford. It would be Russells' pub the first day, Tambo the second.

'Another parliamentarian Dad knew was old Bill Hamilton, the shearer who got into parliament. He had feet that big that he had to get boots made for him. If he bought a pair in a shop he had to get the heels and toes cut out . . .

'Being the son of the most famous man in the west had its responsibilities. When I was a little boy and Dad had the Barcoo hotel I was so proud that when fellows would say what was my name I'd stick my chest out and say, "I'm Henry Howe, the son of Jack Howe the champion shearer of the world," and if any of the kids disputed it, by cripes I'd hop into them, never lost a fight.

'Dad didn't talk about it much. At home at night-time he'd talk about it a bit to us sitting round the fire.

'One thing I do remember him saying and old Harry Dunne too – Harry was his pen-mate that day at Alice Downs – that there was a medal offered for the highest tally in the colonies and this particular day old Harry said that there was some particular fellow there who was going to ring the shed and they used to put his tallies up on the tally board after each run-out. After the first run after dinner they knew Jack Howe was going to break a record today and Harry Dunne told me that when Jack Howe was getting close to the cut-out of the last run the men saw his tally, and his pen was full and he'd got up to 318 or something like that and they said "you've done enough!" and Harry said they threw his sheep back in the pen and tickled Jack Howe and jumped on his back and everything else to stop him, so he knocked off on 7 hours forty minutes shearing time when he got to 321 sheep. If he'd carried on to the end of the time he might have done 350.'

(It was only at Alice Downs that the men playfully grappled with him to stop him shearing, Henry says. Jack Howe had broken the record and that was enough for the men.)

Howe was an enthusiastic member of the Shearers' Union. On 12 January 1890 he seconded a motion for the amalgamation of the Blackall shearers with the Queensland Shearers' Union. In 1891 he was a member of the Queensland Shearers' Union committee which decided on amalgamation with the Queensland Labourers' Union. He later became a committee member in the amalgamated union. In 1901 he bought the Universal hotel, Blackall. In two years time he bought into the Barcoo hotel and later in 1907 returned to the Universal.

He remained a loyal member of the Labor Party. He was President of the Blackall Workers' Political Organisation. It was at this time, in 1909, that he played his active role in the election of T. J. Ryan to parliament.

Jimbour changed to machines

The Howe family bought Sumnervale station in 1919 and Blackall gave their famous shearer the biggest send-off in the history of the town.

The powerful 'Bradman of the Shearing Board' died on 21 July 1920 aged 59, and T. J. Ryan sent a telegram to Mrs Howe: 'I have lost a true and trusted friend and Labor has lost a champion.'

The young shearers now say Jack Howe might not have been able to shear the sheep of today in the time he took that day in 1892. There will be no way we could know. But some old shearers say, 'those wrinkly sheep we shore had a mass of deep wrinkles to be shorn. Any one of them would be the snob today. And the machines in those days were lively; if a sheep kicked it out of your hand and the machine got away it could rip a fellow's arm to bits, Drop it and it snaked across the floor away from you. So, all things being equal, Jack Howe would still be a champion today.'

Machines were already in sheds before Jack Howe shore his 321, but his record sent the blades out in a flashing memorial to 'the glorified scissors'.

25

The Bicycle Brigade

During the 1890s, that decade of boom and bust, one of the things that boomed was the bicycle. It cost most workers a couple of weeks wages and a shearer with a decent tally could afford one in a week. Suddenly, the horse that had been the 'flash' shearers' symbol of superiority over the men who travelled to the sheds on the Hobnailed Express (their feet) must be swapped over quickly for this newer and better motive power. It was not fickleness. The shearer is too professional to permit whims of fashion to dictate his ways. (When motor cars arrived in the country, it was many years before working men could afford them but, as soon as he could, the shearer left the old bike in the back shed and bought a second-hand car. The 1930s saw fleets of tin-buggies rattling over the mulga.) No, the bicycle was more efficient than the horse, or the odd camel that men had ridden, so whole brigades of them took to the bush.

Push-bikes, or bikes as they were always called, required little maintenance, and most, if not all, of that could be done by a rider himself. The deciding factor for the shearer who travelled over arid and semi-arid lands was their ability to survive without food or water. Unlike a horse, they did not have to be hobbled out at night or rounded up in the morning, nor rugged in the bitter night air of the desert. This meant that the shearer could travel lighter.

In no other country have men had to move constantly throughout the year to work as have the shearers, and from the 1890s until the 1930s, the rear-driven, pneumatic-tyred bicycle took these kings of the sheds up, down and across the continent. They were light enough to lift over fences when taking a short cut, they were extraordinarily strong considering the lightness of the frame, the repair kit was small and, like all bush workers, the shearer was an expert in adapting whatever came to hand to get him over a bad moment. Men rode bicycles up to and over one thousand miles a season. One man I spoke to chalked up almost three thousand miles one year. Those who travelled the arid lands had their tyres and tubes shredded by bindi-eyes in the eastern states (also called

On the final leg to Cordilla Downs. The Tasmanian shearers, Bill Greig second from right, have crossed The Cobbler, the worst stretch of the most strenuous, dry journey any shearers undertook. The Cobbler was named after the last sheep in the pen – the awful one that no one wants to tackle

three-cornered jacks in some areas and double-gees in the west) and *Emex australis*, a vicious thorn-like growth that has been estimated to produce 8700 burrs per square metre in heavily infested paddocks. When tubes could no longer be patched, the undeterred riders stuffed them with grass if it was available; if not, they used rope, old clothes, whatever they could lay their hands on. When a tyre was beyond repair they wrapped strips of greenhide round it or cut canvas strips and tied them around the tyre on to the rim. 'I've seen men get to sheds without even tyres,' Chum Watley (Tas.) told me, 'riding on the rim of the wheel that they'd protected with bindings of whatever they could find, greenhide, canvas, old trousers, bits of leather from old boots, anything.' 'Yes,' Ray Watley (Tas.) claims, 'Riding a bike from Bourke my tyre wore out so I killed a tiger snake and wrapped it round the rim, shoved its tail in its mouth and off I pedalled!'

The bike was speedy. A man could do one hundred miles a day, day after day, and it could bear considerable weight. 'On the deserts and arid lands we were careful with weights, but then a man had to carry very little on a horse on those dicey journeys. But on reasonable to bad going, as opposed to bloody hopeless tracking, we could have got away with up to 140 or so pounds if need be. But why would we want to? Mostly we took the bare minimum – a water-bag of course, a quart pot, a change of dungarees and shirt (not always), sometimes a

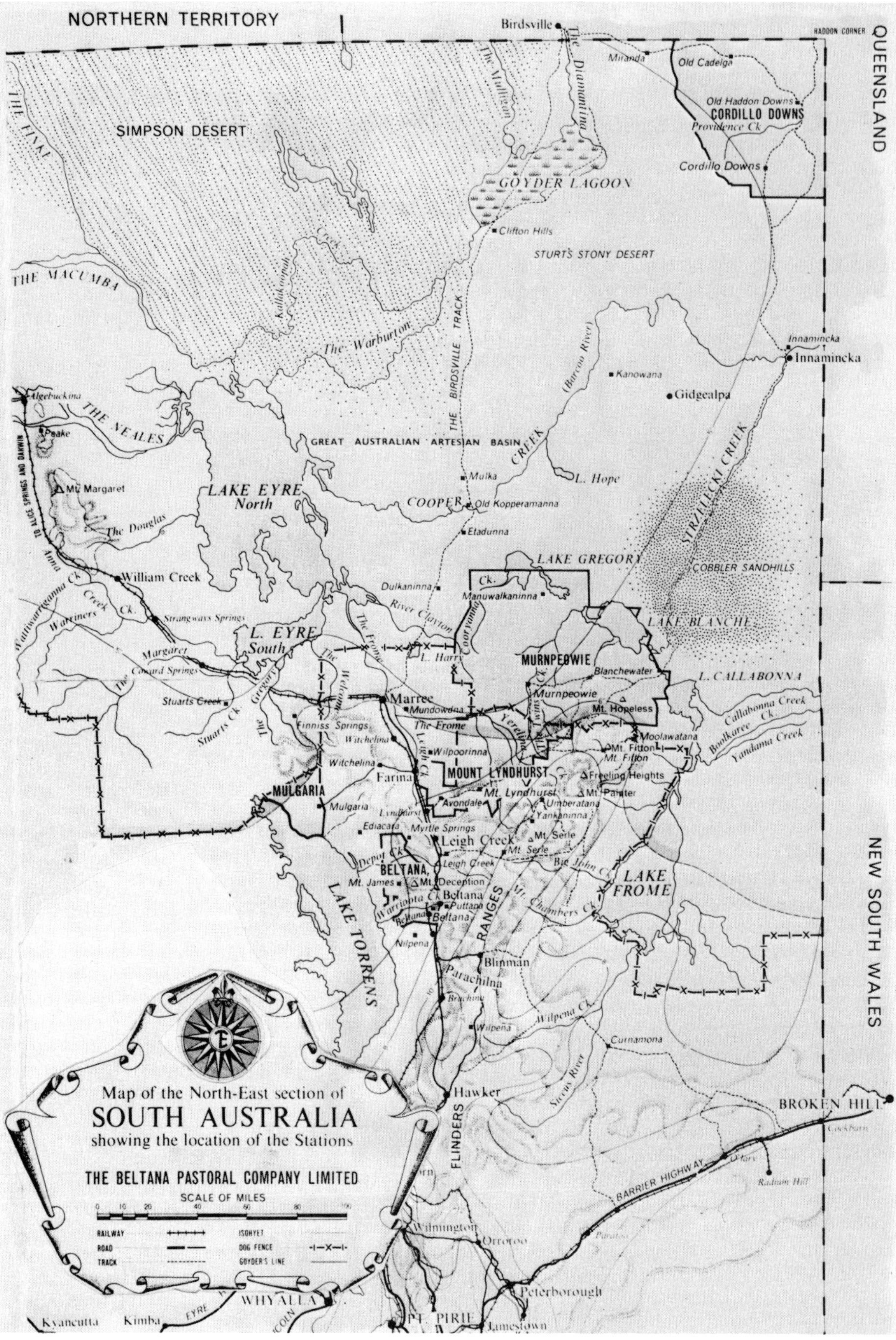

Map of the North-East section of
SOUTH AUSTRALIA
showing the location of the Stations
THE BELTANA PASTORAL COMPANY LIMITED
SCALE OF MILES
RAILWAY
ROAD
TRACK
ISOHYET
DOG FENCE
GOYDER'S LINE
NORTHERN TERRITORY
QUEENSLAND
NEW SOUTH WALES
SIMPSON DESERT
Birdsville
HADDON CORNER
CORDILLO DOWNS
GOYDER LAGOON
STURT'S STONY DESERT
THE MACUMBA
GREAT AUSTRALIAN ARTESIAN BASIN
THE BIRDSVILLE TRACK
LAKE EYRE North
L. EYRE South
COOPER CREEK
STRZELECKI CREEK
COBBLER SANDHILLS
LAKE GREGORY
LAKE BLANCHE
L. CALLABONNA
Innamincka
Gidgealpa
William Creek
Marree
MURNPEOWIE
MOUNT LYNDHURST
MULGARIA
Leigh Creek
BELTANA
LAKE FROME
LAKE TORRENS
FLINDERS RANGES
Blinman
Parachilna
Hawker
BROKEN HILL
BARRIER HIGHWAY
Wilmington
Orroroo
Peterborough
WHYALLA
PT. PIRIE
Jamestown
Kyancutta
Kimba

blanket, and some blokes rolled up a bit of oilskin tight to spread it out over the bindi-eyes at night so he could sleep on it, but that was a bit sissy. We never carried much, you could share things among several of you, such as a pint pot or billy, but each man had his own water-bag in case we ever separated.'

The Tasmanian shearers were among the best known of the cycling shearers. Leaving the island state in March, they put their locally made bicycles (the Cressy bicycle made by Mr Tapley Senior – his son, Ray Tapley, was one of the Longford shearers who rode the Cressy bike) on the boat, and sailed acrosss Bass Strait to Melbourne. Once there, they'd head for 'The Shearers Hut', Batman's Hill Hotel in Spencer Street, opposite the interstate station and near the boat pier. When all had gathered, they set off on the train to the station nearest to where they had been promised sheds.

The men from Longford usually started at Cordilla Downs, one of Australia's most isolated properties. Seventy-five-year-old Ray Watley recalled in 1978; 'If you agreed to shear Cordilla Downs the contractor would say he'd give you easier sheds to follow. Well, I shore Cordilla, but I don't remember any easy sheds.' To reach this place the shearers went by train to Adelaide, then north to the end of the line. The Ghan (the north-west train) at first only went to Marree; later it crept up to Oodnadatta, and in 1927 it reached Alice Springs. But Cordilla Downs was north-east, up in the far corner of South Australia, so they left the train at Farina, took the bikes out of the guard's van and followed the camel pads to Innamincka. 'When the wind blew, which was often, any track marks blew to hell and away. One year two rouseabouts turned up two days before cut-out: they'd been lost in the desert wilderness.'

Lorna Watley, wife – or shearer's widow as they were called in the nine months' absence of their husbands – brought up her family while husband Ray was 'over the water' on the mainland shearing. 'I used to sit in the cold Tasmanian winter and think, "I wonder what run they're cycling off to now?" Sometimes I'd think, "I wonder if they made it alive to Cordilla?" I wondered if they'd got through the Cobbler yet.' (The Cobbler is five miles of sandhills as high as a kitchen table.) 'We'd let our tyres down a bit and try to get over the hills. None of us were fat men I tell you,' Ray said. If they ran short of water on the Cobbler they could die, as men die even now in these waterless wastelands.

'We struck water in the Cooper once and had to be ferried over it, bikes, water-bags and all. This land out here uses any weapons it can lay hold of against trespassers. Bill Greig Jnr mended his tyres all the way to Cordilla once. 'Three-cornered Jacks pierced the tubes the whole way from the time we left the train until we reached Cordilla. It was hard to find the puncture – we could carry water enough for drinking only. The terrain is bad out there, real badlands, sand, rough stones, separately and together. A lot of chaps found it very bad on their eyes, the desert threw up terrible glare, lots of light like arrows from prisms fire up as the sun hits the facets of sand.'

Ray Watley first rode to Cordilla Downs when he was sixteen, and started as a shed hand. 'A shearer got crook with appendicitis and they had to take him

---

The Tassie cyclists put their bikes on the train at Spencer Street and took them off at Farina. From there they followed the camel pads to Innamincka, then it was 'off on the long lead' to Cordilla Downs (top right)

to Farina so the overseer says, "You'll take his pen". "I'm not a shearer," I yelled, scared suddenly. Anyway, I managed 1100 for the run and got £27. I was very proud.

'The whole trip was exciting and new. We saw Innamincka gleaming on the horizon. There was a stack of bottles there that could girdle the world. That sight made us pedal like hell through the sand. "Goat or galah" was on the menu at Innamincka pub. It could have been worse, an old shearer told me. "It could be a Barcoo sandwich: a goanna between two slabs of bark."

'We'd say, "Off on the long lead" – that's what we called the trek to Cordilla. The driver and people would stand by the train watching us go.' Ray Tapley, now an elderly man says wistfully, 'Oh, it was a lovely life. Oh, I loved it all. Hard, tough, lovely.' Ray shore from 1926 to 1946.

'We'd go over on the old *Loongana*,' he reminisces. 'Roll! Could she roll! Then the overnight train from Spencer Street to Adelaide. Then, "Look out Cordilla! Here come the Tassie cyclists!" We were all good riders in those teams. Bill Greig won the Latrobe Gift one year and of course I could ride well; it was my father made the Cressy bikes. And we loved it all. When it was too dark to ride when we were on the long lead to Cordilla we'd dig a hipper in the sand, throw the blanket in and settle down for the night in the Star Hotel.'

The Greigs and Lyons of Longford are cousins. The Lyons tell this story: 'Uncle Bob Greig was riding his bike up to Cordilla and he came on a lonely boundary rider. "How far to Cordilla?" Uncle Bob asked. "Aw . . .", drawled the man who rarely saw another man to converse with, "As far as the crow flies . . . about two hundred miles." "Bugger the crow flying," says Uncle Bob, "If the crow was on this bike how far is Cordilla Downs?"

The bicycle was popular among shearers during the 1930s depression. In those years a man could buy a £6 10s 0d bicycle with a deposit of 10s 0d and pay it off at 4s 0d a week.

In *On the Wooltrack*, C. E. W. Bean tells of the army on wheels when he saw it in 1926, as 'a wave'. 'Some part of Australia is always alive with it, for by the time it dies on the outermost edges of the continent it is beginning again in the centre. It works towards the south and towards the sea like a stir of wind over the surface of a great forest . . . each year from July to November, the shearers come across from New South Wales. And the sign of them these days is their bicycles.

'It is true some still come on horseback, some in sulkies, some on foot. But of late years the bicycle has spread through the country as fast as the rabbit. Their tyre tracks struggle across the very centre of Australia . . . in paddocks as lonely and bare as the Sahara.

'They are ridden, or driven, or ploughed or dragged wherever men can go and not infrequently where man cannot go. They shear through mile after mile of red sandhill, over a debris of splinters and burrs which must have crackled under the wheels for twenty-five miles at a stretch. But the bicycle gets through if the man does.'

Bean says the shearer sets off as if for a ride in the park. 'For exactly the same trip the average European would probably requisition a whole colonial outfit, compasses, puggaree, sun-spectacles and field glasses. The Australian shearer didn't even change his hat. If he wore a 'hard hitter' [a bowler] back in his home town or city, well then he rode off with the hard hitter on his head, as many snapshots of that era show.

'During a certain week a little before the spring, there begin to appear – leaning against the huts around the big, bare, shearing sheds that stand out somewhere toward the heart of Australia – bicycles. There is not a bicycle shop within two hundred and fifty miles. But bicycles turn up all the same. Some have brand-new tyres, full of air. Others are full of rope . . . If you look carefully at the bicycles that come in you will notice that some of their tyres appear curiously lumpy. The rider has foreseen trouble to them, and has got on his way a kangaroo hide and slit it into lengths, which he has bound round the rim. Often, when punctures grow too many to mend, they stuff rope or strips of basil inside the tyre.'

On the 600 km from Farina to Cordilla Downs (1921), men must carry only that necessary for survival, the most important being water bags specially made to fit over the bicycle bar

Photographs show us the shearer's bike set up for those epic journeys across an almost unknown area. Doubled and tied over the handles and front fork, sometimes over the back fork, is the swag that reminded Bean of the cavalryman's kit carried over his saddle bow or an infantryman's overcoat slung round him like a horse collar. The swag was covered in a canvas tent-fly ('Handy to sleep on out in three-cornered-jack country,' Dick Burton says, 'although nothing made by man could keep all those thorns out of you.') From the bar hangs a water-bag, usually a triangular shape made especially for bicycles, and hanging from any place where they would not clank too much were a billy, a pannikin and sometimes a quart pot.

A new word entered the shearers' language from the coming of the bicycle: punctured. To be exhausted, done in, beaten, was to be punctured. 'I tell you, by the end of the day I was punctured.' In true bush style they adopted their phrases from the everyday things around them.

Ray Watley and his mates have no more romantic nostalgia for the days of the bike than for the days of the horse: 'The shearers looked for the best means of transport, that's all. Nowadays they've all got cars. When we were young if you had a bike it was as much as you could hope for. And some have planes you say? Well, good luck to them. You work hard enough in a shed without rocking yourself silly on a horse or bike to get there. Still, I suppose they're not allowed to have plane races like we raced horses and bikes, are they?'

The push-bike was even part of bush poetry in the late 1890s:

> I pushed my bike from the Shearers' Strike
> Not wanting a funeral shroud.

recited the bush bard in *Humping the Drum.*

The goal: Cordilla Downs woolshed, in the far north-east corner of South Australia. The team cycled over the Strzelecki track (1909) in four and a half days, The Cobbler, sandhills and all

Monty Hanna from Old Galah on the One Tree Plain out from Hay saw plenty of wheel men looking for work during the 1930s depression. 'I remember one miserable coot. He had bound rope round the wheels of his bike when the tyres had shredded off. Too poor to buy new ones. He had string for laces in his boots. He was a shearer looking for a pen.'

Moses Mallock used to push a bike from Longreach to Winton, 355 kilometres there and back; '216 kilometres he'd pedal to Blackall and 256 to Jundah – and back again. And he wasn't even a shearer.'

Moses Mallock had a mixed goods store in Longreach. 'He sold everything from horse-shoes to the odd bottle of red Ned.' His marathon cycling trips were to drum up business.

In 1882 Moses came to Australia from the Lebanon, aged 15 and speaking no English. He began life in the outback as a hawker, his first step towards starting a small shop. 'Even when I was a kid you'd see Moses pedalling down the street, and he seemed an old man to me even then,' George Douglas recalls.

Moses' son Peter says he was only in hospital once in his ninety-six years, and that because he was knocked off his bicycle. 'It's a wonder it wasn't daily, with him riding down the middle of the main street as though cars had no right of way where bikes were concerned,' George says.

Everyone in Longreach knew Moses. The shearers say, 'He stood by the cocky when times were bad and he stood by us shearers when our hides were cracking. He gave us credit through the strikes and never pushed us or skimped our measure.'

Moses' moustache was famous in the west. All the shearers knew it. 'Great long thing, he waxed the ends and twisted them up into fierce curls each side of his face. One day when he was push-biking out to a shed at Molong drumming up business from the overseer, he got bushed. He got real worried. He couldn't recognise the property he'd landed at. He stood there beside his push-bike twisting his moustache until the wax melted and the ends dangled down. 'Is this Molong?' he asked a shearer on the track. 'Too right mate, too bloody long altogether,' said the passing shearer.

His family took his bike off Moses when he was 82, but he remained active in the shop packing up gear for the shearing each year. One morning after his

Larrikin show-offs, fag in mouth, some wearing their old army jackets, they later drove cars up the track. 'Setting off as if out on a picnic,' wrote C. E. W. Bean

ninety-sixth birthday he sat up in bed and said, 'Bugger it. It's a bit early yet,' and died.

The Cobbler was so tough to cross that it was necessary to harness a team of camels to an old-style mail coach to take over from the mail car in the 1920s. But not even camels could cross when the rivers sent water down

Moving around the wool track outback one hears names repeated over and again: men who have been recognised by the mob. (Mob is the only word to describe it: there is a freemasonry outback that takes in every man who is interested in what is interesting. Bush telegraph does the rest.) Some men are notable because of an idiosyncrasy, 'The Leaning Stump' for instance, the great shearer who appeared to be listing to starboard; cooks are often recognised but usually only if they are worse than awful or if they have uncommonly apt nicknames. In the main, the men who are legends in their time in the outback are the strong, the tough and brave in a workaday way.

Ian McTaggart is such a man. I first heard talk of him out from Port Hedland (northern WA). I went to Adelaide to meet him, but he was on his way to Mundabullangana, one of the oldest properties in the Pilbara (northern WA). He sent me a message, I received it at Quilpie (SW Queensland) and a contractor told me, 'Yes, you'd better meet McTaggart. A remarkable bloke.'

Some months later, at Broken Hill, a shearer said: 'If you really want to know about some aspects of the game you must see McTaggart.'

Now these areas are separated by distances such as Europeans would scarcely travel in a lifetime. I had gone two-thirds of the way round Australia, a matter of thousands of kilometres, and the name of Ian McTaggart kept cropping up.

A surprising thing was that no-one spoke of his age to me and because he was ricochetting backwards and forwards from Adelaide to Port Hedland, about an eight thousand kilometre round trip, I assumed him to be of an age to take such junketings. Actually he *is* of such an age, but few others who are his chronological age could make the journey, let alone grapple with running a property in perhaps the trickiest area for such activity in Australia. 'Flies himself,' the shearer said. 'Had a licence, oh years and years.' Tony Ryan, from the AWC, had known him when Ian had Nonning (SA), out in the arid lands.

He wrote to me once from Mundabullangana where he was weathering a cyclone. I had asked him about push-bikes.

'Push-bikes certainly played a very important part in the early days in the pastoral areas of South Australia; from 1900 onwards, motor transport was practically non-existent.

'Our property, Nonning, was situated about 145 kilometres west of Port Augusta and the drag of driving bush-fed horses over that distance when serious sickness was involved was a nightmare never to be forgotten. I remember well on two occasions when Father drove into Port Augusta, with Mother each time nursing a baby, with us children hanging on to the ricketty buggy lurching over the unmade tracks. Unfortunately, neither of the babies survived.

'The next time there was an emergency – Mother was very ill – telephones were not in use, but they tapped the Overland Telegraph to contact the doctor. He diagnosed the problem, the chemist made up the prescription – but how to get it out to Mother in time? Then a strong station lad spoke up. He would go and get it. He set off on his push-bike and arrived in Port Augusta the same afternoon, and after having a short rest decided he could make it. With the medicine in his pocket he set off without lights to ride the 145 kilometres through the night, and he had the medicine in my father's hands before breakfast next morning – and Mother lived. It was a wonderful trip when you remember there were no roads, only cart and bullock dray tracks that other vehicles followed.

'Bikes helped in the social field too. Strong active lads would save their pennies and buy a bike and they would ride surprising distances into local towns or visit friends on neighbouring sheep stations.

'There was also the sporting side of the push-bike activities: shearers would ride great distances to shear at a sequence of sheds and they, plus the station hands, would gather together at a long weekend for a sporting and social get-together. The station management would play their part and organise the facilities and there would be cycle races run over different distances. Opportunities would be taken to organise foot racing and athletic events and, all in all, it would be regarded as a very well-spent weekend with a great deal of enthusiasm from everyone.

'As time went on, push-bikes were used more and more, particularly in time of severe drought such as in 1914 and again, the worst in my memory, the 1927–30 drought. We were almost completely dependant upon the bikes and whole teams of musterers would go out and save the situation, the point being that horses were so severely handicapped because of the shortage of both feed and water, or again in the 1939–45 war, the situation became acute because of both feed and petrol shortage, making it almost impossible to cope without the bikes. My personal effort was to ride 15 000 miles (24 150 kilometres) during the course of two years of war-time effort to keep the situation both supervised and assisting with the mustering for shearing, etc.

'. . . A good deal of attention was given to making the bikes more efficient for the job in the difficult sandy country [with] the nuisance of numerous punctures to mend in the course of the day's work. The Dunlop Rubber Company came to our aid with the manufacture of the now famous thornproof tyres, the Dunlop Bushman Tyre, which were almost puncture proof; again, we ourselves converted the wheels to take a mini-balloon tyre; this idea was really stolen from the auto-cycles and was of great assistance in the heavy sand country. A great many bikes were fitted with gears [which] made the pushing easier and I did nearly all my push-biking with all these improvements fitted.

'However, a great advantage of using push-bikes was that the cost per mile was so low. In those times people were tough and active, and it was not beneath their dignity to ride push-bikes over long distances rather than face the long drag of using horses slogging along [when] there were not the facilities that are there today with road houses and motels . . . to assist the traveller.

'During the last war I did a pretty good one-day effort in company with an old boundary rider when we were mustering for shearing. I rode my bike out to where the old man lived in his hut the day before, and next morning with his two dogs and my dog, and well set up with water-bags and lunches, we started out across a twelve-mile paddock to get into the paddock we were about to muster.

'After riding along for about five miles we noticed about three hundred sheep that had been missed before. I said, "What a chance to get that little bunch; perhaps I should get them while you go on?" "No," he said very definitely, "The old dog will easily manage that job. In any case he is getting old for the longer trip."

A wool scour was installed at Cordilla Downs in 1885 to wash 40 per cent of grease and sand from the wool to avoid some of the high transport costs. Stacks of wool stand out like white mountains on the desert (1920). Camels took scoured, baled wool down to the railway

'Tom turned to the old dog and said very sternly, "You take them home." The old dog rounded them up and set off for the outcamp yard. We continued on our way into the next paddock and after we had travelled some considerable distance we noticed our dogs sniffing about. "That's settled it," said Tom, "And by the number of tracks I would say we have the main mob as good as in the bag."

'Away went the dogs and we lunched under a big spreading myall tree, then we had a little siesta, and the next thing was we were awakened by the bleating of sheep – they were all around us.

'Tom quickly made arrangements. I was to take the main mob while he would look for more, then we would come together again at a convenient sheep yard where we could yard them for the night. Tom counted them carefully, they were all present and correct. He had found quite a fair mob which were all put together for the night.

'A little wearily perhaps, we peddled along thankfully, it would be a lighter day tomorrow and eventually we came back to the outcamp yard. A pleasant scene: the old dog presumably fast asleep and the sheep peacefully camping inside the yard.

'We peddled on the last half mile with the dogs leading the way, they well knowing that there was plenty of kangaroo meat in the camp, and we well knowing that there was ample salt-bush mutton there also.'

John Howard, who with Wolseley devised the first and best shearing machine, went looking for more sales outlets in 1899 and hit on an idea of a cheap, self-contained plant which would appeal to the small flock-owner. The motive power was to be supplied by an ordinary bicycle, the rear axle being raised on a stand as are motor cycles at rest to-day. The power was taken from the tyre of the rear wheel by a friction pulley with the usual flexible shaft covered by a flexible tube similar to that used up to the early 1890s. One of these outfits was demonstrated at the Sydney Show in 1899 by the shearer James Davidson, with the cyclist Rod Hughes supplying the power. Eye-witnesses of this demonstration asserted that the cyclist usually 'petered out' after two sheep had been shorn. The price of the whole equipment including bicycle and sundries was only £18.

Although there appear to have been few sales of this ingenious plant, Mr Howard had a further plan. What better way to demonstrate the shearing machine to the outback graziers than to let a man pedal his way from station to station with a shearing plant ready to set up in a few minutes wherever required? Naturally enough, after his success of ten years previously as a missionary for the machines, Davidson was selected to give the scheme a trial. And so began one of the most remarkable single-handed efforts of a century.

Early in 1900 Davidson set out for Queensland with his bicycle – actually supplied by himself under the terms of engagement – and some 70 lb of gear, but with no knowledge of the prejudice against machines still raging among shearers and even among pastoralists. That hostility added to his difficulties of getting a heavily loaded push-bike through the sandy stretches or over the deep-rutted tracks which passed for roads in the hinterland of Queensland and New South Wales and even the patience, good temper and resourcefulness of James Davidson must have been sorely tried.

His usual plan was to arrive at the station whilst shearing was in progress, then shear with the machine some of the sheep that had already been shorn with blades. In this way he demonstrated most convincingly the machine's ability to take off half to a full pound more wool than could be removed with blade shears.

The shearers usually treated the demonstration as a joke until they saw the miniature fleece coming off a sheep they had just shorn. This changed their attitude and some even lost their tempers, wagering that they could shear a sheep cleaner than any machine. At this point Davidson would tactfully explain that he was not trying to show them up, only demonstrating the superiority of the machine over the old-fashioned blades. He would further point out that it was against the shearers' own interests to show that they could shear any cleaner with blades than they had been doing, as it would only land them in an embarrassing position with the boss. This line of argument always produced satisfactory results with the more reasonably-minded shearers, except at one station, Malvern Hills, where a bucketful of stones was got ready to 'speed the parting guest'. Fortunately, the Malvern Hills manager had warned Davidson and when morning came he was able to depart by another route, missing the fusillade.

At Eurella, near Jericho, almost the first shed visited, an Aborigine was hired to provide the necessary pedal-power, but the scheme was scarcely a success. After three sheep had been shorn the blackfellow refused to go on.

Davidson met the famous big-gun blade shearer, Jack Howe, at Alice Downs, at this time doing good tallies with machines also. Davidson always recalled with the keenest pleasure this talk with Howe.

At Glenelg, near Stanthorpe, hostility again reared its head. Only an appeal from the manager for the shearers to take a broadminded view and to see that

Donkey teams bring wool to the scour. Station store (background) is built to withstand cyclones

the machines were for their good as much as for anything else enabled Davidson to make an unobtrusive exit. Working south through New South Wales, the situation at Piedmont, Barraba, took a similarly ugly turn, but the owner intervened in time. Through trials and tribulations, Davidson came at last to Tamworth, where he wired Dangar, Gedye & Co.: 'Too early for local sheds – advise.' The laconic reply, 'Please yourself', gave him a welcome opportunity to join the first Sydney-bound train. His three-month journey must be unique in the annals of individual efforts in salesmanship. Every shed visited by the 'bicycle barnstormer' installed Wolseley machines the following year.

The famous bicycle outfit

26

The One Tree Plain

In fact the saying far and wide
Along the Riverina side
Is Hay and Hell and Booligal.

One Tree Plain was the 'Hell' Patterson wrote of in 'Hay and Hell and Booligal'.

It never was a town as we envisage the word.

A patch of grey discoloured sand,
  A fringe of tufty grasses,
A lonely pub in mulga scrub
  Is all the stranger passes.

Yet one could scarcely write of shearing without including the desolate ocean of smoky mirage where the scorching north wind blows, for it has been commemorated by shearers and drovers in verse, song and story for over a century.

Today it is no less desolate and, until one has driven across it, there is no way of understanding the fascination and terror it has held for men since sheep first came to this region. On either side of the unfenced road the sun burns into a grey desert that disappears into the distance where it drops off the flat earth. There are sheep stations here, but one sees no buildings or fences. At rare intervals a few rangy sheep munch what appear to be dead, dry, stunted clumps of grey twigs – the nutrient-rich salt-bush.

During summer, when all things approach the traveller on a mirage, as if moving on a cushion of visible air, the plain has little joy about it.

At the Nine Mile Box hotel between Hay and Booligal, the proprietor of the little inn would cry each night: 'Any of you gents like to see the mail car turn into Booligal?' Not wanting to miss anything in a land where *nothing* happened, most of the men in the bar rushed out. 'Halfa mo! She won't be long now! Just look up the road.' The men would look up the straight track. Then, 'There she goes!' and like a falling star going sideways, was to be seen the bright light from the coach turning in to the Booligal Post Office forty-five miles away.

Dust storm from the plains blots out the town of Hay

In the centre of the One Tree Plain where a solitary tree graced the otherwise empty landscape, partway between Hay and Booligal, was built the One Tree hotel, and in the 1870s and 1880s this little hostelry was 'the centre of progress and settlement of the vast area of land'. The drovers, teamsters, shearers and swagmen, the travellers on Cobb and Co. coaches and the property owners all congregated at the hotel, sometimes to meet the coach and collect mail, goods or a passenger, sometimes just for company or to have a meal (the table d'hôte was 1s 0d) and a drink at the bar. Some arrived in time to save their lives.

There's a run on the Western limit
  Where a man lives like a beast;
And a shanty in the mulga
  That stretches to the East;
And the hopeless men who carry
  Their swags and tramp in pain –
The footmen must not tarry
  Out there on the Great Grey Plain.

Out West, where the stars are brightest,
  Where the scorching north wind blows,
And the bones of the dead seem whitest,
  And the sun on a desert glows –
Out back in the hungry distance
  That brave hearts dare in vain –
Where beggars tramp for existence –
  There lies the Great Grey Plain.

In the 1850s and 1860s a few squatters had acquired enormous holdings comprising most of One Tree Plain and the surrounding country. T. F. Patterson had Merool, now Ulonga, over 200 000 acres of salt-bush that stretched from Booligal to within a few miles of Hay. By 1885 he had 60 000 sheep on the plain. His woolshed, ten kilometres from the One Tree, had fifty-five stands – all bladesmen. For weeks before shearing started men could be seen tramping over the plains and congregating around the woolshed and the single tree growing on a clay pan near a shallow fresh-water well. Some came on horseback, some in sulkies and carts, some on bicycles and some on Shanks's Pony carrying their swags. Many came hundreds of miles to this famous flat land. The only advice a man 'on the tramp' was given was 'do not tarry on the great grey plain'. They were to carry little, and that little was to be water.

They camped and waited, and by the time shearing began up to one hundred men, sometimes more, would apply for work as shearers, shed hands, cooks. For those who were not selected it was out again on the blistering One Tree Plain. They were followed by those who had been selected but whose ability after work started failed to please the boss of the board, and they were sent off

Halfway between Hay and Booligal, the One Tree hotel was built in 1862. Shearers carrying their swag over the great grey plain found a temporary haven

'on the tramp'. A handful of men always hung around with hopes of being taken on when others were sacked.

'It was a long tramp for those poor fellows out on One Tree Plain,' John Adams (Vic.) who had horses out there in the 1890s told me. 'If a man wasn't fit and well he was in dire straits. And many of them had families. It was hard all right. None harder.'

The town of One Tree sits on part of the old Ulonga run but it never consisted of anything but the hotel. The tree was quite a famous landmark and was gazetted as Timber Reserve No 32 880 in 1901, the smallest timber reserve in New South Wales, but the historic gum tree had crashed to the ground during a gale on New Year's night in 1900. Many of the shearers, drovers, teamsters and swagmen were heard to say with genuine regret, 'So, the old tree's gone.' Its welcome shade had been appreciated when the white heatwaves blazed on summer days and the camp fires welcomed them on desert-cold winter nights. Over the years hundreds of men had camped awhile or paused for a short time under the sheltering branches.

Near the hotel, the big government tank watered up to 12 000 sheep at times,

and teams of horses and bullocks carting wool were watered at one penny a head.

Tom Patterson, who came here in 1870, died on 27 October 1902, but his name lives on in Henry Lawson's 'Flash Jack from Gundagai'.

I've shorn at Burrabogie and I've shorn at Toganmain
I've shorn at Big Willandra and on the Colleraine,
But before the shearing was over I wished I was back again,
A-shearing for old Tom Patterson on the One Tree Plain.

All among the wool, boys, all among the wool
Keep your blades full boys, keep your blades full.
I can do a respectable tally myself whenever I like to try,
And they know me around the back blocks as Flash Jack from Gundagai.

I've pinked them with the Wolseley and I've rushed with B-bows too
I've shaved them in the grease lads with the grass-seeds showing through
But I've never slummed a pen me lads whatever it might contain
When shearing for old Tom Patterson on the One Tree Plain.

I've shorn at Big Willandra and I've shorn at Tilbaroo
And once I drew me blade boys upon the famed Barcoo
At Cowan Downs and Triada as far as Moulamein
But I was always glad to get back again on the One Tree Plain.

I've been whalin' up the Lachlan and I dossed on Cooper's Creek,
And once I rung Dadjungee shed and blew it in a week.
But when Gabriel blows his trumpet lads, I'll catch the morning train,
And push for old Tom Patterson on the One Tree Plain.

Shortly before his death Patterson had paid an annual rental of £116 7s 6d for an improvement lease of 14 700 acres from 30 August 1902 to 29 August 1930. (Improvement leases were granted over Crown Lands considered unsuitable for settlement until 'improved'.)

The selectors were brave, valiant and game for anything. On the black soil plains in western New South Wales a family named Eade walked 'from up north' with 'all their worldly goods on a wheelbarrow and took up a selection north of the Lachlan River. In the big drought of 1901 there was only one wet water-hole on the Lachlan bordering their acres. The father and his two sons began to dig for water. Down and down they dug, hauling each other up and lowering down a relief digger. At one time the father looked at his boys and thought they looked tired – and realised they had dug all day, through the night, and now, twenty-four hours later they were still digging. And Eades still hold the land.'[1]

The Sheaffes at Eurugabah out from Booligal have battled as hard as any selectors, right up to this date. Eurugabah was taken up at the same time as Ulonga and borders the Lachlan River. In good seasons the grass is so high that the present owners truthfully boast that 'we have to carry a ladder on the back of the Landrover to see the sheep'. But in November 1980 their woolshed was an oasis in a blackened land, the legacy of the grass fire that ravaged those acres in November 1979. Both Bill Sheaffe and his wife, Jennie, suffered burns in that fire. Then the call came: 'To the woolshed!' and the family and men gathered there, burnt a back fire across the dry grass and hoped.

'We had it terrible hard in 1928–30,' Montgomery Hanna told me. Monty had been in the Light Horse in World War I so he knew the varying degrees of toughness. He had called his selector's block Old Galah: 'for a time it seemed galahs was all it would feed.'

'We were nearly fifty kilometres from the river here at Old Galah. By the time those poor fellows with swags on their backs reached us they were in a bad way. We weren't in a good way ourselves. We had three dust storms in a week. It was hard on the women. We had no water for cleaning up. Dust was inches thick on everything. The wool on the sheep was clogged with it.'

Dust storms are 'big blows' on Old Man Plain. The most ferocious is believed to have been in February 1900. Two days previously the thermometer had registered 104°F., next day 102°F., but by noon on the 27 February the official reading was 113 degrees. By 3.30 p.m. it began to get cooler. A khaki cloud was observed in the south-west. Before 4 p.m. all doors and windows were shut. In Hay the shopkeepers closed their shops, traffic stopped. For two hours dust blew over, red, black, brown, white it appeared in turn, a solid wad of matter. Squinting, the residents 'could not see across the footpath'. As the wind went past the dust fall lingered and as the hours went by 'there began to arise a certain pride in the performance'. It was the worst dust storm they had known. People from 'Wilcannia, Cobar, Broken Hill, Western Australia were here beleaguered and one by one admitted it was the worst day they had ever experienced'.

'An old inhabitant spoke about storms of '68 or '83 until the dust became thicker and finer and more penetrating, until he gave in. A squatter had become lost on his own property, the Booligal coach had to pull up at the Nine Mile Box, the train was held up at Illilliwa. For pertinacity, for grandeur there had been nothing to surpass it. Even the Boer War had to take a back seat.'[2]

When the next day dawned the elation subsided. . . . The scene presented at every home was a sad one. Everything was covered with a very fine dust laid on thickly and evenly. And that included the wool of the sheep.

Old Galah is 4050 hectares (10 000 acres) and shears 3000 sheep. In 1923 Monty's father bought Embah, 4860 hectares (12 000 acres) of salt-bush 'and not a tree on it'. The Hannas were able to hang on but many others could not, and their deserted properties are sad cenotaphs to their foundered hopes.

When he came back from the war Monty planted trees, and there are now great, shady arbours round Old Galah homestead. The only telephone out that way is the line strung along wobbly, knobbly bits of skinny tree trunks that Monty and six neighbours carried there and erected in 1935. Like many properties, there was a pub or shanty house on the boundary fence. 'Mostly there was a pub of sorts every fifteen or twenty miles or so – far enough for a man or horse to make it in ease in a day's march.'

Monty, now aged 84, all his years spent with his hands in wool, is adamant about who did most for wool in Australia. 'Sir Samuel McCaughey, the Peppins, the Falkiners and the Collins of Bungaree (SA), all did more for Australia and its sheep than John Macarthur who did much for himself.'

South Australian sheep are popular in this area. 'South Australian sheep can out-walk any of the other breeds we get here. They can walk to water. Once you cross the Lachlan you run out of SA type sheep.' This is a slow shearing shed. 'A man who shears forty a run elsewhere can only get through thirty on the rough sheep at Old Galah.' They are sheep that depend for much of their lives on salt-bush. 'After good rains there is barley grass, trefoil and cotton bush. But anything, man or beast, to live on the plain grows tough.'

One Tree was known wherever sheep were shorn in the eastern states. A shearing Pyalong (Vic.) poet named Fred Walter wrote:

Said Jack to Joe I think we'll go and start to get things ready
For Tom and Bill will soon be here and so will little Freddy,
The 'Pig' and 'Buck' they are held up, by rain somewhere up country,
And no-one knows when they'll get here, for they're comin' down from
  One Tree.

The 'Pig' was a shearer, J. Pigdon, and the 'Buck' was a local, J. Buchanan. And all knew the great grey plain eight hundred kilometres away.

From the camp, while the rich man's dreaming,
  Come the 'traveller' and his mate,
In the ghastly daybreak seeming
  Like a swagman's ghost out late;
And the horseman blurs in the distance,
  While still the stars remain,
A low, faint dust-cloud haunting
  His track on the Great Grey Plain.

27

Drought

There was a time when all Australians knew what a drought was. Our poems, stories and news items told of it. When radio came we had serials. We wept when Dad gathered Dave and Mabel and Mum together and told them: 'It's beaten us. The stock's all dead, the grass – well, when did we last see grass? We've cut the trees down to feed them, but there's no water to keep them alive. The few that we've kept alive are so poor no shearer worth his salt would shear them.' Mum says, 'So it's come to this Dad?' 'Yes Mum. After all these years. You've been a slave when I've been a slave. And Dave, you slaved from the day you were born. But, it's come to this.' Mum begins to pack. 'No Mum, You can't take any of those things. The mortgagee, the bank, they own the furniture now.' 'But Dad . . .' 'No Mum. Nothing. We take nothing.' Mum draws a breath and then she says bravely, 'You're wrong Dad. We're not going with nothing as you say. We're taking all that matters, our family and our hope and our memories.' Music fades in. 'There's a track winding back, along the road to Gundagai.'

The plain

Yanda, Bourke district (NSW) in drought-parched land

Corny? Corny as hell! But by God it, got to us. We *knew* what a drought was. *We knew* how it affected the land and wrecked the people. Today, if such country matters as droughts get more than a short paragraph, they are aimed at an academic level that makes it an intellectual exercise that reaches only a small part of our mind. Children rarely hear the names of the inland, the litany of rivers, of towns and poems that weren't literature but stirred our blood.

In 1982, leaving Fort Bourke Stockade behind you, you could drive out past Deadmans Creek, over the lignum that sheep don't eat, pass flocks of emus changing colour from bronze to black as they veer in the sunlight away from a big red six-foot kangaroo, close your car windows when the smell gets too bad from the rotting carcasses of kangaroos killed by cars and sheep killed by hunger and disease, and drive in a dust cloud to Yanda. The shed built in 1889 stands – just – and a motley crew of shearers is on the board. 'No reputable skilled shearer would take my sheep on,' the owner says.

Eighty-one-year-old George Cock bought the old property with money earnt the hard way – toiling. He fished the Darling, twenty miles from Bourke. 'I had to be a careful man.' The fish he carted into town helped him buy three blocks, old selections, Redbank, Hamilton Park and the 1888 property, Yanda. He was running one sheep to six acres (2½ ha). He had had 12 000 sheep. They died. Now he is shearing 4000. The big problem for him to shear them now is their weakness. 'They can't walk far. There's not enough feed to risk droving them far.' He had to cart some across the river on a punt.

'It's a hard country.'

Yanda homestead decays away down on the river bank. There is a large picture painted on the wall above the fireplace, mud nests of swallows, and the floor boards are burnt by the fishermen who hurl up to the mud-hole in the river and camp in the house.

In 1917 Yanda had 60 750 hectares (150 000 acres) and 40 stands in the shed. When George Cock bought it in 1941 there were 42 525 hectares (105 000 acres). Now, 1982, there are 34 830 (86 000).

'It may be the last year that shearing takes place in the old shed. It needs repair, and what with everything else . . .

'The land is in a bad way at the moment. Big floods two years ago covered much of it and when it receded it left behind salt that had leached into the river from more land that had been flooded, and so it goes.'

In the shed wool is coming off and the smell isn't too good. The sheep are lousy and the wool is cotted (knotted, tangled). Because of the bad season the sheep are poor and wool suffers. This wool is hard to pull apart as opposed to the free fall of good wool.

'You're seeing it in a bad season.'

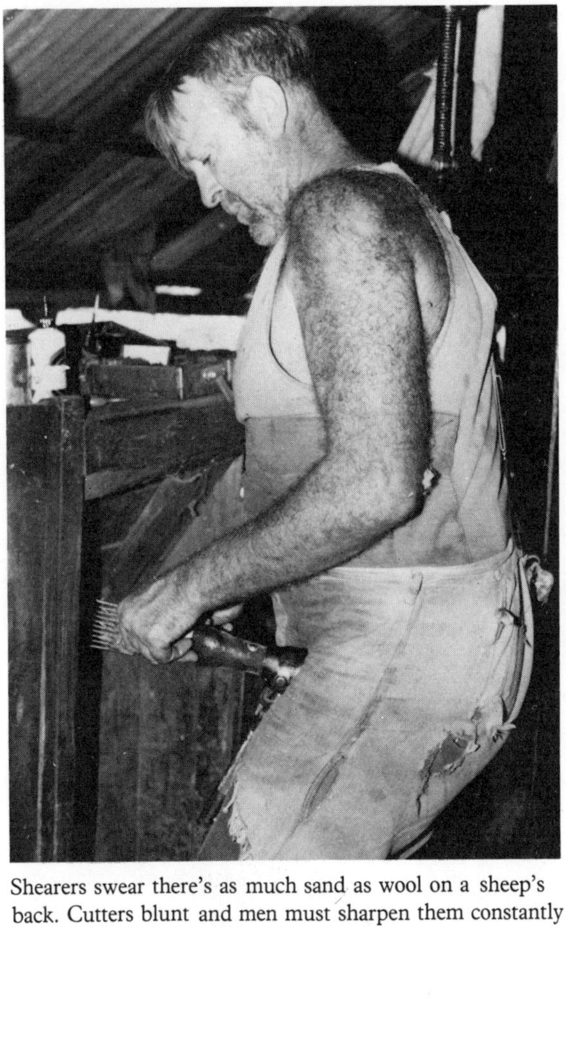

Shearers swear there's as much sand as wool on a sheep's back. Cutters blunt and men must sharpen them constantly

John Mackay, expert-all-rounder, cares for the Yanda machines

28

The Desert Schooner

They were the last beasts of burden to enter the country, but within a decade the prints of their feet were on every track, from the Gulf to the Strait, from Bourke to Meekatharra. They were hung with shells from the Orient, tiny bells tinkled from saddle and cinch and rein, as they plodded

The leader arrives to take on the Onepah (NSW) wool

Leaving on the 600 km trek with Cordilla Downs wool

through the dust and heat to take the wool in to the railheads of Western Australia, New South Wales, South Australia and Queensland. In dry years camels were used regularly, but in wet years their large padded feet slipped too easily on the wet red clay pans and the infamous skids of the black soil plains.

When drought found the weakness in horses and bullocks, camels brought the wool in from the furthest points of the outback. With a bale of wool on either side of an animal, the long strings swayed across the miles of mirage-flooded deserts like Arabian caravans, their turbanned Indian drivers walking beside them over the desolate country from what seemed to be the end of the world. For half a century camels were the major form of transport over one-third of the area of Australia.

Sir Thomas Elder, who was deeply involved in the pastoral industry in South Australia and western New South Wales, brought camels to Australia in 1862. He later set up a stud farm at Muloorina out from Lake Frome. With his first importation of camels Elder brought handlers from India's north-west frontier, near the border with Afghanistan, and camel drivers became known as Afghans – although it would seem from the sketchy records that no more than 10 per cent were other than Pathans, from India (now Pakistan).

Elder has been attributed with designing the modern wool bale – shaped so that one bale can be carried on either side of a camel's hump.

The camel took to the outback and the outback took to it. 'The "Belle of Beltana" rides a camel!' the Tasmanian shearers wrote home.

29

Bullocky O!

Where, in clouds of dust enveloped,
roasted bullock-drivers creep,
Slowly past the sun-dried shepherd
dragged behind his crawling sheep.

– Henry Lawson

Bullocks do not often get ideas. But once their master, the old Bullocky, passed his tongue over them they got the idea of his scorn.

Shearers at smoko, nearing the end of cut-out, used to watch the bullockies moving off to the railhead or the river with their loads of wool. Dick Burton

Bringing the Kilfera wool down to Hay. Through Booligal, across One Tree Plain the caravan trundled to the railhead

(Tas.) told me, 'They were never stuck for a word. I saw a bullock down on its knees one day. It was shamming. "Scarpy" the bullocky walked up to it. "I'll teach you to pray if you don't get off your knees pretty quick!" he said, and up the bullock got, pretty quick.'

A bullocky used his tongue more than his whip. 'Poor beggars,' a bullocky might say in a rare moment about an animal, 'He's a slave and I'm a slave; we're both doing the best we can.' 'I'll give you something to go on with!' the shearers heard one call to a bullock straining to get the huge waggon loads moving from the shed. 'I'll see you in a minute, I will! You see if I don't!'

Some men who knew them well have claimed that the bullocky's reputation for being more foul-mouthed than others is not true, nor are stories that he was very cruel to his bullocks. 'Why would he be cruel? They were his friends, often his only companions for days at a time. If he hated them his work would have been unbearable. As for cursing, I've heard as bad or worse in a city street,' Buddy Wagstaff of Thorn Leigh station (Qld) told me. Hanging over Buddy's front door at her century-old home, Thorn Leigh, out from Blackall, are the great spreading horns of the lead bullock that dragged Buddy's grandfather's waggon over the great grey plains to found this property when no other white man lived here.

The shearer on his way back from shearing a property passed the bullock teams lumbering across the plain. 'You saw it miles away, a great red cloud on the horizon, and you knew that somewhere in that dust was a waggon, a man and his team.'

The teams that brought the wool down from Thargomindah are a fair example of the work and the period. This station is in Queensland, near the borders of New South Wales and South Australia, and the wool often was brought down 390 kilometres to the railway in New South Wales. Over this red land there was an average of one man for every 250 square kilometres – and an average of one waterhole. At two places the waggons must cross depressions and, if

it rains up in Queensland – hundreds of kilometres up that is – these two depressions become roaring rivers, the Warrego and the Paroo. These two maybe-rivers are 145 kilometres apart at this point. If they were running fast, deep or wide, or if the bed was too soft to cross, the teamsters camped. And waited. Sometimes for weeks at a time. Swear? Who wouldn't? When the Namoi River came up teamsters were known to camp for three months: a stretcher of sacking on two poles under the back wheels of the waggon and the tarpaulin that covers his load pulled down as close to the ground as he can get it for shelter. And there he sleeps, with his dog near the camp-fire, until the water goes away.

Freight in 1902 was £4 per ton per 100 miles. If there was grass or feed on the route, a man might be able to make three or four trips back to a shed before the whole clip had left it: quite often the competition was keen and if a man struck a dry patch and couldn't force his bullocks, well then, he missed out. The wool season is not a very long one, and it is the one time of the year a bullocky could make his main profit.

Disasters could befall him. Bullocks eating thistles before they had become used to them killed many a beast. In 1904 a mob of 600 bullocks on the march from 'up north' were camped on the Common near Gunnedah and by morning 585 were dead. With bullocks worth from £7 to £14 a head at the time, a battling bullocky may be forgiven for swearing.

One of the first big land-holders at Bathurst went through the fierce drought of the 1830s, when eight bullock teams died taking his wool to Sydney – 'and

Ready for the long road, Micalago (Qld) 1890

North of Booligal a waggon with wool from Clare station bogged, 1925. Angus Waugh, owner of Clare, brought Reg Sharpless, a young jackeroo with a camera, from Mossgiel station. They left their buggy the other side of the mire; a load from Bundarra station rolled in the treacherous mud. A third waggon has successfully got through the Wilandra Creek crossing

that was only a 150 mile trip.' Teams could range from four to forty bullocks in a team.

The bullocky's manner was different from the quick-witted shearer: more deliberate, his philosophising had the quality of original thought. Sitting on his upturned grocery box reading a paper or a book, smoking his evening pipe in front of his camp-fire: that's the sight many travellers had of the bullocky.

William Gough, the old drover, first came to Hay in 1884 as a five-year-old behind a team of fourteen bullocks and four horses being driven by his father. He later knew many of the well-known teamsters and drivers, 'All tough, fearless men. And the pay wasn't much then. Drivers £12 to £14 a month and found themselves.

'Then came the bullock teams. Fourteen to twenty bullocks. Carriage rates from nine pence to one shilling per ton per mile. A load of eight tons from Mossgiel to Hay was £40. Eight tons was a good load on those roads then. The teamsters used to travel in pairs so if one got bogged the other could pull him out of the bog. When I was a lad I saw four waggons loaded with wool bogged down to the axle beds at Adelaide camp – forty bullocks on one waggon, four drivers, two on each side of the team with fourteen-feet green-hide whips. And those chaps could use those whips, they could make the hair fly with every stroke.'

There is a nice story told in Victoria that Geelong got its name from bullock drivers shouting 'Gee-long!' 'As they swung their whips to get the teams started from the beach with the drays heavy with stores for the inland stations they'd cry "Gee-long!" to get the bullocks pulling steady on their yokes.'

From the same area comes the story that most consignments included rum, and that once a year there was always one keg short. This used to be sunk in a hole in the creek near where Titanga woolshed stands, and during the wool carting season each teamster, after making camp, would haul this keg up the bank and have his tot before carefully replacing the keg for other drivers.

Horse teams were so beautiful – even when they were a common sight – that artists painted them and photographers snapped them by the score.

Westward Ho Hotel, Isisford, as good a place as any to rest a team

There was a high spirit of competition existing among the teamsters. Each man searched for a job that would be more outstanding than the next man, feats that would be spoken of around the camp-fire for many a year.

In the 1920s Tom de Fraine was one of the best known carters at Glen Innes (NSW). He had a team of seventeen horses and it was said that 'his lead grey knew more about team handling than de Fraine himself'.

The old teamsters, like any onlookers, saw most of the game. W. 'Bill' Robertson was a teamster in 1914. 'Before I went off to the war in 1914 I carted timber to Alma station [twenty-three kilometres from Booligal, NSW] for the new twelve-stand shed. The old shed was the last in that district to shear sixteen stands of blade shearers.

'There used to be quite a lot of shearers from the Monaro district come out there with pack-horses before that war and they would have a race meeting for shed hands only. Tarwong station in the Oxley district is a place I saw with a good race meeting too.'

Fred Elliott, a teamster, carted wool by horse team from outback stations to Hughenden (Qld.) His usual load was 100 bales. A 'small' load would be 60 bales. (When motor transport came Fred was jubilant to find the early trucks were able to carry only 20 bales). Fred took misfortune as part of the day's work; he had his ankle broken by a kick from a horse while out on the track, and 'it took the horse three weeks to get him into town', Fred's son Meggs remem-

Near Nelligen (NSW), 1910

bers. 'Later again, Dad had his leg broken out with his teams. Gangrene set in before he could get to help. Cobb and Co. coach came on him and wouldn't turn back to take him to Winton, and as that was closest and they thought he'd be dead before they got him to Hughendon, they left him there with his team. He used to ride with one leg after that.'

Wool was the only produce that could be grown in the inland for export (refrigerated ships were not in use until after 1880). But it was so valuable that the high cost of freighting its weight and bulk by land and sea was well covered by the return it brought to the growers. Those men who have been recorded as the profane teamsters travelled over the terrible terrain to bring this crop towards the ports.

The dust is blowing in Kelpie Land –
  The dust that follows five thousand head
Over the black soil, over the sand,
  Over the ridges red.

– *Anon*

When Charlie Gough was young, his father's sheep dog came to him and rested his head on the little boy's knee. Charlie patted the dog and 'the next thing I knew my ears were ringing. My Dad had given me such a clout on the side of the head. "Don't you go ruining a good sheep dog," he said to me, "Spoiling the best dog I've got".' When Charlie went droving himself he knew that his father was right. 'If it's a pet you want, get yourself some fancy dog and pet it. But if you're droving sheep you get a Kelpie and treat it like a good drover's dog.'

The thing about Gough's story is that on the surface it may appear that the pampered pet is the lucky one; but anyone who has seen a sheep dog at work knows of the satisfaction that a dog has.

When flocks were small in the early days of the sheep industry, Smithfields, bob-tails and Russian collies shepherded them. When the inland was opened up and huge flocks were established, the long-haired dogs were of little use in the heat and rough terrain and over those distances. The sheep men tried many types but all dogs lacked the ability to work 'wide' and silently. Searching for these characteristics brought the Kelpie into being.

In the 1870s Scotch collies were brought in to Australia; not the Border Collie of today but, it is believed, a smooth-haired, black and tan variety of dog. In that decade the King brothers, two men who owned Hanging Rock and Wollongough stations (NSW), began breeding from Scotch collies for sheep dog trials. These trials, which were first held at Forbes (NSW) between 1870 and 1876, were won by one of their bitches called Kelpie. People began to speak of King's Kelpie, and when the first Kelpie had pups they were known far and wide as 'Kelpies'. It was a good name though it came by chance. In gaelic mythology a kelpie was a water-sprite or devil which is to be feared. As more pups were bred from the original bitch Kelpie, they were known as kelpies, the name of the new breed.

There are as many theories and disagreements about the origins of this unique sheep dog as there are about almost every other thing in the bush.

Some claim that the kelpie is largely of dingo extraction, the shape of its head and ears, the lope, colour and working style being all characteristic of the dingo. Many kelpies do resemble dingoes in their manner of working sheep, but

Loading the waggons, Cootlantra 1914; a family day out as it was at most stations

breeders believe that the relentless approach and concentration which all good kelpies share with dingoes is merely an attitude that the dogs have in common.

Other theories maintain that the fox was one of the ancestors of this dog because of the prick ears, the foxy sharp appearance and smooth coat, allied to the fact that there was one red pup in an early litter. But similarity of appearance is no evidence of a throw-back relationship.

Late in the 1870s Mark Tully of the Forbes district mated a smooth-coated, prick-eared black collie with a bitch of the kelpie strain to breed a black dog known as Barb. Barb has been credited with having founded a distinct strain of the kelpie, known as 'barbs', but the *Australian Encyclopaedia* asserts that this is 'one of the myths of Australian dog history'. 'There has never been such a strain. Barb was a kelpie. And although kennel clubs refer to "kelpies or barbs", all dogs so designated are kelpies. It was with dogs of the Kelpie and Barb lines that the most famous line of kelpies was developed, and almost every good modern kelpie can be traced to them.'

Kilfera, out from Ivanhoe, bought an 'Electric road-train' (1913) to move their wool across One Tree, down to the railhead at Hay. Prime-mover (front), is power plant, engine and generator, leaving woolshed

The kelpie is an attractive dog, active, alert, lithe, capable of untiring work. 'They work till they drop,' Mick O'Donnell says, 'but by the time they drop, the drover and the mob of sheep have been dead a week.' It is not a biting dog. The things about it that are indispensable in working with sheep are its ability to control 'wild' merinos – some go so far as to call it 'an uncanny ability' – its toughness and its tractibility. In the inland they claim that for mustering or working sheep in yards, a good kelpie is worth three men.

In the hot, arid regions its smooth, weather resistant coat and clean-cut build along with its determination make it the ideal sheep dog. The hair on its ears, head, legs and feet is shorter, thus protecting it from grass seeds and bindi-eyes. It can be any colour, black, black-and-tan, tan, red-and-tan, red, fawn, chocolate-smoke-blue to almost white – but the lighter colours are not favoured for show purposes.

Down south of One Tree (NSW), it is as eerie to travel the Old Man Plain today as it must have been when the first sheep men came. So much of it is different from anywhere else one has been. One sees a sheep dog tied up at a culvert and slows down, thinking to see a drover round about, then gets to the next culvert some miles on without seeing a living soul and – there is another dog tied up. Once I saw a dog tied to a stake in the middle of a dry red clay pan. Still no drovers. Far on one horizon I thought I saw a fence: the mile width of the road is arbitrary but apart from a few lanky-legged sheep my car was the only thing that moved on the plain.

In Hay, local drovers told me that the dogs are left tied up there to keep the stock from wandering. 'And it's safe enough!' says Mick O'Donnell. Safe? From what? 'Safe. No-one would pinch a dog that side of Hay because it would be a Denny dog – pinched off the street or the back of a stock truck in Deniliquin. No bloody good. No Denny dog is any good. I tell you, no man would leave a good dog tied up like that on the roadside. Swish! He'd be gone in a drover's plant like that.

'Now I'll tell you something else: up at Ivanhoe (NSW) a dog tied to a ramp is a good one, take a bone and smoodge to it first, then undo it and get it on your plant quick and get going. It will be a beauty. The "permanent dogs" are chained along the road to head cattle or sheep back. Not much good for anything, a Denny dog.'

The trail is long and the dams are dry;
The burrs are clinging to raw, red feet;
Under the imminent brazen sky
Hovers a haze of heat.

– *Anon*

30

The Sweat Extractor

The great strikes of the 1890s led to a change approved by shearers and wool-growers alike: the introduction of contractors.

Until this time neither men nor growers were enamoured of the man-between. Now, both sides recognised the value of having the wool harvested by an independent firm. For the pastoralists, it meant that the break in rhythm of station life that shearing caused would be far less; for the shearers it meant they could be sure of work; and for both it meant a buffer between the two still-opposed factions.

A few contractors had set up prior to 1900, but they worked under difficulties; the owners resented paying a little extra to have the organising done for them and the shearers could not believe it was any other than 'a bad principle'. Yet, it was in truth the only manner in which shearers could live reasonably and make a decent living.

Before shearing by contract came in, a man wrote to perhaps half a dozen sheds for his run, sending £1 to each to secure his stand. But he rarely got all six. Sometimes a man might manage to get to less than half of them: the first might not cut out in time, weather or any one of many causes being to blame, he would then miss the next shed, and sit around (with no money coming in) waiting for the next. Like as not he would be held up on that one as well. He would lose his deposits on the other stands because he had been absent at roll call.

Working with a contractor, he could get eight to ten months' work in a year with only one deposit to secure him a pen for the season. And being known to the contractor, he would be sent to sheds that suited his shearing style.

Under the old system a man might travel from his home in Victoria to western Queensland and get a run of only three sheds. In between he would live hard and rough – and hungry, if he was married and sending money home. The contractor gives a man as much shearing as possible in one particular district, thus reducing travelling costs.

All the pastoralists have to do is keep the sheep up to the men; the working of the shed is the contractor's problem.

The hitherto disorganised and intermittent work of the shearers suddenly became organised and more secure. Of all innovations since the coming of machine shearing, this was perhaps the greatest benison to the intinerant shearer.

Among the documents and papers in the office of many a property are bundles of applications from shearers begging for a pen in the days before contractors made life simpler. At Meningoort in Victoria's Western District, the McArthur family have preserved these of the 1880s:

> 'I now write these few lines two you to sea if you would be kind enough to keep me a pen on your bord for shearing has we shore at Miningorett me an my brother and gave good sadfaction their thats too told me two write two you was dennis Riley he told me he shore for you and he new I would suit you in shearing so if you would keep me a pen please write soon and let me know wat time you are going two start I will send you a deposit if you like so I wont disipoint you so I remain you truelly
>
> William Green
> Manfred Station
> Balranald NSW
> Please write soon'

Another, short and to the point:

> 'Mr McCarthy Dear Sir If there is any vaciency for shearers in your shed wil you be kind enought to keep me a pen and oblidge your humble servant J. Ryan
> Please write soon and let me know when you commence. J. Ryan, Chah Sing Station, Via Moulamein NSW'

A typical agreement between owner and contractor is that of Redcliff station (Qld) and contractor Stubbs and Harvie in August 1907: 'The station manager promises to deliver to the yards in front of the shed all the sheep he requires to be shorn. There will be a minimum number of 19 000 and maximum of 26 000 [some indication of the vast mustering job when 7000 sheep are thus 'floating'].'

The contractor promises to have the sheep shorn 'in a good workmanlike manner'. The station shall pay the contractor for carrying out this contract 'at the rate of 5½ pence per head for all sheep and lambs. Rams and stags to count two for one'. (Today the usual arrangement is 'cost plus'.)[1]

In 1907 the rate of shearing was 20s 0d per hundred sheep. At a rough minimum of 20 000 that means £458 for the contractor, from which he must take out the cost of transport, rations, machines etc. and £200 for the shearing team's earnings.

The weekly papers of January 1899 carried the first advertisement of a shearing contractor.[2] It was L. McLean & Co, also trading as the Victorian Shearing Company, who seriously began the new era. Within twenty years, 90 per cent of the sheep in the eastern states were being shorn by contract. The land would not again be sprinkled with men in groups, often in twos and threes and many as loners, tramping up the great shearing tracks from shed to shed, looking for work. For men who began shearing during and after the 1920s, working for a contractor has been part of the life.

Jim Carew says the contractor is 'the meat in the sandwich' off the board, and the overseer has that title when on the board. 'By 1934 when I contracted we had to use the harness shed at Tinnenburra to shear in. It was good, a fine lignum roof and a damned sight cooler than the famous old shed of James Tyson's day when the 101 stands were in the shed with iron roof and walls. Of course we had only four stands but the shearers were pleased, they thought I was a great bloke getting them such a cool shed!

'The wise contractor gives his men a good shed, then a bad shed with the bait of a good shed to follow. You have to use your head, such as knowing to ease the men over periods when their tally will be low because of the sheep. For instance, wethers' wool is dense and that brings the tallies down quite a bit.

'Contractors are used to bodgy names – there are always some who don't travel with their own names: they had their own reasons. You met plenty of "wife starvers" out west, cleared out to get out of paying maintenance for wife and kids. The police would come sometimes after "wife-starvers" and next day the man would have gone. He'd tell me, "I'm off."

'Of course, it worked both ways. In the old days when men were away from home for eight months of the year they headed back at the end of the season with their cheque and found the wife had cleared out.

'One thing that always amazed me, though, was the cocky. No matter how many days of the shearing there are, the cocky will be doing something when you start that he should have done yesterday. Always.

'It's regarded as the coming of the plague when the shearers are at the station. The cocky is glad to see them come and glad to see them go. All is hustle and bustle when the shearing is on and in some ways it excites the cocky and his family, but in other ways they're glad to see it end.'

When Laurie Walsh of Broken Hill is contacted to put a team in a shed, the first thing he looks at is the accommodation. 'Most faulty conditions are due to lack of maintenance, a nail here, a lick of paint there that would keep a place in shape.

'Men are mostly concerned about having a good bed. The work is too physically gruelling for them to have to lie on a sway-backed bed with a lumpy mattress. They need plenty of hot water – the day when the shearer . . . waits until Saturday for a bath is long gone. And a fly-proofed room to eat his meals in,

Sunrise, and Laurie Walsh arrives in his Cherokee Piper at Netley (NSW)

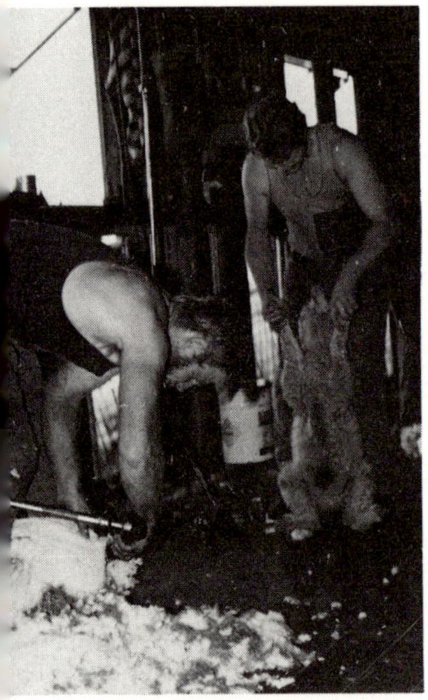

The Welling brothers, Lindsay and Colin, fly their own plane to sheds out from Broken Hill

and good, hygienic toilets. It doesn't sound like much but you'd be surprised how many properties don't have these things.

'In 1947 at Mt Hesse when I was shearing you had a palliasse with the choice of chaff or straw from the stable as filling. Some growers have tried to belittle the shearer, demean him and say he is used to no better: the trouble is such a man has usually never been in a shearer's home; he'd get a surprise if he did. They're the same as everyone else's home who has an income the same as he has. First class.'

Some workers in wool believe it has been the grower's tactic to demean the shearer. 'It justifies the poor conditions they made him work under in the past. Like they fail to accept what a fine worker he is, skilful, careful, quick and remarkably tolerant given his aggravating working stance.'

Laurie Walsh can understand much of the antipathy. 'In 1950 in New South Wales there was a lot of trouble. Properties were cut up. Graziers didn't have reasonable houses themselves and yet were forced to erect "reasonable" quarters for shearers. They hated the shearer.

'In 1974 the great rains came. The whole of the country was like an inland sea. There was not a sheep shorn from mid-March to mid-April. Westward Downs was isolated for three weeks and the Julnie team for 4½ weeks. We were dropping supplies to them from the air. You looked down and it was like they were marooned on an island. We kept tobacco and tucker up to them but couldn't drop booze. Boulia could only get tucker in, no booze. The fellows suggested getting two ten gallon kegs and dropping them in the big water tank! Skilled pilots dropped tons of food and rarely was a box broken. No-one in these areas had seen such floods. Yancanninnie Creek area was like a raging inland sea; they had a ten-foot high levee and the whole bank swept away. In four days six inches fell at Julnie. It's fairly flat land and the water can't get away. Homesteads, sheds, camps stick up out of it. When the water went down sheep were found drowned in the branches of trees.

'It's always the end of August that the cry goes out that there is a shortage of shearers. Every station wants to shear August-September. Of course, there's a shortage of men then: you can't shear thirty million sheep in two months.

'There would be no shortage of shearers if wool-growers would stagger their sheds. Contractor's phones ring non-stop in December with fellows looking for runs and there is no shearing. You phone up the growers, "No, gets too close to Christmas and New Year". In the Broken Hill area shearing moves along mid-February.

'During the last war and for a while after it there was food rationing. You had coupons to take to the store. It was funny. You took your own butter in a bowl with a crock lid and put it in the coolgardie after every meal!'

In 1974 Walsh's teams shore 494 000 sheep. He had eleven teams, nine shearing and two crutching, 120 men moving through sheds across an area as large as Scotland. He contracted in a 325-kilometre radius and travelled – still does – 80 000 kilometres each year, keeping an eye on the running.

'The only way a team can run successfully is to have a topnotch overseer. The moment a man on the board gets the upper hand it's the end of the team. The overseer must act and keep control. He should have been a shearer first. Only an ex-shearer can handle shearers. And cooks.'

He now has a licence and flies his own Cherokee Piper – 'I got sick of open-

The squatters called the arrival of the shearers, 'The Coming of the Plague'! A well-laden vehicle in the Charleville area of Queensland

ing gates, and talking to myself' – and can get round five teams in one day. 'Anything going wrong, such as wool classer or cook missing, I can get a replacement out quick.

'These days fresh meat and booze is taken out to the sheds once a week. The booze is put on the end of the verandah. It has to be issued by the overseer. In sixteen years I've never seen the men abuse it. If I don't have it there at the huts for them they'll go somewhere else for it. And that's when the trouble starts. They earn their booze more than any man alive, and I mean that.'

He does some of the biggest sheds left in Australia, twelve and eight stands, places such as Netley, the old Kidman property that used to run for over ninety-five kilometres. Mostly he works western New South Wales where the sheep have long legs. They need them for walking the distances they do to water.

'Mostly South Australian sheep here. As fly-proof as you'll get any sheep. Imagine if we had Vermonts up here: walking blowfly traps with their small corrugated wrinkles . . .'

The middle man, the contractor, stands between two adversaries. By the very nature of things what wool-growers want cannot be what shearers want, and vice versa. But the contractors attempt to be fair to both. They are in a position to see both sides of the matter and it is because of them that the industry flows so smoothly today. It isn't that alone that excites admiration, however; it is their total involvement. Perhaps they love the work they are doing – there seems no other reason for their killing hours, hard, lonely travels on bad roads, on call twenty-four hours of the day. On top of that they leap in to involvements that a cautious man on a wage could never do.

In July 1976 Frank Doecke and Ron Schultz did a journey that was reported by the press as 'Shear Agony', and they did it of their own volition. Working for Stockowners' Shearing Co. they learnt that Milo station out from Adavale (Qld) changed from sheep to cattle and had eighteen Moffat Virtue stands lying waste. The two drove from Adelaide in a car and trailer over 'indescribable roads' for 1850 km to Adavale on the Bulloo River, 210 km west of Charleville, loaded the machines and gear and began the trip home, averaging 80 km an hour because of the trailer. From Cunnamulla to Cobar the road is unsealed, as it is from Adavale to Charleville and Charleville to the South Australian border. They drove for six days. 'The old gear was renovated and was welcome on farms and stations down here.'

'We had a lot of fun,' said Fred Cotterill (on donkey) with his shearing mate Jim Edgar (later killed in World War II). They were shearing Coonamoona in the Flinders Range (SA)

Stockowners' Shearing Co. is directed by wool-growers whose names are well known in South Australia. They employ as their managers and contracting managers men who have been shearers, as do most contracting firms. In the early days, the manager travelled to his distant customers by train, then onwards by mail coach. Some properties have had their sheep shorn with Moffat Virtue machines by the company since its inception, because this agency came with the take-over in 1917.

In its first fifty years the agency averaged 1 200 000 sheep shorn annually, 813 000 being for 42 operations in the first year. Their lowest figure was in 1960 at 346 000 for 26 operations. By 1967, the firm's fiftieth anniversary, the figure was 950 000 for 90. Until 1932 shearing was only from June to December, whereas autumn shearing constitutes a major portion of today's shearing. Their records show that in 1923 the average cost for shearing was 9.5 pence per head, which included administration charges, travelling and insurance, 9.75 pence when wool classing charges were added in 1925. In 1931 the average cost per head was 7.65 pence, when there was a decrease in shearing rates for all hands.

In the 1930s this shed in the Hughenden district (Qld) was considered crude, but many men today say they prefer a 'bough' shed in those northern latitudes

A portable plant at McCulloch out from Billilla station, Wilcannia (NSW), 1926

One of the interesting 'sidelines' of the company cared for by the present manager, Frank Doecke, is the hand-piece repair section, which has a turn round of 1300 to 1400 hand-pieces annually. Some of their men, including Frank Doecke, shore in some of the best known and toughest sheds in the state. Mr G. Burton, who was with the company for 46 years, began as a shearer at Cordillo Downs in June 1910 and was shed overseer until 1939.

'You don't see the funny men any more,' Tony Ryan and Kevin Fiebig told me. 'The sheds have changed . . . there aren't the con-men, the characters, real funny men that had you laughing before they spoke. The Tasmanians had a lot of funny men with them always.

'I was a lad when "old" Bill Greig came into Calpernum station mess. "Who's the ringer here?" old Bill says. Well, no ringer ever tells, you know, you don't boast. But someone points the ringer out. Then old Bill Greig says, "I'll relieve you of that position first thing Monday morning".'

At dawn one day I flew to Netley station in western New South Wales with Laurie Walsh the contractor in his Cherokee and lanky 'Lofty' Ian Gore, a 47-year-old shearer. Control was contacted: 'There's another for Netley ahead of you.' 'Yes,' said Laurie, 'It's rush hour.' The other plane was the Welling brothers, Lindsay and Colin, heading out as they do each morning from their homes in Broken Hill to whatever shed they are working. 'Suburban shearing' it is called.

Laurie carts his own $4000 power press with him and hires it to smaller growers. 'An average shed in the Broken Hill area is four or five stands. If you've got 10 000 sheep with four shearers snagging away at it you're busy enough.' When he was shearing in the 1950s it was forty stands. 'Boonoke had 32 machines, 8 blades; Toganmain had 23 shearers; Corona, a blade shed with 50 men on the board. They changed over and put in head gear for twenty machines.'

They may not wander as they used to do, but today shearing is still a different life from that most Australians lead. I tried several times to meet Tony Bourke, shearing contractor of Broken Hill. Once, by telephone, I met his wife. 'He's not home.' 'If I came up from Melbourne could I meet him?' 'Well, you'd have

Penshurst was a shearers town. Peter Archbold lived to 86 after a lifetime of shearing, both blades and machines. His three sons followed him into the sheds, and his grandson holds his 1891 and '92 ASU tickets

Penshurst shearers setting off for Cunnamulla, 1937; two of Peter Archbold's sons, Laurie and Jack are second and third from left

'A fair cow of a trip,' they wrote home

a fair way to go. He's out in a shed.' 'Oh, I don't mind. How far?' 'Several hundred kilometres I think.'

Looking for Peter Manoil was similar. I rang his number and a child answered, 'Dad's out bush.'

Bill Schuller's wife pressured him 'to give the game away'. 'I used to call him Midnight Bill. At the end of a run he'd come home, weary, sometimes gone again next morning. So I said, "Enough".

'Of course accommodation wasn't much in those days. We had thirty-two men camped in one room in three-tier bunks. Up the middle of the room was a plank laid on stumps where you ate. Dirt floor of course. A big fireplace at one end and when the cook lit the fire in the morning we got up – or be smoked out. At first we weren't given palliassess, just bags, and stuffed them with grass or chaff or anything we could get. I always grabbed the top bunk of the three. If some dirty coot let go, well . . .

'On our way to the shed we used to begin collecting empty kerosine tins. We'd need them. For washing in, and washing our clothes and dishes, and boiling up water. There was never anything at the shed for those things. You could have heard us shearers arrive at a shed by the rattle of the tin cans we carried. In my

Where there were no shearers' quarters contractors put up tents such as these at Glen Innes near Beta (Qld)

day we travelled by truck, sitting on our swags. At nights it would be a tarpaulin muster: all asleep on the tarp with feet to the centre and waggas over us [a wagga can range from sacks sewn together and padded with 'spare' wool to the rather folksy waggas made by mothers from scraps of old warm material sewn together and padded with less respectable looking scraps of woollen odds and ends, like old socks].'

Kevin Fiebig, now rural manager for Sunbeam (SA), had a legendary background for a shearer. He was born on the Murray River, not beside it but on it, as his mother was being taken across river by boat. An Aboriginal woman helped at the birth, between Mannum and Swan Reach. He lived at Wongulla on a farm until he went shearing in the Murray Mallee, 'Lower and Upper North' as South Australians say. Clare, Jamestown to Hawker, and 'over Goyders Line', the line dividing a certain rainfall area: outside Goyders Line is marginal land.

Kevin couldn't get started as a shearer; no shed was taking on beginners. In 1939 his chance came when, in hospital for a minor operation, he helped a sick boy. The lad's father asked what he could do to repay the kindness. 'Teach me to shear sheep,' Kevin said, knowing the man was a gun. He shore for seventeen years.

A Uardry ram appeared on Australian shillings from 1932 until decimal currency came in 1966. Mr Roward Jameson MBE (left), Manager, on the Board of Directors of Uardry for many years, lines up rams with the help of Owen Capper (right) and young jackeroos. 'Shearers? Wonderful men. You wouldn't trust such sheep to any but the best. I've always found them to be splendid.'

Frank Doecke began shearing in 1936, with blades. 'There are more stud sheep in proportion to sheep numbers in South Australia than other states, so many of my generation over here learnt with blades.' In 1938 he bought a Cooper Small Flock air-cooled motor, 'a petrol driven job', and taught himself to shear, from a book. He still has the book.

Charles Flavel of Broken Hill is a contractor-shearer. 'I started my run usually at the corner, up Tibooburra way, and worked down this side, Pindara, Clare, a

Uardry sheep have been shorn by the same contractors for thirty years. Mervyn Shung (left) has classed wool here as long as has Neville Wade the presser (seated right). Tom Smith (second from left) and his brother Ron (far right)

south-east run.' All tough runs in dry lands, Kidman country. Charlie had been mining at Broken Hill until he turned thirty. 'Then I got divers' cramp, I was down working the air lock, building Murray Bridge.' He went wood cutting for 5s 0d a ton, then began as a shed hand. The shearers taught the ex-miner to shear and within a few years he set up as a contractor with a T-model Ford and a mobile two-stand shearing plant. 'Most of the places I got were small selectors, cockies struggling to establish themselves, and they couldn't afford shearing sheds. I shore in 78 sheds and put through 503 000 sheep. Shore some stations for forty-three years – started and finished my work contracting with them. And I had some shearers six years straight, presser, classer and expert longer.'

Charlie, aged 89, is still a big, raw-boned, tough man. 'Did you ever have to use your fists?' I asked him. He roars with laughter. 'Not often, but I liked it to be thought I was a Holy Terror.' In the 1930s a policeman, by chance, told his team that the ceiling in their quarters was six inches lower than the award had decided on. 'The men had a vote. Eight decided to stay, twelve left. We went on shearing. A month later when we'd cut out I was driving along the dusty road and there, camped on the spinifex, were the twelve strikers. I looked up to the sky. "Got enough head room now!" I yelled as I passed them.

'Those Kidman places were big before they were cut up. Netley on the river had 60 000 sheep and I'd put in twelve shearers.

Charlie contracted in western New South Wales's arid lands. He says all contractors are crawlers. 'If you get a board of hopeless chaps you crawl to the owner. "Give them a go, the're not a bad bunch." And if the owner is a pig you crawl to the shearers. "Give him a go, he's had a bad run." No-one worried about what sort of run I'd had. At some places I was directed to my hammock – a sheep skin hammock with the legs still attached.' Charlie engaged his men from a corner of the bar. 'Sat with my back to the wall. Engaged five teams of twelve to thirty men a team, "You go here, you go to that station"; did it without pen or paper.'

Meggs Elliott 'went through the wool game' after starting as shed hand, mainly so he could learn to shear at Tattoo, his father's property. When Tattoo was sold Meggs bought it back. Tattoo is one of the few bough sheds left working today. 'The shearers call, "Water Meggs!" and I out and puts the hose on top of the old bough shed.

'You rig these sheds up yourself. Put in posts. Put wire netting over the top of the posts and cover it with spinifex. If a hole appears you pile on more spinifex until over the years the roof becomes six feet high. It keeps the sun and heat out, I can tell you. Then, put up a bag floor – you sew the bags together yourself. A tent for yourself and the cook. Rats are bad in some areas. Our bag toilet blew down once and the shearers reckoned the rats pushed it.

31

Clancy

Have you ever been droving out west
Where the flies are a terrible pest
And mosquitoes at night, by the Jesus they bite,
And bulldog ants crawl up your britches in fright?
Where there's no bloody bird, not even a crow,
By the Jesus you know you are earning your dough
When you take on for droving out west.

– *Anon.*

There were not so many songs or folk-style poems written about the squatters as about the men who worked for them, for the simple reason that workers outnumbered squatters twenty to one. In 1857 the census for the Broken River and Lower Goulburn Squatting Districts shows only twenty squatters and settlers, although the total number of fencers, shepherds, drovers, bullock drivers etc., brought the number to 400.

The squatter's lifestyle did not excite the imagination of the emerging Australian. It was stories of men who were living with the wild land that the new patriots wanted to hear, tales like that of George Booth, a shepherd of Nathalia Shire (Vic.) who lived a lonely and isolated life at Ulupna station. In 1842 he had three months alone with 1147 sheep out on the run before he was able to come in to the head station 'for the excitement of shearing'.

There were stories filtering down from the far north of Western Australia where everything was bigger than Texas – truly – of sheep stations so huge that the men mustered by fire, setting alight to the spinifex as a signal to their colleagues that they had come on a big mob of sheep in the half a million hectares (one-million-plus acres) of the run. There were camel drivers, shepherds, boundary riders, rabbiters and teamsters whose stories hinted at a land larger than any migrant had known before, and of drovers who fought drought, flood, fire and starvation to bring a 'mob' through the stock routes that could stretch a thousand kilometres or more.

'Clancy's gone a 'droving, and we don't know where he are,' replied the bushmen in A. B. Paterson's poem 'Clancy of the Overflow', and that's just the way it is

Of all the song-poems written about these battlers, 'Clancy of the Overflow' speaks up most eloquently for them. There have been many explanations about the who, where and why of the Clancy that Banjo Paterson wrote, but that is all too silly – we see and hear of a man like Clancy even today, and they certainly heard of him often in Paterson's day.

'At the Blow-hard Sand-hill, on the night of the 10th I camped with a party of six sons of Belial, bound for Deniliquin with 3000 Boolka wethers off the shears. Now, anyone who has listened for four hours to the conversation of a group of sheep drovers, named respectively Splodger, Rabbit, Parson, Bottler,

The dusty stock route, Hay (NSW)

Dingo and Hairy-Toothed Ike, will agree with me as to the impossibility of getting the dialogue into anything like printable form. The bullock drivers were bad enough but these fellows are out of the question.'[1]

The 'thin red line of adjective' was not used in literature in those days, and no-one had yet written of the kanga-bloody-roos at Tumba-bloody-rumba, but the men in the bush were saying it: it's just that the Mother (and Father) Grundys tidied them up before presenting their stories to the public. Stories such as the damned dog and the tuckerbox. Could we really have believed that a driver or teamster would think that it was the end of a day of bad news if 'the dog sat on the tucker box'? He'd been bogged to his axles, the offside poler had slipped his yoke, the wool had tumbled off his load and – the dog *shat* on the tucker box nine miles from Gundagai!

Ted O'Reilly, veteran drover from Horsham (Vic.)

Dogs, horses, mules, donkeys, bullocks and camels had to be worked in sheep country, even goat teams brought in wood and water, and no man or woman worked them without having a fair command of the language.

At first, drovers used encircling fires to keep the sheep 'yarded' at night, then a type of fence made of strips of calico stretched from tree to tree or fastened on pegs driven into the ground. Sheep for some reason treated the calico with great respect and never pushed up against it as they did with wire-netting when it was introduced. (Some say it was the brumby runners out rounding up the wild horses who first thought of using calico as drives up to the yards they had made in the bush.)

The 'drovers plant' didn't change from the earliest times until quite recent years. The 'plant' usually consisted of a drover's cart and a few broken-in horses to pull it, twenty-five to thirty horses for saddle work, saddles, tents, cooking gear, tucker, swags, shoeing tools (including a small anvil) and some means of making a yard for the sheep at night. It was an economical set-up, developed from the need to take necessities only, but making sure that nothing was left out.

The number of men accompanying a plant differed according to the number of sheep and the distance to be travelled. For 10 000 sheep moving 500 kilometres or more, a cook, horse-tailer and five drovers were needed. (The horse-tailer must care for the horses that were not being ridden and bring in the hobbled horses at daybreak.) Some boss-drovers used pack-horses in place of carts in areas where a cart might bog, but it meant caring for twenty or so horses. In comparison, some mobs of sheep have been 'droved by two men and a dog'.

The splendid early drover was Oscar de Satge, a Frenchman from Clermont in France (Queensland's Clermont district was named for his birthplace). Although he later took up land, including almost 1¼ million hectares (3 000 000 acres) on the Georgina south of Cloncurry, on arrival in Australia he began as a drover. 'Droving is undoubtedly a fine, healthy life to lead for a time and an excellent introduction to a thorough knowledge of country, livestock and men you employ.'[2]

In 1858 he took a job with a drover moving 20 000 wethers about 1600 kilometres, from Dalby (Qld) to the Campaspe River (Vic.). It took around seven months. This was before strips of calico were used to keep sheep yarded at night; his men would light fires at intervals round the mob and the night-watch had to keep these fires stoked. Watching the sheep at night meant that every man did half a night's watch every third night. He had nine drovers for the 20 000 sheep. In the summer months their half-night-watch would be short, but in winter it might well be six hours a man.

De Satge is the example of what a good man in the bush could do when he set his mind to it. By 1862 he had a place of his own, Wolfang, near Clermont, and had driven 16 000 sheep there. By 1872 these had bred up to 100 000, and that count was after 15 000 were lost in the 1871 flood. He sold up for £70,000, bought what is now Corena Downs (near Aramac), bred up again, and once more sold for 'a fortune'. He had good backers, and why not? His droving feats proved he had the staying powers that financiers look for.

In 1863 George and John Sutherland out-did many an exploration when they drove 8000 sheep from Rockhampton in the east of Queensland to Rocklands on the border of the Northern Territory. To have come through alive in the decade that had claimed Burke and Wills was one thing, but to have brought their sheep through was another. For most of the journey they were stalked by Aborigines; at all times Sutherland had to do 'flying' trips to search for water and at one stage, when tucker ran out, he had to take a dray to Burketown for food, where he found little, and that spoiled by sea water. Clothes and boots wore out and they used blankets in their stead.

It was a fearful trip. Of the eleven men who began, only five finished. When they reached the limestone hills area, with its covering of spiky spinifex and very little water, the men had to walk. For 160 kilometres they struggled through rough country until they reached the Barkley tableland, 2093 miles from their starting point. It had taken seven months. It may have been called 'a droving trip', but when one realises that rather than head for a known place, they were beating their own track to where they reckoned land might be found suitable to settle, it was as great a feat as any of the explorers had achieved.

Some men went droving to work their way north or south, as the need might be.

Most drovers had a hammock – sometimes several – made of old sacks suspended under the cart in the shade. When the cook moved on with the cart the dog that was not needed at the time hopped into its hammock and travelled in cool comfort, fresh and ready when called for work. 'In really hot areas the short-haired kelpie can work in the heat better than any other dog,' men say.

Among drovers' trials are the plagues of rodents, rats or mice, that sweep across the plains like a rolling grey blanket. 'At night when we'd be sleeping on the ground rats would bite our toenails and chew our hair. But they left the rest of our body alone.'

Some claim that drovers' cooks are the worst cooks of all, but H. M. Barker knew one who was supreme. 'He made yeast bread in camp ovens all the time. Everything he made was good. He actually made sausages while travelling, using sheep's entrails for casing and an affair made of hollow bamboo for forcing the sausage meat into it. He said he learnt that from a chinaman in Charters Towers, also recipes for the various kinds of sausages....

'The best of the old-time owners and managers always stressed the benefits of serving some time at droving. They maintained that no man should be allowed to manage a station unless he had done at least two years' droving.'

It may be that 'The drover's life has pleasures that the townsfolk never know', as Paterson wrote. And Henry Lawson wrote right back to tell him that the drover's life has hardships that the townsfolk never know, either.

These superlative bushmen perfected the art of survival for themselves and their flocks and many a shearer has lifted his hat to them. 'We'd go rattling along in the cool of dusk on the dusty track towards wherever we had got a shed and there, far over the horizon, we'd see the coloured streaks in the sky that reflected the dust of a mob being driven in the distance. And many a time we've said, poor bastard,' Jim Carew remembers.

Drovers from the beginning of outback settlement have been entirely dedi-

Ted's plant, with his netting ready to fold his charges for the night

cated: 'Our job is to deliver sheep in good condition and account for all of them at the final tally.'

Jim and Pat Ryan of Darraweit Guim (Vic.) tell of an old drover of forty years ago in that district. 'He'd set off with a big mob, train his dog to lead from the head of the mob and he'd come on behind. Huge mobs he used to take and we'd watch the cavalcade pass: first his dog, then the sheep, then the drover with his horses and covered cart. It was a lonely, terribly hard life, but we never knew him to lose a sheep through carelessness.'

Sheep on the move are usually limited to flocks of three thousand. Sometimes if a larger number is on the move the mob will be split and two men with their dogs go off in charge of each flock. By law, sheep travelling on stock routes must travel a minimum distance each day – in most parts of Australia it is six miles (around 10 kilometres), but in drought conditions when sheep are poor, permits are often given for the reduction of a few kilometres or so.

Wool-growers had to put an escort on mobs being driven through their property to see they didn't stray and eat too much: record books at Uardry (NSW) refer constantly to 'seeing mob through run'. There were upwards of 18 000 sheep in one mob mentioned.

In the *Riverina Grazier's* office in Hay, is William Charles Gough's handwritten story of his working life. Born 23 July 1879; nine months schooling; drover by necessity owing to having been 'born with crippled feet'.

'Dad had to drove sheep. He couldn't work with cattle because he couldn't get on his horse quick, he couldn't have got away if cattle charged,' Charlie Gough says. 'He was always reading. He read good books. He had a bullock dray, and he took me out when I was five or six. I remember our first stopping point – everyone's – Mark Tree Point [One Tree Plain]. Forty teamsters were sometimes there.

'There was no P.P. [Pastures Protection Board] those days. All the T.S. [Travelling Stock Routes] were held under lease by the stations, and their sheep ran all over them, and a drover in charge of sheep had to give the lessee not less

than 24, or more than 48 hours verbal notice (or by letter) of his approach and entry on to his part of the T.S.R. under his lease. It meant that a drover had to employ a reporter to ride ahead to give notice – the drover-in-charge was held responsible for any breach of the act and deemed to be the owner of the sheep whilst in charge.

'A drover had to sign an agreement to pay the sale value of any sheep that he could not account for, if dead and shorn to carry the wool and the ears of the dead sheep. There was no loafing in those days, the squatters' boundary riders saw to that, saw that the drover did his six miles every day, sometimes more to get to a good camp. Where there was timber we drovers made a break out of limbs from the trees. The stations' men would burn them on the plea that breaks harboured rabbits so then the drovers had to watch the sheep all night, 2½ to 3 hours watch, as rope fence-breaks as used now was not thought of then.

'A capable drover had to be able to count and draft and have a knowledge of sheep in general, he had to sign delivery notes on behalf of the owner – I myself have signed for thousands of sheep and I have never had any complaints from the owners. Nowadays a Stock Agent will drive out and count the sheep on departure and another Agent counts the sheep on delivery and pays the drover at their office.

'There are now yards or camping places every six miles and nearly all the routs and reserves are under P.P.B. control. Little or no notice is given and that can be done by phone – not like the old days.

'There were long spans between watering places, only three between Booligal and Hay – fifty miles – and unless the sheep was very strong there was unavoidable losses in the summertime.

'The drovers' rates of pay was low: small mobs up to 1200, £1 per day, up to 3000, £1 10s 0d per day and 4000 and over from £2 10s 0d to £3 per thousand per week. The drover had to find plant – horses, saddles, waggonette and all gear and pay and "found" men to assist him. Their rates of pay was £1 5s 0d per week and found. Each man had to have his own sheep dog or dogs. The rations then was rough. Mostly mutton, damper (some bakers bread when passing through towns), jam, no milk or butter, and a Brownie now and again when the cook was on a good camp.

'I don't claim to be a super drover but I have had as much experience as the next man. I was one of the first drovers to use rope fences to hold the sheep at night. I learnt the game in the hard school, watching sheep at night in the rain. I do claim to be the first drover to travel 3600 sheep from Robe to Menindee (on the contractors' train from Ivanhoe to near Menindee. Plant and all loaded, eight horses, waggon and gear on a big train. Unloaded at end of the trip, no yard, so got sheep out on a sand hill beside the line in a 10 000 acre paddock with Station sheep running with them). I travelled those sheep on near Sterling North near Port Augusta in South Australia with the loss of eleven sheep for the whole trip. The sheep was then railed from Stirling North to Leonora (WA) up past Kalgoorlie. They was the property of Mr Roy McCaughey of Coree near Jerilderie (NSW).'

The 'drover's right hand'

The distances travelled by Gough and the sheep in his charge are immense even by today's reckoning. 'I have travelled 5500 four-tooth two-year-old sheep from Barringun on the Queensland border to Coree near Jerilderie, started from station 15 May 1931, rout via Burke, Cobar, Hillston, and Carrathool. Held up by '31 flood and had sheep shorn at Clover Downs shed. Lost nine sheep that time from poison weed.

'I was one of the three drovers (C. Chapman from Kiandra and J. Oddy from Tumut) that brought 9000 sheep out of the snow on the high plain in 1915. This was a very difficult job – the ordinary rout, Rules Point, was snowbound and we had to bring the sheep through Lobbs Hole – now known as Ravine. It used to be an old copper mine. Very rough country. No road, only a blazed trail (marks on trees). We had to cross eleven running creeks by making bridges out of saplings and bark and we had to swim the sheep over the Yarrangobilly River, about a twenty feet swim in fast running water. We lost 29 sheep in this, which took a whole day to do. We had to watch the sheep all night to keep dingoes and foxes away (they were very hungry and daring when the snow was falling in the mountains). We trucked those sheep at Tumut, to Hay, and then by road to Clare Station. [In this detailed account of rescuing sheep in snow and snowstorms, he not once mentions the hardship it must have been for himself or the other men, wet, freezing, unable to rest, totally on their own with the whole responsibility of the 9000 sheep, only 29 of which they lost on that forced swim.]

'One year I took delivery of 4500 one-year-old wethers in the wool at Mossgiel

Station and travelled them to Barratha (It was the Back to Hay Year). Those sheep was sold when I reached Pretty Pine and I drove them on to Jerilderie via Conargo. There they were again sold and I was kept in charge of them to take them to Kea Lodge in the hills, Berridale via Narrandera, Wagga, Tumut, Kiandra, Jindabyne and Adaminaby and had those sheep shorn at Lake Medion near Narrandera. (I remember coming down to Hay on the train for one day while those sheep was shorn.) The whole trip took 18 weeks with the loss of 5 sheep. I employed 3 men, 2 shepherds and a cook.

'That was one of my best trips. Of course I had some very tough trips. I took delivery of 6000 old ewes, very poor, right off the shears. And run into a cold snap – it killed over 70 of them. But I managed to get the rest of them to Kiandra in the mountains.

'In the years I was droving I lost some sheep in storms at night but I was always able to account for them. I had to pay for 18 sheep only once. They swam the river when it was low into a flock of station sheep the opposite side of the river and the owner of those sheep wouldn't allow them to be mustered so that I could get my sheep back, so I had to pay for them.

'In my experience I've found the best type of sheep to handle is merino wethers and the worst is crossbred weaners. It's hard to travel sheep with pink-eye (a disease that blinds). A good idea of Hell – a drover in charge of a mob of pink-eyed sheep.

'I think droving was the worst paid of all contract work considering the responsibility the drover had to undertake. There were hundreds, many other drovers, that did as much as I did, but there are not many of them alive now.'*

In the back country the stock may spread out as wide as the stock route allows. In western New South Wales and Queensland this can mean a kilometre and more – it usually depends on whether times are bad and the drover is trying to get the best he can from any feed for his animals. When it's not possible, it is a case of 'punch them along, keep them moving'.

At night, while on the road, the sheep are kept in temporary enclosures known as 'brakes'. Originally a semicircle of fallen trees was used as a half-yard in which to keep the sheep at night, the open space in front being occupied by the cook's dray, the men's swags and wood fires. A man on foot was kept on watch. Wire-netting was tried but it lacked stability and it was difficult to handle, so the drovers still took it in turns at night-watching. A solution was found with a rope brake and rope yards are one of the few changes in a drovers' plant. The McMaster Brake has five strands of light rope, laced at intervals to form a large-meshed netting. It makes a perfect temporary fence, is easily carried, durable and cheap, and can be quickly erected on stakes. With the evolution of this type of brake, night-watching ('The Drovers Death') came to an end. Sheep are let out of the brake about sunrise.

Sheep which have been on the road any length of time lose much of their timidity. They quickly become accustomed to new sights and acquire a cunning which makes them difficult to handle other than at their own pace, which they decide themselves. 'They get bloody cheeky, that's the trouble with them.'

When rail and road transport began to cater for moving sheep it was thought

*Mr Gough died in 1965, one year after recording this story. He writes – and no doubt spoke – of 'routs', the expression used outback for what others might term 'routes'.

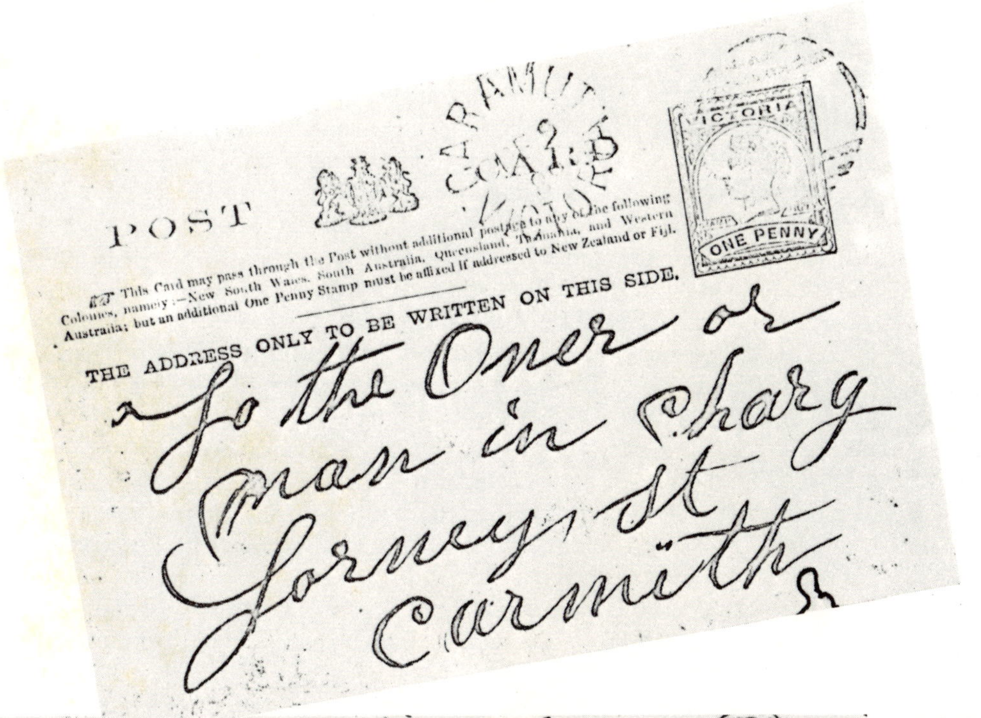
POST

This Card may pass through the Post without additional postage to any of the following Colonies, namely:—New South Wales, South Australia, Queensland, Tasmania, and Western Australia; but an additional One Penny Stamp must be affixed if addressed to New Zealand or Fiji.

THE ADDRESS ONLY TO BE WRITTEN ON THIS SIDE.

VICTORIA

ONE PENNY

To the Oner or
man in charg
Gorney St
Carmith

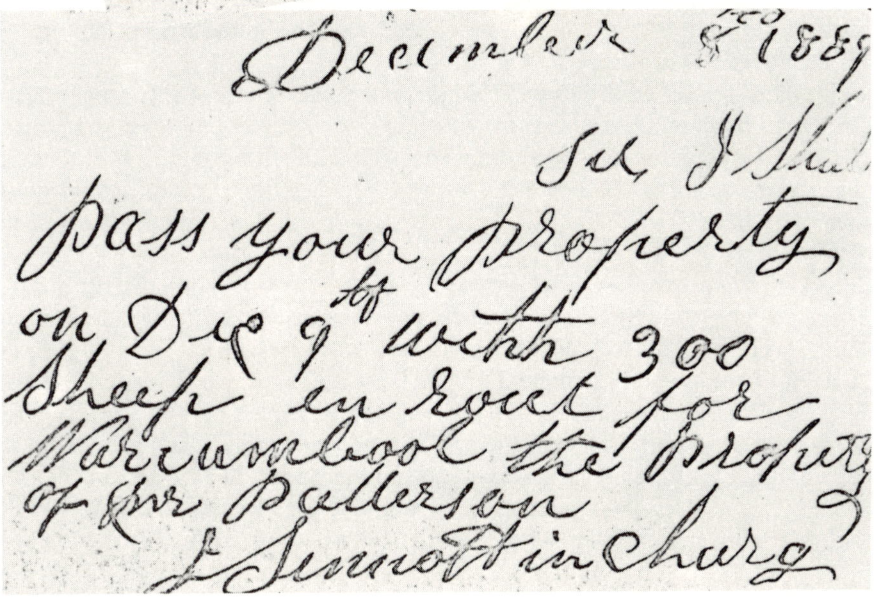
December 8th 1889

Sir, I Shal

pass your property
on Dec 9th with 300
sheep en rout for
Warrnambool the property
of Mr Patterson
J Sinnott in charg

Drovers must warn an owner when a flock will pass through a property. In 1889 this was done by hand, today by telephone

to be the end of the drover, but there is still a call for a good man, and there are still plenty of good drovers. Sometimes a pastoralist may buy sheep at market but, needing time to move other animals around on his land, he sends them off with a drover. If the property is a hundred kilometres away he will have ten days in which to prepare for them. In times of drought the routes carry heavy traffic as sheep are sent off to agistment 'down south' – and sometimes go down and back without agistment: if the drover is an 'old dog for a hard road' he can save his employer much money. 'You've got to keep your eyes and ears all over the place.' The stock inspectors are as wily as the drovers, and moving stock creates a constant battle of wits.

Mick O'Donnell of Hay was droving by the time he was twelve years old. 'In a brown and white topped waggon. A good droving plant, I remember. Horses, dogs, drays, camp ovens, all a man needs. We were off to Dirranbandi on the Queensland border. We got back to Hay six months to the day that we left.'

In the flood year, 1931, they had to drove their sheep down the main street. It was a red metal road, but with so much rain cars were bogged in front of the shops. 'You could bog a duck there in those days....

'Of course I've been out on One Tree Plain. What drover hasn't? There used to be a tree there. The drovers called it the Chloroform tree – they'd have their lunch under it and go right out to it – lying on their backs as if they were asleep.'

One pub and one tree were on this plain stretching from the Murrumbidgee to the Lachlan. And no city dweller could appreciate what both meant to the traveller. 'The drovers and stockmen always lit fires under it. There was usually a bottle of black O.P. Rum therabouts.'

Charlie Gough sometimes went droving with his father. 'Like in 1931 to Barringun in the big floods. Six months droving sheep. Had my bulldog when we were pushing them through Cobar.

'What I remember most about my fathers' plant was his dogs. He had perfectly trained dogs... four dogs and a breeder bitch. Some drovers had up to forty dogs, but Dad's were real good.

'Hot days it was bad droving out on the One Tree. There was two pannikins of water to wash in then give it to your dog to drink. My dad used to say two days at most for sheep to go without water. "You knock the tripe out of them if they have to go longer." Sheep can survive longer without food than without water. You have to travel six miles a day by regulation but sometimes you push them to get them to water.'

Some drovers say they have pushed sheep about thirty kilometres a day, 'When we have to. They can go five-six days without water if they are not too poor. You don't like to do it but this is hard country here.'

Mick O'Donnell and his mate Tim Gash are still droving. 'We used to have 20-50 horses. Tim was sudden when driving the waggon. He'd give the harness horses a burst of parrot shot in the arse if they were sulky.' And they still use horses. 'Hobble them out at night; keep one tied near the waggon so we can round them up in the morning.' Tim's parents were drovers and he was born on the droving track. 'That's why I can't read or write.

32

The Water-ways

Along with the shepherds, drovers, musterers, teamsters and bullock drivers, the river-men shared the life of the inland bushman and shearer. In September 1853, the *Lady Augusta* pulled in to Poon Boon on the Murray to take on 200 bales of wool and the ceremony of shipping the first bale was performed. The first woolpack was rolled down the bank of the river to the plank communicating with the vessel, and then passed on board jointly by Mrs Younghusband, Mrs Finniss, and Miss Younghusband.

The coming of the river boats opened up the country west of the Darling for sheep. Kinchega, Silistria, Balaclava and Kars sheep runs were established. Stations had their own landing stages on the rivers, homesteads had their front doors facing the water.

The big grey iron shed of Dunlop at Louth (NSW) would keep a steamer

Loading wool at Billilla out from Wilcannia on the Darling River. 'Each season', Ian Mann of Billilla says, 'we cleaned out the cutting with pick and shovel and got a plank out for the landing.'

(usually the *Jandra*) going for two months, up and back each week with the cargo of wool from the factory.

In some ways the river traffic was as important to the wool industry as it was to cotton in the Mississippi-Missouri waterways, but unlike those rivers, the inland streams of Australia are unreliable. The uncertainty of the water level was such that the *Jane Eliza* was stranded high and dry for fifteen months, 'although her draft was such she could have steamed off on a light dew, even on a damped down wool bale if someone had thrown it in front of her'. But there was no dew at all, and the ship had to be constantly tended to keep her from disintegrating in the kiln-like air of the empty river bed.

Yet when the floods spread out from the rivers, some of the river boats set out on amazingly adventurous voyages. In 1870, one boat skipper 'navigated the Paroo River right up to the Queensland border and on his way back he steamed cautiously to within a mile of the edge of the floodwater that surrounded Newlands Warloo woolshed, and they ferried the bales out to him one by one on the two small station boats lashed together to make a makeshift raft'.

A river-boat captain would watch out for an old, scarred T scored into a gnarled tree-trunk – the signpost for Toorale – and ease his steamer in to take on wool valued at thousands of pounds. Sometimes the clip was worth £30 000, more money than many seaports had brought in for the year, yet the only evidence of traffic was the smooth, shallow cut in the bank of the river made by the thousands of bales of wool sliding down onto the boats over the years.

The signpost for a station, which may be half a million or so hectares, might be a sheepskin on the lower bough of a gum tree. 'Put out a flag' might mean a wool sack thrown carelessly on a log, a scarred trunk on the bank, an old paling

Billilla wool heading down river, 1926

The barge *Kulmine* (towed by the famous *Pevensey*), 1900 bales on board, August, 1931. Heading down the Murrumbidgee they will steam into the Murray and on to the railhead at Echuca

stuck sideways in the mud, or a dinghy drawn up under a tree. 'It takes some seamanship to find one mouldy sheepskin in 3000 miles of river bank,' C. E. W. Bean wrote in 1911.

Some of the river boats took over the work of drovers, carrying up to a thousand sheep in their barges up, down or across the rivers.

Within a few years of the coming of the steamers, upwards of one hundred stations were taken up along the Darling, and as many paddle-steamers were plying the river. There was even a missionary boat, the *Etona*, to bring The Word to the people beside the rivers. Back in England, Eton College had provided subscriptions for the vessel and she served the area well. As late as the floods of 1956 the old *Etona* brought shearers and stores up to Nap Nap, but only on condition that she was supplied with firewood on arrival, as the whole of the country was under water.

Nap Nap sent wool out by steamer until 1936 but by then only to Mildura, for Echuca had almost ceased to be a port. In 1939 the *Pevensey* came up and took back Nap Nap, Toopuntal and Murrum wool, about 900 bales, the last to go regularly down river by steamer.

The ports of the inland, Port Bourke, Wilcannia, Louth, Tilpa, Menindee, Pooncarie and Wentworth, were as well known in the 1860s as were those of Sydney, Melbourne or Hobart.

The ships brought stores, tobacco, grog and visitors to the stations and took away the wool. They were fuelled with timber being cleared by settlers from new pastures. Enormous quantities of firewood were needed, and great stacks were piled up on the river banks waiting for the scream of the whistle of the river boat. If the woodcutter wasn't about, the captain 'wooded up'; with one look at his stack the cutter would know how much had been taken and, by the

Echuca wharf in the wool season

sound of the whistle, by which ship. The captain would pay up next trip or the one after.

The river had a whole life of its own. 'The Darling whalers' were known to the shearers for fifty years: swagmen in canoes, they bartered their catch of fish to the bladesmen waiting on the banks up and down the rivers.

The newspapers had a Weekly Shipping Intelligence column such as that in the *Riverina Grazier* at Hay. The column for 20 October 1880 gives an idea of the traffic on the rivers and the stations which depended on river boats:

'River at Hay 9 feet 3 inches, falling fast.

'Wm. McCulloch & Co. report – Since last report *Rob Ray* and barge *Swan* arrived from Echuca and has been loading wool from here; will clear 1500 bales.

'*Freetrader* and two barges arrived . . . 1500 bales. *Princess* with barge *Namoi* . . . 1500 bales, topping up at Mungadel and Toogimbie. *Hero* expected to-morrow with full load from up river . . . *Riverina* and barge *Benduck*, and the *Pioneer* and barge *Trader*, also *Agnes* and barge *Scottish Chief*, from Echuca and the two former loaded up with wool for Echuca and the latter went on with a large cargo for Narrandera.

'*Enterprise* and barge *Belubla* arrived this morning from Tubbo . . . leaves Echuca this day with over 1000 bales of wool. . .'

Arriving in Sydney, 1920. En route by truck to ships waiting at the wharf it's a long way from the lonely stations and the romantic transport ways of the inland

Charles Dunstan was aged 17 when the Victorian Railways sent him off to Echuca in charge of the railway wharf. It was 1888 and the start of the wool season. 'I knew this was Victoria's second busiest port, but I hadn't expected so much traffic, steamers and barges tied up along the 1000 foot long wharf waiting to be unloaded and others waiting for a berth ... The fleet had sailed down from the Darling, Murrumbidgee and from sheds along the Murray itself to bring the wool to the railhead. There were 5000 bales waiting for me.

'I took on forty casual labourers, two tally clerks and a foreman ... They worked from five in the morning till ten at night. The casuals were paid one shilling an hour – nothing for overtime ...

'The railways collected stevedoring charges of twopence a bale of wool. Each steamer carried about forty or fifty bales and hauled a barge carrying 1000 bales. There was a crew of five men on the steamer and one on the barge. Some of them had been 1000 miles up the rivers, as far as Bourke.

'We unloaded about 60 000 bales of wool a year from the Murrumbidgee area and about 20 000 bales each from the Edwards and Darling rivers.

'Wool from the Darling River stations was not very popular with the men. No-one liked handling it because much of it had been brought from back-country runs and stank of camels that had carried it on their backs all those miles in the heat till they reached the river steamers.'

33

'Wet' Card or 'Dry'

On 11 May 1908 a conference opened in Melbourne to investigate the term 'wet' sheep. R. O. Blackwood of the Federal Pastoralists' Association was President, W. G. Spence, Vice-President.

They decided first to advertise for inventions that might help settle the problem, and that brought many strange devices to the fore with 'wet wool determinators and detectors' and 'chemical compounds' prominent.

By August the conference had settled on Richardson's and Gillespie's Wet Wool Detector, a moisture-estimator that worked on the principle of an acetylene generator, calcium chloride mixed with a sample of wool. Quite forgetful – or unaware – that shearers claimed that it was the sheep and the humid atmosphere that troubled them when wool was 'wet', not necessarily the wool itself, 'scientific and medical' men were sent off 'to ascertain to what extent the shearing of wet sheep affected the health of shearers, and especially as to what precise amount of moisture in the wool is harmful'.[1]

The problem was considered of such urgency that when the Prime Minister, the Rt. Hon. Alfred Deakin, was approached and told of 'the National Importance' of the question, he made £500 available for the cost of the investigation.

Two scientists visited 78 sheds in four states and New Zealand, and spoke to 1030 shearers with an average of sixteen to a shed. On receiving their report the conference resolved that: 'In cases of dispute the instrument – the Wet Wool Detector – was to be used and if no more than 8 per cent moisture in Merino or Come-back wool and 10 per cent in First Cross or British breeds, the sheep were to be considered fit to shear. Tests on the wool are to be carried out by the manager or deputy and a representative of the shearers.'

It was all very wonderful, seeing that shearers for generations had a deep belief that they *knew* when sheep were wet. They even told the scientists with the calcium carbide detector: 'Wet sheep give off an unpleasant vapour', or, 'the hand in the wool method is infallible'. Others said the legs of their trousers became cold and clammy from the mixture of water and wool grease absorbed from the wool, as the man holds the sheep between his legs during shearing.

Many complained that wet, 'hot' sheep are the worst. All said a shearer gets hot very quickly and sweats much more freely than he would otherwise, that the atmosphere about him is much more humid, his sweat evaporates much more slowly and his body temperature rises in consequence so he becomes greatly overheated and more liable to catch cold on leaving the shed.

Yet the calcium carbide test could not comment on these beliefs. An almost universal one among shearers used to be that the steam from the sheep enters the human body through the pores of the skin and does the man material harm, they could not pretend to understand in what way. Some complained they got sore eyes in consequence.

Doctors who examined the men believed that the pace at which the shearer works plays a great part in causing the troubles ascribed to shearing wet sheep. For two years, 1908 and 1909, the five men, doctors and 'scientists' travelled around sheds.

In 1909 a farmer told C. E. W. Bean that although he had shorn sheep wet or dry for forty years, he had never had anything wrong with him. 'I've an idea that when a man shears a wet sheep it is all right so long as the sheep is cool. Even if the sheep's hot it doesn't hurt if a fellow goes about it easily . . . I've seen men shear wet sheep scores of times, all the same – especially towards the end of shearing, when they have booked stands in another shed, and want to get away to it, the sheep are never wet then.

'It's when the shed is near a town or an hotel that the sheep get wet. Someone hangs up his machine, and no-one likes to stand out . . . I've seen them knock off three days for a point or two of rain when a shed started. At the end of the same shearing it had rained heavily, but one day was enough to dry them then. It all means that a fellow who doesn't want to work gets his way, and the fellow who's there to knock out a big cheque loses two or three days. You may reckon this as a safe thing, anyway – that if the wool is dry enough to press, it is dry enough to shear.'

A 'horny-handed shearer' told Bean: 'I think wet sheep are a myth. I don't think anyone suffers through shearing them. I don't think most shearers believe in them. They do when they want a spell. Two doctors followed the question through from Riverina to New Zealand last year, and said that every complaint of illness from wet sheep reported to them could be traced to some previous trouble. There's hundreds of men besides shearers has rheumatism, aren't there? If a man is inclined to rheumatism he naturally gets it after getting wet or even before if there's rain in the air . . . There might be thirty men in a shed who could all shear wet sheep without any danger, and one whom the wet would affect. Well, they all had to knock off for him. That's what "wet sheep" is.'

If wet wool is pressed into bales it may heat and burn by 'spontaneous combustion', the growers claim. Many men have given a scientific decision as to whether 'wet' wool would cause rheumatism. But as to his decision being accepted, that is another thing.

'There was never many wet sheep in my day,' Tom Duckmanton, rouseabout, says about his twenty years in the sheds. His mate, the old shearer Percy Riddle, says, 'I never voted them wet more than five times in fifty years.'

'Mark you,' says Tom, 'It wasn't for want of trying . . . Nobby Templeton and I were rousies at Mt Sturgeon once and when the Melbourne Cup was near we wanted a holiday. So we could go to Melbourne. Nobby and I sneaked over to the shed in the dark that night with a bucket and a watering can and sprinkled the sheep in the pens. "Holiday tomorrow!" Nobby assured me. But it never came off... the shearers voted them dry! The few times they voted sheep wet in my day us rousies were put to work on other jobs.'

Percy had reason not to vote sheep wet. 'There was no point voting them wet at that time because before the day of motor cars we couldn't get off the property. I've shorn hundreds wet. They do affect your eyes and your head feels as if it is blowing up. When you shear wet sheep you get pains and aches all over. But we needed the money. We worked in flannels and long johns and they kept us cool in summer and warm and dry in winter. That's what saved us when we shore wet sheep.' Percy got yoke boils. 'Very sore things. A man can be in agony with yoke boils.'

John Adams got yoke boils. His wife Bridget remembered them 'to her dying day'. 'Filthy things. So painful he couldn't bear any cloth even to touch them once he got out of the shed. They returned every time he went shearing.'

Herbert Hutchins believes that 'Shearing wet would damage a man's health. Because I work alone I can usually avoid it. But yes, a man feels bad shearing wet.'

Before the union was founded, voting 'wet' was one of the few retaliations against the squatter that the men had. When sheep were 'raddled' and therefore not paid for, a little 'wet' eased off the frustration of the men. They could get their own back in a small way. Except that, as Jim Pugsley remembers, the boss also often prayed for rain.

'You get a full board of good shearers, racing for a cheque, and it's damn hard to keep the sheep up to them. I've seen squatters look almost relieved when the fellows voted wet.'

After the need for retaliation was past, the use of a 'wet' vote was still handy for the odd days off. Bob Pulsford, Bierbank (Qld), had the shed as full as he could safely do without smothering them. There were enough sheep to keep the boys shearing all day. He thought there was no way they could get out of it. Then an old hand called out to him, 'Don't you think you could shove a couple more in Bob? They might sweat wet then.'

Only one vote a day is permitted on the question of 'wet' sheep, to enable the men to return to their quarters or home, as the case may be. The shearers say, 'The cocky would be having us vote every time the rain eases'. They believe shearing wet sheep "brings on" rheumatics, and every team can tell you of a shearer they know who 'is crippled up' with arthritis or rheumatics after shearing 'wet'. But both contractors and owners believe it to be a quite iniquitous condition, the cause of much unnecessary loss of time and a great trouble to the grower in the movement of his sheep.

On the other hand, the classer is entitled to stop the shearing if the wool definitely is wet. (The men vote by placing cards marked Wet or Dry in a box.)

Even the shearers admit that there are an inordinate number of 'wet' sheep at the beginning of a season. They get very dry towards the end, particularly if 'wet' would mean bringing the men back for one day after a weekend off.

Meggs Elliot has seen it all. 'The shearers want to go to the races this Saturday so they voted the sheep wet – six to two. The cocky put lambs in. Lambs dry quickly so they should be all right he thinks. No. Eight voted them wet. "They can't be", the cocky said. "Lambs dry quicker than grown sheep."

"No," said the shearers, "They're littler. They're further away from the sun."'

There are signs that let the wool growers know which way the men are going to vote. 'If they have a shower at lunch-time you know they are going to vote wet,' some say. And once I watched men file back into the shed after lunch and the owner told me they would vote wet. 'How do you know?' 'See the ringer, the big gun? He hasn't got his cap on. I've never seen him shear without his cap on.'

Like the 'eastern shearers' the Western Australian growers are apt to agree that wet sheep do affect a man. David Fitzgerald and Don Murray, both sheep-owners, say: 'Headaches, a dopey feeling, cold, difficulty in breathing, swollen joints – these things do seem to affect some men.'

Another Fall of Rain

The weather has been sultry for a fortnight's time or more,
And the shearers had been driving might and main,
And some had got the century that never had before
But now all hands were wishing for the rain.

Chorus

For the Boss is getting rusty and the ringer's caving in,
His bandaged wrist is aching with the pain,
And the second man, I fear, will make it hot for him
Unless we have another fall of rain.

A few had taken quarters and were coiling in their bunks
When we shore the six-tooth wethers from the plain,
And if the sheep get harder then a few more men will funk
Unless we get another fall of rain.

But the sky is clouding over, and the thunder's muttering loud,
And the clouds are driving eastward o'er the plain,
And I see the lightning flashing from the edge of yon black cloud
And I hear the gentle patter of the rain.

So lads, put on your stoppers, and let us to the hut,
We'll gather round and have a friendly game,
While some are playing music, and some play ante-up,
And some are gazing outward at the rain.

But now the rain is over, let the pressers spin the screw,
Let the teamsters back their waggons in again.
And we'll block the classer's table with the way we put them through,
For everything is merry since the rain.

Final chorus

And the Boss, he won't be rusty when his sheep they are all shorn,
And the ringer's wrist won't ache much with the pain
Of pocketing a season's cheque for fifty pounds or more,
And the second man will press him hard again.

34

Barcoo Rot

Like men in most industries, shearers believe they are prey to illnesses peculiar to their work. They have suffered in the past from the old bushman's disease they called 'Barcoo Rot', a form of scurvy caused from lack of fresh food and vegetables on their long treks and in their camps between sheds and, in the old days, even in the station quarters.

Jack Rose (left) and his mates spent Sundays at their quarters 'trying to clean up'

Since 1970 small numbers of men in central and western Queensland have suffered from what they call 'Q' fever. A dengue-type heat and delirium builds up until the man must be hospitalised. Some take months to recover. Shearers believe doctors are mystified as to its cause or even to specifically what it is, but other observers believe it to be just another manifestation of the Barcoo Rot.

There have been fewer accidents with machinery than may have been expected. Carl Pulsford and others remember a shearer being battered and eventually dying when his coat sleeve – he was putting the coat on at the end of a run and it flew up in the air – wrapped round the spindle and smashing him up in the saplings supporting the roof.

Long hair was not popular in the sheds with a man bending and moving round exposed machinery. 'There was one bloke years ago had a lump of long hair dangling down from under his hat – he always wore his hat on the board, everywhere, and the hair was a strip of kangaroo fur falling down his forehead.' It was his sad, gallant attempt to conceal the fact that he had no hair *anywhere* on his body, said Carl.

Jim Carew, who spent his working life in the sheds, knew the hardships. 'There was scabby mouth, a dermatitis that used to affect the sheep around the mouth. Infectious Labial Dermatitis. Two fellows got it in a shed I was at. One died. Since then there's been changes.

'Some, of course, got up to their tricks and invented it, took a cigarette butt to their lips to bring up a few blisters, or got a scrubbing brush to the sheeps mouth and when they came out of the pen shouted "Scabby mouth!"

'For years the only first aid kit we had was a bottle of kerosene. Just whack it on a cut. It was good. I never lost a patient with it.

'In lush country you'd see a green strip around the men's armpits from the green pasture off the sheep's feet, see men come in to tea after a wash with a green swathe around the armpit, their trousers thick with yolk, blood, grease and sweat. And all they had for cleaning up was kerosene tins: one for boiling clothes in, one cut down the side for rinsing.

'The old boys used to get yolk boils, all over them, some of them were a horrible mess. It might have been because of diet, I don't know. It was an awful hard life. It was hard to keep clean if you were a shearer. Now they shower every night and daily have a fresh change of clothes. If a man isn't clean today his mates would get stuck into him.

The wood-fire copper, a luxury few stations provided

'They used to wear a breast plate of basil [untreated sheep skin] across their chest for where they'd pull the sheep's head up. Nowadays they use denim strapping instead of basil.

'Board bunks were there for most of my time and straw palliasses, sometimes no windows. Later they brought in fly screening. The union made them flyproof the doors and windows but they didn't bother about the air gap of a foot or more up the top of the walls.'

'Scabby mouth', a disease in sheep, is one of the conditions that alarm shearers as much from a type of folk belief as from any concrete evidence that it affects the human body. 'Several men died of it,' Jimmy Pugsley believes. 'I was always afraid I'd catch it. We all were.

'I had Q fever. Had eleven days in hospital. Lost 28 pounds in weight. Sweated, shivered, shook. Terrible temperatures. Caught from animals. Some say from ticks. I was delirious. I was the thirteenth man to catch it at Tambo in 1960.'

Among the diseases, typhoid was most prevalent, most apt to kill a man.

Figures on actual shearers' deaths are not available, but in 1888, between April and June, 52 deaths from typhoid were recorded in Broken Hill.

Out in the huts on the stations the smitten shearer either drank his pain away in his dirt-floored humpy and was certificated 'died of alcoholic intemperance' or hung, shot or otherwise removed himself from the disease. (As late as September 1980 I saw 'pit dunnies', blowfly-filled holes in the ground covered with a rough wooden board with a hole for the backside to squat on, no cover, nothing to prevent the flies from flying, as they did, into the kitchen and men's mess that in the earlier days had no fly wire.)

There is heat. I was at Netley station out from Broken Hill on 17 November 1980 where the temperature was 46°C. The shearers often say, 'Hot! It was 46 in the waterbag!' Well, it was nearly that this day. The iron shed; eight shearers going like tigers; the heat of 1500 sheep in the shed and holding pens; sun glaring in the portholes of the walls facing west. It wasn't good, but the tallies stayed up in the 140-160 region.

There is Ngora burr, a native clover, the burr that comes in the sheep's wool off the creek banks and rips the shearers' lanoline-soft hands, causing festering and deep infections. Roly-poly is even worse, the big, dried-out pieces of stick matting the wool together – sometimes binding the sheep's legs together – rip into the shearers' flesh. And 'Crows Foot', a corkscrew-type of thing, rips skin off the hand. One rarely sees a glove today, but a cut-away fingers-off sort of mitten often is used.

There are other 'foreign bodies'. Bathurst burr (sheep can be black with them) is a very hard burr, and Trefoil burr will lock the cutter, causing machines to leap from a man's grasp and rip flesh. 'The cutter is going across the teeth at 3400 revs and if a burr jams it, it can lock. Something dangerous,' Ern Holmes says. When Stonk Lewis was learning to shear, his hand was badly burnt by the hot box in the hand-piece. The old-timers told him to urinate on it. 'I did every hour. It didn't get any worse. But I can't say it made it any better.'

35

The Longest, Strongest Scab Train

Twas early in the New Year
From platform number nine
Steamed the longest, strongest scab train
That ever graced the line

It was 1931 and once again the wheel was turning full circle. It was hard times again, a depression worse than any before. The pastoralists were again being pressed by the banks and harassing the shearers. The shearers were forced to strike at the worst possible time.

Working men, often honest and good toilers, tramped the roads of the outback taking any work they could get – scab or otherwise. And the trainloads of 'free labour' steamed out of capital cities 'with double engine haulage' because of the vast number of carriages and truckloads of desperate men. 'The stink of shit was heavy,' goes a piece of doggerel of the day, referring to the fear of the scabs who knew the battlefield they were steaming to. 'But they were well protected by one hundred Queensland police.'

The broadsheet carrying this poem was circulated during the 1931 strike along with 'blacklists' of men who were shearing non-union. This and other such sheets were prepared by a man known only as 'Shorty' or, 'The Ghost of the North', a sort of Robin Hood of the depression who used also to publish train time-tables for north Queensland for bagmen who were 'jumping the rattler'. The hatred of working man for scab working man was deep and terrible. One version of this poem ends:

When at last you're buggered
And a homeless helpless wreck
May you slip back through your arse-hole
And break your bloody neck.

Alf Kain AWU Secretary for the Central and Western districts of Queensland, 'old shearer, old battler'

Another says:

And when I hear you're buggered
And your life they will not save
I'll slip across the border
And piss upon your grave.

Alf F. Kain is AWU secretary for the central and western district of Queensland, a very large territory. He started in the sheds when he was aged fourteen, rouseabout at Augathella, and by the time he was fifteen he was having a go at wool pressing and shearing. He shore for two contractors in the Longreach area for seventeen years, the Select Shearing Company and Bob Wedgwood. 'Things were bad in the Depression so, in 1930 I migrated out west – fifty-two years ago.' He was in time for the 1931 strike.

'In 1931, about four hundred men were camped at the old shearing works in Emerald. Some people were very good to us. [A Chinese] storekeeper supplied our camp with everything we needed. He knew we had no money. When we came good later we paid him but he couldn't be sure of that when we asked for tick. His sons are still at Emerald.

'They were lively days. We had nothing to lose . . .

'Meteor Downs brought out almost sixty scabs, and forty policemen, led by Inspector Battersby, had them at a big old blacksmith's shop with the cars that would take them to the station [the police remained in town]. They were locals' and cockies' cars. When they were on their way back to town we stoned them, broke windscreens, let them know our feelings.

'When they were transporting another big mob from Emerald to Barcaldine Downs a row started, and the police attacked the strikers and bashed them over the skulls, and fifteen strikers had to get attention at the Emerald Hospital.

'In 1929 and 1930 there would be at least 100 men tramping between Longreach and Barcaldine and another 100 between Longreach and Isisford . . . They were forced by law to move on . . . I was an old bushie and I never went without meat. But for some it was bad. There were desperate men in 1931.

'There were shearing quarters at Emerald Depot and the striking shearers were let live there, sleeping on the boards.

'But feelings were very high. There was a shot fired at "the shearers' hotel", the Railway hotel, and young Pluman . . . lost his eye.

'The Police were nearly run into the ground chasing us and raiding us, orders falling on them from Brisbane who in turn had been ordered by the cockies to "protect" them and their scab labour.

'I used to get information from Herb Lewis, a policeman . . . my cousin. He used to tell me when the trains were coming and everything, and one day he told me to . . . get out of the camp . . .

'That night I told them in the camp that the police were coming down and I thought we all ought to get out for a bit. Just then a Sergeant Languish arrived in charge of the police and we all went, hell, west and crooked.

'At the Blue shed, twenty miles from Longreach, Charlie Hulmes was the contractor, he had a team out there and the police was there to arrest us when we arrived so we beat it.

'Things got lively. There was a chap they chased up at the works and the young fellows got him, they were going to hang him. We sent him back, paid his fare back home. Old Tom Kerr was contracting Crossmore station. They had ten shearers on, but after they finished there they just kept going. Didn't wait for any more.

'Lots of stations shore with scab labour. Northampton Downs, Inneskillen . . . They collected chaps from the cities who'd shorn no more than a pet lamb; gaol birds, and sometimes legitimate hard-up family men who were willing to forget mateship or honour to feed their kids. You couldn't blame some of the men but you could despise the cockies for using them this way, for knowing the men wouldn't have done it if times had been better. Trains were leaving Number 9 platform in the city almost every day, full of scabs. It was bad times, Mad, Bad.

Meeting the Scab train at Barcaldine, 1931

'We lived on tick. There were four or five cooks working, no pay; the butcher supplied us on tick . . . and all paid for when we got back into work nine months after. The town looked after us, they were in sympathy with us. Mary Nolte who had the White Hart hotel, she looked after us. Some other publicans did too. Railway men looked after us too. My cousin, the policeman, kept me in tobacco. Some of the townspeople would come to our camp at nights and listen to our sing-songs round the big campfire.

'It was a long, hard nine months though. When it was over I couldn't get a job under my own name, so I took the name of Green, and Billy Simmons, a cook, and I came to Longreach and worked at Hill View.'

One of the few working class men to write memoirs of his involvement in the far-left-wing of the Labor struggle was Tom Brislan. Throughout his story one senses the isolation of the classic loner. He neither counts nor is cared for by employer or employee. The pastoralist reading his maniscript will declare that his Communist philosophy is that of all unionists; the average worker will as readily disown him as being not only a thorn in their side but a lure for the opposition to whip all unionists.

'This is the way I saw it,' Tom says in his introduction.[1]

'When the depression settled on the land and the 1930 decision of the Arbitration Court to cut the shearing rate from £2 1s 0d a hundred in NSW to 32s 6d, the shearers asked the union for strike action. By the end of 1930, the Queensland shearing rate had been cut by 9s 0d from £2 5s 0d. Later it was to sink to £1 7s 3d a hundred. The militants began calling the AWU "The scabby WU" and the "Spewy U" . . . Dozens of sheds were due to start in New Year so the Queensland men were faced with a new challenge. The *Worker* had come out with a front page banner headline saying We Stand for Arbitration, so there was no doubt about official policy towards strike action. Nevertheless, meetings in all the main centres declared for a strike and set up rank and file strike committees.'

The strike fell in February. 'Scabs were brought in by train from Victoria and NSW guarded by policemen with .303 rifles. One train from Rockhampton to Longreach was preceded by a pilot engine with a crew briefed to derail the train. At Barcaldine, women and children joined in a demonstration against the Scabs.

'In the Sou-west the famous "Border Patrol" was organised: teams of shearers who used their cars to travel thousands of miles to intercept would-be scabs and send them back over the border. In one legendary operation led by Trucker Brown, all the scabs were taken from several sheds and escorted in a long convoy out of the state.' They were taken as far into New South Wales as Moree and told not to cross the border again.

'At a station at Surat, shots were exchanged between the beseiged scabs and the patrol before the scabs surrendered. The patrol finally stormed the shearers' quarters . . . The only casualty was the shearer "Cabby" de Moleyns, who received a head wound when a scab struck him with a piece of wood . . .'

According to Tom Brislan, the strike was broken by the Labor Government of the day and victimisation of the strikers then became 'vicious'.

After the strike, Alf Kain got a job with the Select Shearing Company 'under my own name and shore with them for nine years. Then I was Experting and Shed overseer on £9 a week, employed fifty-two weeks of the year. But I fell out with a cocky who wanted me to sack a man who in my opinion was in his rights. I never sacked a man.

'I went back shearing, with Wedgwood and Company. In ten years I shore 332 000 sheep. I used to be in the first five of the men on the board. One shearer shore 380 000 in ten years. He was one of "the Guards" as we called the guns, the cream. The Guards always came back to the same sheds. One man was given a silver cup by the grazier for coming back for sixteen years. These men were consistent, reliable, clean shearers.

'Up until recent years shearers could work starting in January and finishing in December each year, New Zealanders, Victorians and Tasmanians were up here (Qld) by the thousands for the winter.'

Alf Kain became a union official on 1 August 1947. 'My father, Jim Kain, an old shearer, had held a union ticket always . . . I was brought up on unionism.' In 1956 the shearers went on strike on 1 January and the dispute lasted until October of that year. As union official Alf visited Isisford (Qld) and went into a shed. 'Only one man would talk to me. Ted Wilson. He said, "Alf, come down here. Those scabby bastards won't even talk to me." He had been there nine weeks by then. That shed went for months – all bad shearers.

'I went to Bilbah Downs and drove straight up, talked to the shed overseer – he didn't know who I was – and I looked at the four shearers on the board. But the owner came down and to get me away he invited me up to the station. I had a scotch and a feed up at the homestead. It was all done to get me away from the shed but it was wasted because I'd seen what I came to see.

'At Ferrick Stud, Blackall, the gates were locked, I couldn't get in. At Parella, outside Aramac, the manager and Geoff Still stopped me from going in. We summoned Ferrick and they got fined by the Industrial Court in Brisbane for obstructing a union official.

'There were nine shearers working at Duneira, Blackall. I got there about 7 o'clock one night. The bosses were quite friendly to me [Alf begins to chuckle at this stage] but next morning when they woke up to who I was, the rouseabouts, classers and all, there were no shearers on the place. I'd bundled them in my utility and rolled into town. We looked after them and sent them all back to their homes.

'I got summoned in the Industrial Court for taking shearers away from the station. I appeared at the court in Brisbane and they kept me there for ten days before they said I could go home. They had adjourned the case. When I went back they dismissed the charge, as there was no evidence against me. Those nine men were real good chaps.

'Manneroo gates were locked. "Trepassers will be prosecuted" signs up. Instead of going in, we charged them with obstruction and they were fined.

'The union itself was fined during that strike. Joe Bukowski as Branch Secretary of the AWU was fined £1000. Joe waited for them to come and collect the money. He had it all in the strong room in cash – 240 000 pennies. Joe knew the bloke quite well who came to collect it and he felt real sorry for him. Now we're all good mates again.

'People forget how brave women were during the shearers' strikes. In 1956 many wives would have gone to work themselves rather than have their men give in. The Old Raters put in 20s 0d in the £1 to help. . . . Wives would come to collect it from the union through the strike committees. There would be a long line up the footpath on Monday afternoons of wives waiting for strike assistance. And you must remember that things were different in those days. Today people in need easily accept assistance, but for some in the past they felt there was a stigma attached, as though they were accepting charity. But not those wives! They knew it was help given by mates of their husbands in recognition of their stand for what they all knew was right and proper.

'Head office had put in £12 000 and the levy helped us see it out.

'There weren't as many scabs came up in 1956 as there were in 1931 or 1945.

'Some scabs were flown in from Western Australia, South Australia, Victoria, and a few Enzedders. Their fares were paid out and if they stayed three months they got their fare back paid. Shearers from New South were rushed up here, leaving old blokes to shear the sheep down there.

'Each town had a scabs' hotel and if the Old Raters went in there and caused a blue the cops would pick them up. And vice versa, if a scab came to clean hotels. The strikers declared Tattersalls hotel [Blackall] black because it catered for scabs. Some publicans used to short-change the New Raters. They used to do that quite openly in Muttaburra. The police had to call one pub black because of the brawling. There were 500 men in a big brawl once.

'It is a fool who says fool things about a man and his union. I've seen men walk fourteen miles in the rain to vote. I've seen 160 crowd in this little union room at Longreach to get help and assistance.'

Alf Kain remembers Tom Kewley, Jack Howe's friend. 'And as for union officials, you would walk a long way to find another man like Kewley. He and his horse and sulky were known throughout western Queensland. It wasn't only his interest that was with the men, it was his heart. He drove hundreds of miles to attend a delegates' meeting. When Kewley died, a stack of people followed the hearse up the street. Mourners came from everywhere.

'The 1945 strike was for a forty hour week. We were out for eight weeks and went back because the court was sitting then on the forty hour week probe. Norman Potts was a good man then. A cook. He was strike chairman in 1945. A man remembers some wonderful mates from the strikes.'

36

A Hundred Thousand Strong

So stir yourselves, you penners-up,
And shove the sheep along
The musterers are fetching them
A hundred thousand strong.
– A. B. Paterson

The logistics of bringing in sheep from a large property, ready for shearing, would daunt an army commander. A 'hundred thousand strong'? Musterers have brought them in 500 000 strong and more from paddocks each of which could swallow a county, from properties that could swallow the British Isles.

Mustering on Belltrees station (NSW) 1932

Since shepherds became redundant, sheep have wandered free within the vast fenced areas called paddocks in the inland. Finding them is a job for an expert. On the 'never ending plain' they blend in with the grey salt-bush and a mirage of heat, but a wisp of dust on the horizon may betray their presence. If there is scrub, then the work is harder. Today some properties have aeroplanes – not as many as newspapers would have us believe – but they do not muster as such, merely find the sheep.

On Eurugabah out from Booligal (NSW) both the owner, Bill Sheaffe, and his wife Jennie have licences and use their Cessna 172 for mustering, but it isn't mustering in the old sense of the word. On their vast, flat plains they spot the mobs and hustle them towards musterers with the horns on either side of the plane. 'They go like hell when they get a blast from that horn.'

The pilot is in touch by two-way radio with the men mustering on the ground and can direct them to the sheep 'lost' in scrub. Manpower today being what it is, the old Eurugabah shed with its fourteen stands is too big: 'It would be hard to keep the sheep up to fourteen shearers today. It was all right in the old days when you had a team of musterers bringing them along.'

Shearing time at Ensay. Musterers bring in the stragglers as the first load of wool goes out, 1909

Until recently, a property would shear at the same time each year, usually within a period of a few weeks. Let the military strategist ponder the tactics necessary for 'fetching them in, a hundred thousand strong'. With some properties running one sheep to five acres because of the scarcity of feed and water, the limit of ten to twelve kilometres a day for walking sheep if they are not to lose condition, the reserving of paddocks and routes to return shorn sheep to distant

Every pool of water on a property is tattooed in the musterer's mind as he brings sheep in over the miles of outback 'paddocks'. Buggies were often used closer to the homestead. 'More comfortable, able to cart more gear.' Scene on Dalkeith, Cassiles (NSW) during 1941 shearing

paddocks, and the almost universal problem of finding enough feed in paddocks close to the shed to hold the numbers of sheep, are a few of the difficulties. Add to these the preparation of troughs and tanks for dipping shorn sheep against lice, gear to attend to foot-rot, pens and drafting yards for culling those that will return to paddocks from those that will be sold, and box (get together) those selected for other purposes such as breeding or for later sale or to fatten. Then add to that all the things that can go wrong with such a vast operation: it can rain. It can rain in the desert where it hasn't rained for nine years, then the sheep are banking up as the musterers bring them along. In 1978 at Thorn

Billilla (NSW) sheep kick up the dust as they are let out of the shed near the river bank, while those still 'in the grease' are brought into the shed

'Off shears' and away on the long journey to distant paddocks; water and feed determine the route

Leigh, Buddy Wagstaff's property towards the Channel Country (south-west Queensland) the woolshed was stranded in six inches of water for weeks on end, because in that area, where the land slopes so imperceptibly towards the dry river beds of Central Australia, the grade is measured in inches per mile. Most sheds provide cover only for enough sheep for the next day's shearing. If everything else goes well, there can be industrial problems: 'wet' sheep, for instance, or fast shearers. (If a farmer gets a team of gun shearers he will be hard pushed to keep the sheep up to them and still perform the score of other activities involved with shearing.)

Mustering on some properties has been so difficult that unique systems have been evolved. In the north of Western Australia at Boodarie station, the Hardy family muster by fire. They have mustered this way since the days when the black boys mustered here'. It is a method suited to the huge sprawling property where spinifex is the main ground cover, and it is fired as a signal when mobs of sheep are found. Properties between Mildura and Broken Hill muster by water. They shut off the water or shut the bores down in different paddocks, leave the gates open, and thus force the sheep to head to the water the musterers have chosen.

At Clare (NSW) mustering used to be done by flags. Using the principle of semaphore, signalling from point to point, they would haul up flags which let shepherds know which flocks to bring in. 'You could send the message over miles this way, each shepherd on the way out would hoist a flag on the line until it reached the man needed.'

On the hard-hit-by-drought stations, mustering is different again. In October 1980, Henry Cock of Yanda (NSW), the old station out from Bourke, found it difficult to get his sheep to walk from the outer paddocks to the yards. In their

Isis Downs (Qld) and 1930 flock in for shearing

weak state 'it took a long time to get them in'. Many had to be brought from an out-station across the Darling River on a punt.

Frank Ayres (Riddells Creek, Vic.) was on the parade ground at Fremantle (WA) in 1943 when the troops were told the outback wool-growers were in trouble getting the clip off because so many of the shearers had gone to the war. Frank could shear, so off he went in an old truck, inland from Geraldton. 'Rumbled on for a few hundred miles and then they said, here you are.'

The soldiers stared out over the flat miles. 'They were mustering by water we were told. In such huge areas they had Buckley's chance of finding their sheep, so they cut down tea tree and built brush fences around the water-holes and left only a small opening. The sheep would go in to water but wouldn't find their way out and so – they were mustered.'

Jenny and Bill Sheaffe of Eurugabah on the One Tree Plain (NSW) fly a plane to muster sheep today. Both remember days when there were no such aids and a steam engine drove the shearing machines

37

'Ducks on the Pond!'

'The ladies are coming!' the super says
To the shearers sweltering there,
And 'the ladies' means in the shearing shed
'Don't cut them too bad. Don't swear.'
– *Henry Lawson*

'I never knew shearers to be touchy about women in the shed,' Jim Carew says. 'I remember well enough years ago hearing them yell "Ducks on the Pond!" when a sheila hove into sight but that was more to warn a man to watch his tongue. Tempers flare a bit when a man's racing to lift his tally . . . Women in those days would have got a great surprise to hear the general talk of a shed, even today it might surprise some women, so the blokes tone it down when a woman comes in.

'Yes, they showed off a bit when a girl was around. When I was a presser I noticed it: they'd get such steam up the wool would be flying at me like feathers from a duck. I'd think it was a good thing when women came in the shed. Gave a better atmosphere altogether. You don't hear the blokes complaining about little things when the girls are there.'

'I think it was at Lorne, main shed of Terrick, that Lord and Lady Lamington came on the board to watch the shearing. My father [Jack Howe] told me about it,' says Henry Howe.

'My father didn't keep quiet, or bowed. He knew that Lady Lamington had a penalty to pay and he put his hand-piece down, crossed the board, bent down and lifted the skirt off her shoe and raddled Her Ladyship's foot. [Raddle is the dye-stuff used to mark a sheep.] By tradition any woman who entered the shed should have her foot raddled red.' Lady Lamington asked what it was all about and Jack Howe told her that the penalty was for the lady raddled to shout a round of drinks for the board. 'She turned to his nibs and asked if he had any money and he said he was sorry that he only had half a sovereign. But that went a long way in those days.'

Patsy Adam-Smith gets her hands in wool with classer Alan Bates at Meningoort (Vic.). 'In five years of visiting sheds and talking to shearers I've found no resentment to women being on the board.'

The signal that women are on the board comes in many ways. 'Lady!' some call, or 'The ducks have landed'. 'Woman on Board!' 'Feathers Flying', or just 'Watch it!' But 'Ducks on the Pond!' or 'They're on the Pond!' is the usual. Usually the wool presser calls first, because most visitors to a shed come through the wool room. 'Ducks on the Pond!' he calls, and the warning is taken up again down along the shearing board. It is one of many warning signals given to stop swearing or 'any toilet activity' as the handbook puts it. 'Hawker!' call the South Australians.

'Woman on the Board!' is the only way my approach was heralded in over 180 sheds I entered, and that no more than two or three times.

These calls are not to intimidate women; indeed, shearers have no more antipathy to women being in the shed than they do to men. (Less so, perhaps. I've seen records broken when a visiting woman or girl appears to break the monotony of hard yakka.) Any visitors are somewhat of a nuisance, for scarcely a corner is not part of the traffic area. And if they are crass enough to march down the board they are a danger to everyone in the shed and to themselves. A shearer's work is done in isolated areas and few have seen him at it. Like most skilled men he is proud and happy to demonstrate his art, but only if the visitors are controlled. In 1980 the American Ambassador to Australia and his grandson were visiting Longreach (Qld), and I drove out to White Hill with them one hot, dusty day, when the shearers were going like steam to get through a big mob. The ambassador stood quietly in a corner of a catching pen observing and photographing everything. Each shearer said 'good-day' as he caught his eye and one asked if the grandson would care to 'have a go'. The shearer gave him his stand and helped the lad shear a sheep – it seems a small enough gesture but it did cost the giver money, about three sheep less on his tally for the day. Did he mind? 'Hell no. Kids love to brag they've shorn sheep on a board.' Well, I don't know about that, but this one did boast, all the long miles back to Longreach.

'Everything that could possibly give offence is strictly taboo while the rustle of skirts is heard. The man that might do anything to disturb the visit would be roughly handled afterwards by his mates,' Jack Howe said.

I have found the same etiquette on board a cargo ship. No profanity for profanity's sake, but a woman should keep out of the way when something goes wrong.

> 'Shearers yell for tar and needle
> Ringers roaring like a bull!

– the big ram has kicked the hand-piece from the shearer's grasp and all hell is let loose. If a woman hangs around at such a time it's her own look out.

E. S. Sorenson explained why women are sometimes unwelcome on the board, in *Life in the Australian Backblocks.*

'While the work is in full swing it is one of the principal sights of the district, and the board is often graced with the presence of ladies from town . . . The board is narrow. There is barely room for the skirts to pass along without brushing against the long row of stooping shearers; and their presence interferes with the pickers-up . . . The shearers . . . try desperately to hide their bare feet, or bag moccasins, and the gaps in their shearing togs, any kind of clothing being worn

Women shearers and rouseabouts are employed in many sheds. 'Reliable, hard-working, sensible,' contractor John Harris says, 'Just like you employ men – you keep them on if they're good and put them off if they're not.'

Stephanie Jones, wool classer, is among the professional women now entering – and accepted – in the industry. Below, she tramps down a bale at Yeadon (NSW)

at the greasy work on the board. They are not seen to the best advantage, and they don't like it, expecially those who "act the swell" in town. They are covered with grease and blood, and reeking with perspiration. Particularly unpleasant is it to the man by whom they happen to stand, criticising the cut and his operations generally, and saying "Poor thing!" when the nervously-held blades nip the skin . . .'

'We never forbade the board to women but it is not the place for outsiders when the pace is on,' said Jim Carew. 'There isn't much space . . . and of all toil it is the one where cursing comes natural as you wrestle an animal noted for its stupidity. The belief has been fostered that women are not permitted on the board. It is a handy lever for keeping pests away. But as these skilled professionals like to show off their art, a woman standing watching and out of the way is always welcome.'

They are girls from the city. (Our hearts rebel
As we squint at their dainty feet)
And they gush and say in a girly way
That 'the dear little lambs' are 'sweet'.
And Bill, the ringer, who'd scorn the use
Of a childish word like 'damn',
Would give a pound that his tongue were loose
As he tackles a lively lamb.[1]

– *Henry Lawson*

38

The Silly Season

How silly can you be! cried the editorial of the Brisbane *Courier Mail*.[1] 'The shearers now are receiving more in relation to the price of wool than they were a few years ago when there was no strike. The new shearing rate is £6 19s 6d per 100; the average price for greasy wool at the first five sales of 1955–6 was 63.14 pence per pound. In 1949–50, the shearers were being paid £5 per 100; the average wool price then was 8.49 pence per pound.

'The wool storemen now going out on strike in sympathy might compare their ordinary weekly wage with the fact that good shearers still can make up to £40 a week.

'The communist disruptors themselves are now not even sure of their own direction. Their hierarchy in the Kremlin have just pulled down their own old Red god, Stalin.'

It was the silly season in Australian politics and just the time to foment a strike that would proceed with as much rancour and hatred as the political parties who were slashing into the under-belly of what today was a friend and tomorrow the bitterest enemy. The times were awry, and this dreariest decade in modern history was further soured by a political war that split families, put children against parents, brother against brother.

1956 was the year of the Red under the Bed; a year that blasted a schism in the community as surely as the 1890s had done.

The Labor Premier of Queensland at that time was Vincent Gair, shortly to leave the party he had successfully divided and enter the Senate for the DLP (Democratic Labor Party, as the new Right Wing political body was named). R. J. J. Bukowski, Queensland Secretary of the AWU, was also a Catholic and there was total, personal animosity between the two men. So bitter was their antipathy that there was no way they could parley in the sort of unofficial negotiations that had virtually kept the shearers out of strikes for fifty years.

Gair followed the advice of his church and helped split the Labor Party; Bukowski ignored the bishops and, knowing that the AWU had always – even

Picketing the wool stores, Brisbane, 1956

in W. G. Spence's pioneering days – been anti-communist, continued with the union. The Brisbane newspapers might shout 'Communist Reds!' but it was Christian sectarianism that smeared the 1950s with a bile that is only now receding. It was a complex time with allegiances changing rapidly – a rather dodgy time to strike; in that, the 1956 shearers' strike during the Cold War had one thing in common with the strike of the 1890s.

In time the political quicksands dragged the original argument out of sight and all that was left were the old catch-cries of 'Reds' and 'Bolshies'. 'Communists have moved in to take control of the Shearers' Strike,' declared W. A. Gunn, the millionaire President of the United Graziers' Association. But the shearers believed that the fall in the price of wool after the Korean War ended and the high unemployment in the country had made the squatters confident that they could lower the price of shearing. And the rate was lowered by ten per cent.

The difference between this strike and that of the 1890s was that although the growers used 'scab' labour to get the wool off, the shearers were able to hold out for ten months because they now had effective union organisation which had been built up over fifty years.

'1956 was one hell of a strike,' veteran union secretary Alf Kain of Longreach knows. 'We won because the squatters didn't stand together as the unionists did – many paid the old rate, many paid the new.

'Our men stuck together, firm, loyal. Those shearers who got work at the old rates levied themselves heavily to help the men who refused the "new" rates. They were tremendous men. Some of the best men this country has known were out in 1956.

'And behind us, there were thirty-two unions refusing to handle "black" wool, backing us in all ways because if our rates went down these men knew it would encourage other employers to lower the wages of all workers.'

It could fairly be said that shearers are among the least militant of all union groups. There have only been two serious, long strikes in the history of shearing in this country: 1890 and 1956. The industry, for the type of aggravating situation it can engender, has been free of industrial trouble. One reason has been that these small factories in the bush usually can sort out grievances right there.

'I don't believe there is anything you can't negotiate in this industry,' contractor/shearer Laurie Walsh states. 'You get in the shed, find them daggy,well, you yarn with the owner, and most graziers when it's put to them right see it fairly. Those who are hard-headed, well, they find it hard to get a team back once they up and head off. Shearers are better organised and support one another better than other industries.

'We were in a shed, the sheep were full of sand. A man was lucky to get twenty-five of them shorn in a run and might do in fourteen combs on the way. It was keeping men broke buying tools. We spoke to the grower, he agreed we had a legitimate complaint and adjusted things for us.'

Laurie Walsh went north to Queensland to shear in the 1950s and has this to say about Queensland shearers: 'Accommodation was pathetic up there. Queensland shearers were said to be shit-stirrers out of their own state, but on their home ground they'd do anything so long as they had a bore drain and a carbide light!'

Although the accommodation act by 1917 was quite specific, economic conditions forced men to accept less than they had fought for for so long. Many shearers were men who had taken up soldier settlement blocks after the war and been driven off much of the unusable land; others had lost out to fate during the depression. 'The union couldn't enforce the act to any great extent during the depression and things were pretty rough, in some places accommodation was as rugged as any in the nineteenth-century shepherds' camps. But when the squatters have the upper hand a man with kids can't complain about anything or he's out on the tramp again with nothing in his kick to send home to the missus.'

Meggs Elliott is a tough man. In November 1946 a bad-tempered horse he was riding took off. The saddle girth snapped, and Meggs flew off and hit hard. He lay for four hours in the sun with his thigh broken. He went back contracting, with a gammy leg. 'I'm not afraid of nothing.'

'We went on shearing in the strike. Free labourers. The union men put broken glass bottles on the road so shearers going out would puncture.

'It was hard to get a team together. I got to one station and the manager said, "Have you got a full team Meggs?" "They're as full as I can get them!" I told him, as the drunks rolled off the back of the truck.

'I got life – with my Uncle Alex Meekin – for scabbing on shearers that

Again the old problems were advertised, this time on city walls time,' Jimmy Pugsley told me: 'Later, all suspensions and disqualifications were lifted, all except Pugsley. That's how it goes.

'They turn your photo to the wall if you play up.'

Meggs Elliott was in the 1956 strike up to his ears. We know he was when he shows the furrow down his forehead. 'They got me with a tomahawk. But that wasn't what felled me. It was the log on the back of my skull did that.' Doddy, Meggs' wife remains silent, neither verifying nor denying it.

Arthur 'Stonk' Lewis played a large part in the fight for a forty hour week. 'It had to come.' He lost a lot of time and a lot of work over that fight. 'The wool-growers were against it on principle. The stupid thing was that many wool-growers were and are flat out keeping the sheep up to a team and are as happy as the men when there is a bit of rain, because it gives them a chance to get the mustering done better. But they were just bloody minded as they have been all along.

'Mr Hargreaves of United Graziers' was Boss of the Board at Oakwood and he said a shearer, Byfield, hit a lamb, so he sacked him. He said Byfield shouldn't shear 200 a day. Said it was damaging to sheep to shear 200 a day. Then he told the next man, Coppick, that he shore too rough. So he was for the chop too.

'We cut three or four strands of wire with pliers and tied up the count-out gate so they couldn't release Coppick's sheep. They were all shorn cleanly and we wanted that to be seen. We were told that if we refused to work they'd get twenty-four more shearers and rousies out to replace us. Mr James, the rep. from Charleville was coming out. We go on working for two and a half hours and he hasn't arrived. So I think, I know what's coming for me, so I joined the army.'

Stonk Lewis knew the police well. 'I only used to fight with police when I was drunk. They knew that. They didn't hold it against me. Sergeant Jim Palmer at Charleville in the 1956 strike was on duty at the pictures. He pulled me up when I came out. "You haven't been drinking?" "No. Not for several days." "Well," he said, "Don't start it tonight. We're going to grab all the rate shearers tonight." I was frightened to tell anyone too much. They'd say I was on with the cops.

'That sergeant once let me out of the cooler so I could raise my bail. I needed £2 and he said, "Be back in two hours with it," and I was.

'As for strikes, they got progressively less violent but in 1956 one bugger burnt a shed down. Stupid bastard,' says Stonk Lewis.

'Men who went on strike in 1956 could remember 1951 when wool-growers were getting boom prices for wool and not paying shearers a penny more to harvest it for them. In 1956, when the price dropped, they cried poor to the Arbitration Court and were awarded a reduction in shearing rate.' It is Laurie Walsh speaking.

'It was a strike that wasn't well handled. I believed everyone should have stopped work, but the Federal Executive of the AWU declared that if owners were prepared to pay old rates the men should shear: 90 per cent of the shearers went out, 10 per cent worked. Every time you went into a pub there was a blue on. I was out at Ingomar, fifty miles below Coober Pedy. The strike came on in March. We told him that if he didn't pay us the old rates, we were out. He said all right. We insisted that he gave us all new agreements.

'We had to have this, then he couldn't claim we shore at new rates: some graziers did this, paid old rates but boasted they paid new. Imagine the ruckus that caused when shearers got together!

'As I said, show your face in a pub and there was a blue. I remember this time we went into Kingoonya. There was a team from Koondambo in the pub, more of them than we had in our team, so we were wary. Not one of us put a foot wrong! Police didn't break up fights because they had nowhere to lock you up.'

Bill Schuller (Broken Hill) had what he calls, 'a poor schooling but a mighty education'. From being a battler he became a contractor for eight years, from 1951–9. 'When the 1956 strike was on, lots carried guns.' Did he? 'Well... anyway, it was a nasty time. There was this lively chap agitating up here, going around the sheds. So someone pointed him out to the police as a strike leader and the old sergeant said he had two young blokes up from the city "spoiling for a fight". "Don't mark him." Well, they put him in a padded cell and didn't mark him but he died six months later. I think they roughed him up too much. The Commos are behind all the strikes,' Bill believes.

Charles Flavel carried a gun. 'A man had to, sometimes people from the city don't know what's happening out in the bush. We're often a long way from law enforcers. We've got to be our own law and order often, when things blaze up.'

The taint of 'scabbing' lives forever, it seems. Sharon Lewis, working at Cunnamulla as a barmaid, knew this. 'A pack rapist and a murderer, I've seen blokes drink with them at the bar. But they won't even sit at the same end of the bar with a scab.' Sharron's father, Stonk Lewis, watched a mother and her kids walking by a pub during the strike. 'The scabs were sitting on the front verandah without shoes. "Jump on their feet as you go by!" the mother yells to her kids on the way up the street. "Scabs! The lot of them!"

'We knew which pubs put up scabs. We never trusted them.

'The stigma is such, a scab will never unload it. It is the greatest moral crime in history. The graziers used scabs but after every strike they didn't want to know them. They weren't good shearers and they weren't good men.'

Good men or not, many shearers were afraid that the men they called scabs might leave them jobless when the strike ended.

The Sober Six

Said the striker to his comrade,
Let us sink a drop more booze,
For they've shorn the Yalco-weaners,
Now they're on the Aven-ewes.
Now shearing through a strike's a thing,
I really do not like,
For how are we to make a deener,
When at last we end the strike.
They have shorn the Yalco-weaners,
Now they're on the Aven-ewes,
And they haven't lost a blow yet,
Call them hungry if you choose.
And from the sheds up north in Queensland,
Down to Tassie spreads the news,
They have shorn the Yalco-weaners,

Now they're on the Aven-ewes.
Shearing sheep's their occupation,
While the strikers get the kicks,
And there's not a town committee,
That can stop the Sober Six.
And in rainy Western District,
Or wherever shearers booze,
You will hear them speaking of the team,
That shore the Aven-ewes.

Tactics reminiscent of the 1890s were used on both sides. Once again the graziers advertised in New Zealand for men, promising 'at least 8000 sheep per man', and claiming that they were flying them over in great numbers.

'No shearers have so far left New Zealand for Australia,' wrote A. Walsh, the Secretary of the NZ Federation of Labour in the *Dominion* newspaper. 'I feel confident that New Zealand shearers will refuse to be used as strike-breakers against Australian shearers in their fight to retain wages and conditions of employment.'

Camps had been set up again for the strikers and for any 'free' labour they encouraged to leave the sheds. Once again the courts were busy. Many of the charges laid in this strike were against shearers who 'contravened an order', or, in other words, left the shed after having signed a contract. Bukowski of the AWU claimed that such men were not on strike. 'Shearers have the right to refuse their service if payment is inadequate,' he told the Industrial Court. His plea was not accepted. From Mona station, the first shed to be so charged, the men said they had become discontented after shearing began because of 'the size and character of the sheep'.[2]

For the information of union members the *Worker* constantly ran lists of Queensland sheds shearing at the 1955 (old) rates. Some of these lists noted 82 sheds, another 117, all seemingly operating union. At the same time the United Graziers' Association constantly denied that certain sheds listed by the *Worker* were indeed shearing union. And so it continued for ten months. One of the few lively comments to come out in court was regarding the shearers leaving Mona shed. 'The manager took his false teeth out and offered to fight me,' the union rep. said in his evidence. From such snippets of levity does sanity return.

'We stayed out ten months until we won,' says union secretary Alf Kain. 'Here at Longreach, where the railway used to be so busy carting wool that many men were employed in the yard in the season, the wool was black, so the railways wouldn't cart it. The railways have never picked up from that time.'

39

All among the Wool

Mervyn Shung, classer

Jim Carew went wool classing in 1929. In his long career in the sheds, he travelled from Julia Creek in the north of Queensland to Tasmanian sheds. When he first went out, twenty-eight stands was the average; when he left the work it was down to eight and nowadays five is common.

Jim's father, James Carew, worked for Sir Samuel McCaughey on South Yanco in the Riverina and wrote a text book on shearing that stands to this day. 'I always knew I'd follow the shearing. When I began at Wellshot with its forty stands I knew I was in among the wool. I stood at my classer's table, in the great and famous sheds, and I've stood on a cowhide at flimsy bush timber contraptions under bough roofs, classing for two stands out from Hughendon. A man can see a lot from the classer's table. On an old timber floor you could see the boards bleached at every stand from the sweat running off the shearers. Later, because there was a great need in small sheds which did not justify a large team, I went off as overseer-classer.

'The overseer-classer stayed at the homestead or at staff quarters. Sometimes he'd bunk in with the men but that's not so good: the men need to chew on about him a bit, and the overseer needs to talk to the owner and is sometimes late in at night. At meals it's no good if some bloke says "You chipped me today". It's better if you stay apart a bit.'

The wool classer is the man who sees both sides of the game. He is a trained, skilled man trusted both by shearers and the wool-grower. Ultimately he is trusted by buyers. A classer's skill is unquestioned. There is only one classer in a shed, no matter how many shearers. 'Sometimes the old song, "All among the wool boys" was too true to be funny,' Jim Carew recalls. 'It would be piled high around a man while he had to calmly continue to assess the quality.'

Wool classers move from state to state. Most know Tasmania well. 'We usually end up there to get away from the heat.' Peter Sams, one of the best known classers in the country, is also in demand as judge at shearing competitions.

The pressers, wool rollers and classers in 'burry' country wear leather gloves to protect their hands from the vicious Ngora burr. 'My wife likes me to look

The old wool press on Vansittart Island in Bass Strait reputedly was loaded with kangaroo skins to make up the weight on shipments to England last century. Some suggest a more sinister use, when skeletons of the now-dead Tasmanian Aborigines were shipped out in bales to the British Museum which paid large sums of money for these rare bones

after my hand,' is the corny comment, but the truth is that Ngora burr can rip lumps out of the lanolin-soft hands of men working in wool.

'Pressers', says Jack Nicholes, '*never* go on until they are old. You will only find young pressers. It is the most strenuous work known. It is comparable to playing football for eight hours a day. Dozens of wool pressers have died with strained hearts.

'In the depression we were all in the same boat. We carried an unemployment insurance book and there was a sixpenny stamp stuck in it each week for as long as you were working. After cut-out, when you came to Brisbane, you applied for work, they called your name, and if you answered, "No work", they paid you 15s 0d per week until your stamps ran out. Fifteen weeks was allowed – less, if your stamps ran out. Then, when that happened, the government gave a man a travel warrant out to the shearing centres. Men would arrive out west,

A classer's decision is accepted without question by wool-growers and buyers at city wool sales. James Carew (left) was sought after in Queensland sheds over many years

A fleece is thrown in such a way as to have it fall evenly over the classer's table

cream trousers and a blazer and tan and white shoes, and they'd be camping in tents, some camping under houses. In those days we used to call Yeppoon, "The Rouseabout's Bondi".

'Ernie Bradford always said he'd like to die shearing and he did just that; finished the last blow and straightened up and fell dead, at Dirrinbandi. Ernie was a good man. In 1931, when the strike was on, he mortgaged his home to feed women and children of the strikers. He didn't say much about it. We were sitting out there picketting the pub, he and me, and I asked him about it, that's how I know.' John Nicholes saw the last of the steam engine-powered sheds and presses. At Historic House, Charleville, there is an old stationary steam engine, the last of the mighty power drivers, and one of the few preserved from the shearing areas. And a ninety-year-old in Charleville, Billy Munt, says he could get it going again in no time at all. Billy had kept it going for half a century at Oakwood station.

The men who coaxed and cajoled the machinery a thousand miles from qualified mechanics or engineers were the peers of the shearers they worked with. Carl Pulsford knew Darky Peard. He was on the steam engine at Cowley. 'He'd ride his pushbike the 105 miles home to Charleville on a Saturday after lunch and get home by Saturday night. On the following evening he'd ride back again.

'His was heavy work, like all the early experts, and he used to write into Fitzwalters to send out "two black shirts" a fortnight. His wife was bewildered as to what happened to all those shirts on his store account. "Oh," said Darky, "They get so dirty and greasy with the oil and work on the engine I have to burn them." (A black shirt, as translated by the storekeeper, was a bottle of rum sold by the gallon only.)

Jimmy Pugsley says, 'You'll have seen the flat-nosed pressers? Got them when the handle slips and swings back.' And yes, one has seen them.

The classer, 1920. 'If they wore a bowler outside, they wore it inside the shed as well.'

There were some big guns among the pressers. A Victorian, Bill Chapman, has done 70 bales a day. Jim Carew helped him do the last of 68 bales out of Longreach some years back. 'You need to be strong and active to be a good presser. They develop a system and their hands are doing something all the time.

'Some of the steam presses needed three men to operate them. They had to be a good team to work well. They were the aristocrats. Guns.'

'Arming' or carrying wool to the press. Tupra, 1980

Piece-pickers 1870

Rouseabout
  up at sparrow fart
Shirt hanging out
Eight stone
Fly blown
Poor bloody rouseabout.

'Rousies are terrible religious,' they say. 'After shearing's been going on two weeks you find them down on their hands and knees praying. For rain.'

Tom Duckmanton, being a rouseabout, neither ate nor slept in the same quarters as his shearing mate, Percy Riddle. 'I was getting 12s 6d a week at Devon Park (Vic.) in 1925. Didn't have showers in any shed those days. We'd boil water in a tin over the fire and wash in it and then wash our clothes in it.

'There was always a fire outside and some huts had a fireplace inside. Mt Sturgeon had an open fireplace eight feet wide. Three men would carry a tree trunk in each night and put it on to burn, and we'd dry our clothes each night.

'Pressers don't live to be old men,' they say. Mel Margrie, pressing at Meningoort, is about to haul the upper half of the press up as he will do for each bale.

T. ROBINSON & CO.,

MANUFACTURERS OF

Agricultural Implements and Machinery.

T. ROBINSON & CO.'S

PATENT IMPROVED

SPEEDY WOOLPRESS,

Worked with Chains. Direct Action. No Friction.
With Tumble-over Box.

PRICE, £30. WEIGHT, 12CWT.

Over 650 of the Old Pattern in Use, and these are very much Superior.

GILSENAN'S Patent Continuous Lever Woolpress, with double wrought Tee Iron Racks and Tumble-over Box. Guaranteed the Quickest and Most Powerful Manual Press Made. PRICE, £45.

T. R. and Co.'s Steel Skid Earth Scoops.
T. R. and Co.'s Two and Three Wheel Earth Scoops.
T. R. and Co.'s Dam and Draining Ploughs.

NOVELTIES IN ALL KINDS OF AGRICULTURAL AND PASTORAL MACHINERY.

SOLE AGENTS FOR

MARSHALL'S ENGINES,
CENTRIFUGAL and FORCE PUMPS.

T. ROBINSON and CO, 247 ELIZABETH-ST., MELBOURNE.

Nive Downs pressing room, 1892

'No, there wasn't so much gambling among the rousies – they didn't have the money the shearers earned, and we were too tired. We ran all day. In bad times they'd only put on two rousies to pick up for ten shearers. We were only young, but after work we'd wash our clothes, have our tea and fall into bed, that's how done-in we were.

'There were no big pay packets for rousies. Old Jimmy Brown once gave me £1 at the end of the shed. Another owner gave me a packet of cigarettes.

'They shore until dinnertime Saturday when I started. Rousies went back to the shed after dinner to wash the board. If you didn't do it to the boss's satisfaction you went back and did it again.

'"Broom there!" they'd shake you up. Most of the old sheds were built over a creek or drain and that washed the manure away as you swept.

'I was working for W. Thompson at Corea for six weeks in 1925 . . . rode my pony out and my mate his push-bike. We rode home at weekends after washing the board. You must remember what a big area the board covers. Some of the sheds had thirty shearers . . . then there's the wool sorting end to clean up.

'There was usually a shed-hands' cook on over the weekends, but we liked to get home if we could, because usually we couldn't. I'd most years be away by July and come home for Christmas. In that time I'd been around 20 000 or more sheep.

'But I loved shed work. I liked watching them shear. I was offered a job piece-picking and wouldn't take it. Standing there looking into space all day wouldn't suit me.

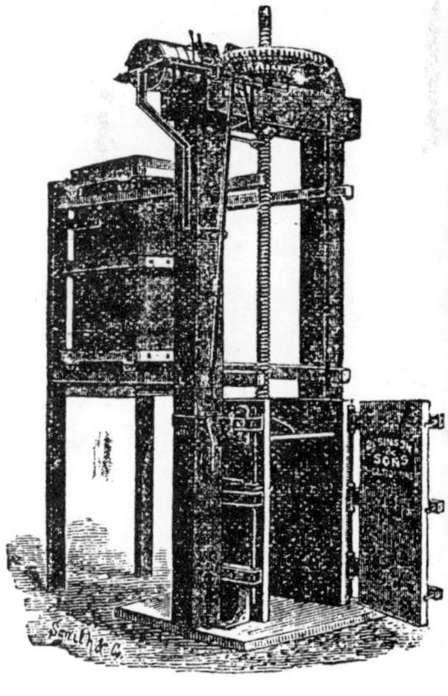

'A rousie becomes pretty nippy pretty quick. You develop a style all your own. "Shake that broom up boy!" You throw the fleece on the table so it opens up, run back down the board, pick fleeces up, run, sweep. "Shake a leg there!" Run, bend, sweep. "Wool away!" throw a fleece, pick-up. At Fulham shed you ran, cleared the fleeces, dumped them against the wall, ran, threw the fleeces on the table when you got a chance. "Sheep O!" More sheep needed in the pen.

'They educated you when you were young. They taught you what you could and couldn't do. You got a smack under the ear and you learnt fast. I was high spirited, played tricks on the shearers, and they returned the compliment! I respected them. They educated you as far as your manners went too. At the table, or anywhere. You stuck your nose in and they'd say "You keep out of this and speak when you're spoken to."

'Andy Riddle taught me to make moccasins. He used to put a bit of wool in them and it finished up like lining. Shearers were older than us boys. They helped us.

'Like most jobs, there was the chance of danger and that excited a young boy. The governor belt at Toganmain broke one day and everyone threw down their machines and bolted. I followed them, fast. The whole place could have begun to fly around in pieces.

'Still, it was the greatest life for a boy.'

Little bare-footed broomie at Talgai West, 1895, and other rousies would be known to the cook as 'brownie-gorgers' – big eaters

Stencilling the brand, Tupra

40

The Baitlayer

They call him the water burner, cookoo, Silly Look, baitlayer, babbling brook. The oldest, corniest joke in a shearers' cookhouse is, 'Who called the cook a bastard?' The reply comes, 'Who called the bastard a cook?'

> Mutton stewed or chops for breakfast, dry and tasteless, boiled in fat;
> Bread or brownie, tea or coffee – two hours' graft in front of that;
> Legs of mutton boiled for dinner – mutton greasy-warm for tea –
> Mutton curried (gave my order, beef and plenty greens for me.)
> Breakfast, curried rice and mutton till your innards sacrifice,
> And you sicken at the colour and the smell of curried rice.
> All day long with living mutton – bits and belly-wool and fleece;
> Blinded by the yolk of wool, and shirt and trousers stiff with grease,
> Till you long for sight of verdure, cabbage-plots and water clear,
> And you crave for beef and butter as a boozer craves for beer.

Ern Barnes was at Urangaline for the start of the 1907 season and after drawing lots for a stand he joined the men balloting for a cook. 'There were half a dozen men who were nominated for the job of cooking for the shearers . . . Some of them were known far and wide – from Bourke to Barcaldine as the saying goes – for their cooking and, what was of as great interest, for their efficiency in running a cookhouse and so bringing down the cost per man to a minimum. Although the others, not including the wool-pressers, but the rouseabouts, wool-rollers, shed hands, and the like, have their 'keep' included in their wages, the shearer has to keep himself in the way of food, and a bad cook can be the means of the messing bill being half as much again as it should be . . . When we voted by writing on a piece of paper the name of the cook we preferred and placing it in a hat, I voted for [Dick]. When the votes were counted it was found that Dick was elected.

'He turned out to be an excellent cook; as well he might, for I found out later that when he was not cooking for shearers, he was chef at a fashionable hotel at a popular seaside resort near Melbourne. As his takings for the season at Urangaline would be in the region of twelve pounds a week (a big salary for those days, made up as it was at four shillings per man per week for about ninety men), with extras in the way of 'fat money', it was better payment than a member of parliament would receive in those days.'

'Life wasn't easy for a cook,' reminisced Stumpy McMahon: 'We had a half-blind cook once and he used to cart water from the outside tank and we used to cut the bottom out of his dipper. We were high spirited with the lack of thought that kids have. The cook had a wood copper to cook in or camp-ovens in coals out in the sand, and we didn't notice how hard it was for him when we were so young. A cook's reputation goes ahead of him. "He's greasy, that cook," or "He pinches the best rations for himself," or "makes weavily burgoo".'

But like most things in shearing, there is another side. The ever observant Rachel Henning wrote to England in 1864 about a cook in Queensland. 'I have just been hearing the exploits of Ned, the shearers' cook: how he borrowed our three camp-ovens and sat up all night cooking for today . . . The productions of his genius were beheld this morning by the wondering eyes of Mr Taylor. Two great cakes, six jam tarts, small cakes and unlimited plum puddings and beef. I think it is fortunate that the weather is so wet, and that they will not be required to shear tomorrow.' And the cook would have the shearers under his feet all day and night, 'whinging about the tucker, wanting to use my kitchen table for a game of poker'.

'All shearers' cooks are mad,' David Fitzgerald says. 'It's the continual heat on their guts that does it.'

Richard Cottersill knew some cooks before he set off to 'go it alone' shearing for himself. 'There was "the Whinger", out from Esperance, "the Long-handled Shovel" – a woman that one – and another woman cook called "the

Shearers' dining table, Claverton station (Qld), November 1905. The cook appears to have prepared a special repast this day

Roo Shooter". There was a woman cook in one shed got on the lemon essence and my mate saw her making pancakes. Something was wrong [but] he couldn't see properly through the fly wire door. He went in, and she was trying to get the pot holder in the pancake for filling.

'But you get some good cooks. Ossie Potter was a chef at one of Perth's best hotels and he could cook bread and pastries in any old oven they put up to him. Many of the stations provided brick baking ovens, and the cook had to bake all the bread, cakes, biscuits and scones.'

Bill Greig took a cook out to a shed he was shearing north of Tarcoola (SA). 'It was 117 degrees when we left Adelaide. Imagine the heat north of Tarcoola! Well, the cook went mad. He drank all the rum in the huts before shearing started. "Never again," he wailed. But what a mess! You know, you've got an enormous amount to sort out without the cook chasing things up the wall. That was the trouble often in those way-out sheds: you couldn't take beer because it was too bulky so they took high-alcohol-content spirits.'

Ray Watley is scathing about cooks. 'Many of the cooks I saw were the sort that made a man say, "The next time I come up here I'll stay at home." We had a cook once who was always boasting he'd shoot us a duck. "Him bring down a duck!" said my mate, "He couldn't hit a bull in the backside with a bucket of wheat!"'

'They're a much maligned race of men, shearers' cooks. How'd you like to toil from daylight to dark and see the whole of your work demolished in eight minutes – if it's good. If it's not, you're likely to be wearing it. I've seen men let the cook have it, plate and all.' It's Jim Carew speaking.

Others say, 'When you get too old and greasy to shear you take on cooking. Cooks, crooks and cuckoos.'

They're a race apart. They have to be, to contend with the hard-working shearers who work at a furious speed all day and finish their last run tired, sore, nerves jangling: annoyance of any kind will upset them. Only 'non-found' members of the mess can vote to sack the cook, that is, shearers and pickers-up.

Carl Pulsford, Charleville: 'At Milo, Jacky Lynn cooked for sixty-six men; shearers, shed hands and all. He baked his bread and cakes and put on beaut

Firewood for the shearers' cook, Isis Downs, 1927

meals, yet he never went to bed before 1 a.m. He was a big gambler, really big at it, never miss a night, but truly fantastic smoke-ohs came up to the shed following a good breakfast, no matter what time he went to bed.'

Jimmy Pugsley told me, 'There was this cook outside Longreach. There was an awful stove and he complained until he got a new Crown stove. Oh, how he loved it. He used to talk to it. "New stove," he'd say. "Are you going to cook well for me this morning?" Always talking to it. Billy, one of the Nine-Fingered Cassidys, was in this team and he hid behind the Crown stove and when the cook came in and asked, "New stove, are you going to cook well for me this morning?", Billy Cassidy said in a deep voice, "My bloody oath I am." The cook took off like a bullet from a gun.

'There was always a fire going outside the quarters. Nights are cold in the west. Gidgea logs make good coals, and the cook used to fill a wash-tub with [them], put a piece of wire netting over the top and grill chops for our breakfast.

'There have been some good women cooks. Mrs Munro was one of the greatest old-time cooks. Jackie Lynn was good too. Ian Smiles was one of the best chefs to come out west but what an experience for him! Straight from Edinburgh, Scotland out to Nardoo in Queensland. A big wood stove. Hungry men bawling out for a feed. Ian couldn't leave the ice chest, when he wasn't drinking iced water he was tipping it over his head to try to cool himself. He didn't have a clue about the bush, didn't know to put meat under cover and it got blown – but he was a top chef at tourist resorts.

'At a shed on the other side of Quilpie this cook was always running out of flour. We couldn't work it out, he would get a load in and next we knew he was out again. When the shearers would be going to work he always had his bread setting, but that was not enough to run him out of flour. We didn't know the answer until one day we saw this huge mushroom rising out of the ground. As we watched and the sun got higher the mushroom flowed up till it was huge. The blokes went over and prodded it. "It smells like yeast," they said, and so it did, and the mystery was solved: the cook was a rotten bread-maker and was tipping the hard dough down a hole out the back and this day, up it come!'

At some sheds you can see a paragon of a cook such as Vince McKenna, cooking at Gunnedorah when I met him. In his white shirt and apron like all cooks are now required to wear, Vince could carry the wood to the stove, a big double oven black iron stove, and remain spotless.

That is not to say that all cooks today are paragons. Contractor's agent Trevor Caldwell of Charleville tells the story of Barry, who 'took a dislike to a bloke in the team and poured a full bottle of senna pod pills in the stew to get even with him. And the whole crew went down with the shits'.

Lorelei ('Doddy'), Meggs Elliott's wife from Winton, cooked at eight different sheds. 'She was always worried that her cooking wasn't good enough,' Meggs says.

'Laurie Wren was cooking and the shearers went off after ducks . . . Laurie had this big camp-oven – three feet across. We came back and thirty-nine wood ducks went in that one camp-oven. That's how big it was,' Mick O'Donnell says. Mr Roward Jamieson, manager of Uardry station cannot remember the ducks but declares that there was a camp-oven 'as big as a 44-gallon drum'.

Jim Carew heard many things said to and about cooks. 'One of the shearers I knew was caught in Singapore by the Japs and thrown into a POW camp. The

When the men shore close to home, womenfolk brought meals from home kitchens. At Lyndhurst, Kilmore (Vic.), 1935. Shearers Dave and Lee Mairs and Vic Lowden ready for lunch prepared by Truda Mairs and Ethel Lowden. (The Lowden shed served as a district shearing depot)

Japs gave them muck to eat. Men were dying of it. A bloke says to this shearer, "You don't seem to mind this muck?" "Why should I?" the shearer said. "I've lived in shearing sheds half my life."

'Of course, a cook's life is rotten and he slaves over a red-hot wood stove in a tin shed in the heat of a northern day and turns a meal out only to see some fellows come in – any amount of them – and scoff the lot down in less than eight minutes and leave the table. No wonder cooks are temperamental. I found women cooks not near so temperamental but they seemed to find it harder to handle large numbers.

'Cooks are always given three days to prove themselves at the beginning. After that it's three seconds. The best cooks often come drunk to the shed. If you give them a hand, help them to straighten themselves up, then you're right for life with them.

'I've seen them all. There was Benny the Dutchman who went through nine bottles of beer a day, and "Tasteless George", who could take the taste out of bacon. "Frying Murphy" hit the boss over the head with a frying pan full of fried eggs. The "Busted Oven", he'd thrown a bucket of cold water into an old brick oven to bust it – there were some dreadful old ovens and no-one blamed him.

'Gympie Howard when he retired got his own back when some shearers found him farming pigs. "Cooking for pigs now, Gympie?" they said. "No gentlemen", said Gympie, "I spent years cooking for pigs." Gympie when he was younger, used to stop anyone who complained with a shovel.

'Coolgardie safes came in during the 1930s and they were life-savers. Until then it was hard for the cooks to keep meat or any other thing in the high temperatures. Employers were not obliged to supply butter between November and March because of the heat.

'If you have men together for a long time you need to change the cook; men get tired of the same type of tucker and the cook gets sick of the same faces.

'At Alice Downs the cook went mad on the third day. Chops were flying everywhere in the meat house while he brandished the cleaver. They'd drink anything. Metho, essence of lemon, anything.

'Some cooks could cater for a big mob. Jules Innocence with two young sons cooked with a bush oven for eighty men in an open-walled building with a bricked oven at the end and little fires around it for warming in the morning. You could stoke a brick oven with coals overnight. In the morning this cook would put the red coals into warmers around the mess room and in pots to keep sausages warm. In the hot oven he'd cook in various stages of temperatures through the day – bread, then meat, etc. For breakfast he had rissoles, chops, steaks. When you're away out there cooking with mutton there's not much variety, fatty too, but he managed.

The cook's gear arrives at Erudina station (SA)

'Men get under a cook's feet if they're allowed in a galley. I saw a cook ask the fellows to clear the way one day and they just lounged around in the galley and he put his hand in his apron pocket and hauled out a handful of 12 guage cartridges and threw them on the fire and that soon cleared the galley. Some cooks threw black pepper on the stove.

'Oh but the meat they waste! You can't make a shearer eat twice cooked food. No use trying to use up yesterday's roast leg for shepherds pie. We had a cook called the Holy Ghost: he resurrected yesterday's dishes. The moment anyone annoyed him he'd say, "All right! Step outside!" He'd whip his false teeth out and tear his specs off and get them in his pocket before the shearer knew what was going on. But if a shearer did take him at his word he'd lock him outside the door ever after.

'There was a fellow near Hay in 1956 who sent out marmite sandwiches for smoko. They all went back. We used to strike a triangle or a gong for meals and expected men to come in promptly. Some cooks would let a man drink on after work and come anytime. But a good cook is usually up by 5 a.m. and he wants the men out by 7 p.m. He has the odd respite in the day but it's a hard days work he's got.'

Some kitchens are small, cramped, hot-boxes in the century-high temperatures of the inland; some are spacious and well planned such as the one at Netley station in Western New South Wales, a beautiful big kitchen with two huge iron stoves, large walk-in pantry and spacious dining-room with open fireplace for the cold inland nights. And outside, the cook, Yvonne Hanslow, has a big wood boiler for day long hot water and showers.

I had breakfast here at 6.30 a.m. on 20 November 1980. When the thermometer went up to 44°C in the shade, Yvonne Hanslow merely pushed the big table back to the far wall away from the stove and went on cooking: chops, bacon, tomatoes, eggs and toast with twenty bottles and jars of condiments on the tables.

In an attempt to keep women cooks, this station, after consultation with the contractor Laurie Walsh, built a separate, motel-type unit with bedroom, bathroom, lavatory and wash trough. Both the manager, Doug McCrae, and Laurie believe a woman not only does not give trouble but 'there is a different, better atmosphere in a shearers' mess with a woman cook about'. Doug likes women workers on his Kidman's property. 'Women use their heads. They work well; they're usually happy and that's infectious in a group.'

Bill Schuller had a cook who was 'on the metho, on anything'. 'This day he races over to the shed. "I'm off! You can keep your job! Seen one snake too many!" I didn't mind. He was in the DTs and no use to us, but he kept yelling about the snakes he'd seen. "Saw one in the dunny just as I sat down, by hell I went! Another in the meat room and after that I needed a drink so I went to get some metho from the pressure lamp and there's another. That was one too many snakes for me!" We let him go pelting down the track and when we knocked off, went over to the camp – and found three snakes, big beauties.

'I was in a shed once that had a real dirty cook. One day the shearers bawled out, "Cook! There's maggots on this meat!" and the cook bawled back, "There aint! I swear I wiped them all off".'

Veteran Frank Doecke believes – and with 45 years in the industry he should know – that cooking for shearers is so different from restaurant cooking that there should be a training course available. 'Shearers must have meals on time, right on time, because time is money, and rest is gathering strength to shear more and so earn enough to make the hard work worthwhile. If a shearer is in the habit of a small rest after lunch, he needs his meal on the table as he walks in. With the unnatural posture he is in for hours a day he needs careful cooking, and because his type of work makes for exhaustion, and therefore a tendency to want no emotional upset, he needs variety and style.

'But then, a shearers' cook must be a good caterer. He is miles from stores. He should not waste food, but he must never be short. If it rains and the shearers sit around for a week eating their heads off, he must have food for them and God help him if he hasn't.

'A quarter of the sheep of South Australia are shorn away from towns with hundreds of men to be catered for in virtual isolation at times. Some cooks are splendid. Bob Snooks is one of Australia's best. He balances his buying, can

Artist Will Dyson said of cooks, World War I: 'He is at his noblest when he has graduated in the shearers' sheds. He stands apart in his little niche of fame eternal and unchangeable, as God and the democracy of the shearing shed made him.'

make his own bread if supplies don't arrive, always carries a few tins in case of rain, flood, or other hold-ups. At Wilgena it's a seven-week stay. It's not easy for a shearers' cook.'

John Harris is another contractor who believes a woman cook is good for a team. Dot Spiers has cooked for him over the past eight years. At Eurugabah, out from the One Tree Plain, the owners' children go down to her kitchen daily for delicacies. In 1980 I saw her. She was 'filling in spare time' knitting booties for her tenth grandchild. 'That gives a warm, homely feeling to a team,' John Harris says.

Dot Harvey, a shearer's wife from Cowra (NSW), cooked at Lisdillon on the east coast of Tasmania. She wrote the following letter home to Miss Lorraine Harvey in what appears to be the 1960s. 'Tell Bill he'd rather keep these fellows a day instead of a week. I've never seen anyone eat like they do, near enough to a sheep a day for meat . . . At least three different sweets and same in vegetables, one of those big black pots of soup, the same size pot of stew or curry, legs of meat and the biggest pie-dish of savoury things, plus two shoulders cold meat, spuds, tomatoes, lettuce, hard-boiled eggs, puddings, plus bread and toast. It doesn't matter how much extras they have they still eat a lot of bread. At Smoko it's a loaf and a quarter cut into sandwiches, cakes and hot scones. Damned lucky they can't chew china or there'd be no plates left.

'I think they are the nicest team of men that have come here since I've been cooking at Lis. They are clean both in themselves, their speech and dress and at the table . . . And can they shear! . . . They have seconds and third helpings of everything that can be chewed. Nothing is left on their plates, and left-over stew or curry etc., is cleaned up at breakfast along with two or three chops; same of eggs plus bacon and any left-over vegetables. They are poor cornflake eaters and the poorest tea drinkers I know . . .

'I am writing this at 4 a.m. No time for reading or loafing about or letter writing during the day . . .'

Stonk Lewis says, 'The better the cook, the cheaper was the mess bill. First thing you did at a shed was to look at the rubbish bucket. I've seen rubbish buckets at sheds where the cook was so good that a rat couldn't get a feed.

'Freddie Johnston was good. Always a four-course meal. He had offsided for a wonderful cook but he couldn't keep an offsider himself. He worked too hard and expected the offsider to do the same.

'He was always dressed in starched whites, cap and all. There were sixty-five men in the mess when I shore with him as cook. On Sundays I've seen him peel 130 boiled eggs. He'd make 60–70 lb of pastry at a time. He'd stand at the table pumping it for hours.

'There were two plum puddings during the week and two on Sundays. He'd have them made early and hung from the ceilings in their cloths to dry out. "Have you seen them mugs punch my pudding?" he said to me. (I'd punched them myself, it was too much of a temptation to resist). "I'm going to fix them!" So he hung up a big stone in a cloth. Bang! went the first fist to contact it. That fixed them.

'If people ate what he considered to be inferior foods – hash, or potato pie – when out and he heard about it he'd say loudly, "One man taking a risk!" or, "Another bloody sport! Gamble on anything they will!" He said to me, "How am I going to feed these bastards if they don't know good food when they see

Many contractors now prefer women cooks in the shearers' mess. 'A woman has a good effect on the men. And she's less inclined to tantrums.' Yvonne Hanslow (above) cooking at Netley station out from Broken Hill (NSW)

Smoke-o! at Darraweit Guim (Vic.); Leonie Ryan pours the tea at this depot shed it?" He'd stand at the dining-room door and watch them. He made sure they all ate fresh vegetables, the lot. He was darned dedicated to the business.

'Fit young men hit the front of the line to the cook-house. Few had eaten such food... There were a terrific lot of bagmen about those days, in 1938. They'd come to the shed for a hand-out. Cooks generally gave them something to eat there and packed them a bit to carry them to the next shed.

'The over-told story of the one-armed cook making rissoles was Wingie Evans. A bit silly, I suppose, to put around that he flattened his rissoles under his arm-pit by flapping his arm up and down. Funny enough, it was usually the townies who made up the tall tales.

'We didn't need tall tales. The characters out there were unbelievable without going into fiction...'

Shearers have told me dozens of stories about cooks. There was one not so long ago who couldn't read or write so he identified the contents of a can by the picture on the label. But one night he served up mixed vegetables for fruit salad, and another time the men complained that the gravy was terrible. 'Where's the can?' Cocoa. All the labels were missing: the supplier had sent out a cheap lot.

'Many shearers' cooks were called the Black Snake. And you know Wally Austin, the Wood Duck, a clean, excellent cook but as quick off the mark as they came. There was Pete Hall, they called him 1080. The union constitution does not permit a reference to be given to a cook regarding his competency or otherwise. But the word goes ahead of him if he's a weed killer like 1080.

'Everyone knows Speed Balls, who served rissoles twice a day, and the Burgoo King who gave us porridge laced with strychnine for breakfast.'

Cooks received more nicknames than any other group. In the history of their company compiled for its fiftieth anniversary, Stockowners Shearing Ltd (SA) listed 'Top Class Men who have worked for the company', and in the thirteen names listed were five cooks: Len Cunningham; Ted Conlan, the Man with the Bowler Hat; Dinkum Dave Gurr; Stuttering Sam Woods; and Beaut Bob Snooks, 'The Prince of Cooks'.

A Queensland shearer who wished to be anonymous told me: 'Old Maggotty English! Gawd! The meat he gave us got off the plate and walked over the table. Gawd we hated old maggotty! When we complained, he'd grind out between his teeth: "If it were old times I'd be having youse all thrashed!"'

Les Frost (Frosty, of course) weighs 21 stone in his snowy, stiffly starched apron on Clover Downs station (Qld). On the big black stove he cooks six vegetables, corned mutton and white sauce, three sweets, custard, and hauls ice-cream and a bucket of cold cordial from the fridge when the men come in from the shed. I've met 'Screw Jack' ('He'll lift anything!'). The Wood Duck, Wally Austin, gave me Queensland Blue pumpkin seeds to take home to my father. Gympie Howard, 'The Sitting Shot', was known for his address to each new man on the team. 'There's mutton on the plate, stew in the boiler and damper by the billy. You can have a go at it or at the cook.'

'I knew a cook once,' says Meggs Elliott. 'We called him The Bent Axle – he had a crooked leg.' Meggs Elliott knows cooks. 'You see some comical things in shearing sheds. At Riddell one night, the cook comes in quietly. "Mr Elliot. I think the huts are on fire." I sits up and takes a gander. "My Christ! They're on fire all right!" A big shearer calls, "Don't panic! Don't panic!" and falls over the stool. That night we camped in the woolshed. At about midnight the cook looks in and sees blokes dossed down on wool bales all over the place. "Look at all those poor shearers," he says. Gawd Struth!

'Rat plagues are terrible for the cooks. They move in like a thick river and you have to wait till they travel on! There's millions of them on the move.'

Meggs cooked for shearers himself once, at Tattoo, when it belonged to his father. 'I had a big skinny black kangaroo dog and I was cooking in the little tent. It was no fun. I cooked chops and put them on top of the kerosene tin, and this black mongrel dingo bastard dog ate them. My brother dated me up the arse for that. When it rained that tent would fall on top of you . . . Yes, they were the good old days.

'Sometimes you tramped to the sheds. You couldn't carry a port. In the desert you got along with a swag that looked like a cigarette. The 1930s were bad. In days before properties got sheds, they'd pull down the shed on one property and move it to another. There were no quarters for men. Shearers lived in tents. A man worked up an appetite I can tell you. And God help the cook if he couldn't rustle up food quick.'

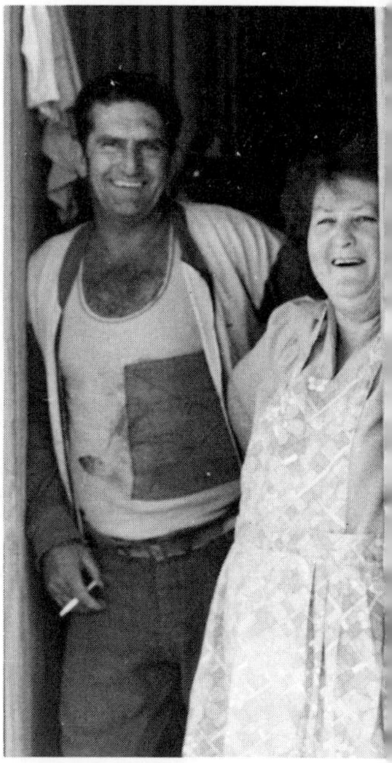

'It's the ideal life,' for cook Dolour Hart and Pat, her Overseer-in-the-shed husband

Fred Cotterill: 'At Toganmain there was a cook on the lemon essence. Lots of cooks got on the lemon essence, it made them madder than any grog. He turned nasty and wouldn't give the tramps a feed of bread so they knocked the back out of his oven and when the cook looked in his oven and saw the men shearing sheep he committed suicide.

'Tramps were always worth their tucker. They would come and sit around the fire at night, or in the shed and tell you news of other sheds, who's shearing where, who's ringing the board and so on. They were very welcome.

'The rep. can tell the cook to clean himself up. He must always say *we*, not "the men want it." Sometimes two men go to the cook, the rep. and another shearer. Cooks are apt to get nasty, cantankerous, dangerous.'

When Syd Ashbury was contracting in the 1950s he took his wife Martha cooking. The men paid £6 a week for the cook then. 'I used to take lots of bits and pieces from my own kitchen, a fry-pan, pressure cooker. I saw brick ovens at some of the places I cooked. A good cook would have to take her own bits and pieces. I cooked at dozens of places. Cooked for twenty men at Geelong (Vic.). There was a father and five sons shearing on the board there. I started during the strike. Cooked at Mondolia near Port Augusta, Poolamak and fourteen miles from Yanko out from Broken Hill.

'Worst place I cooked was Willawa. We arrived after dark. No lights. Sand so deep you scarcely push your feet through it. The overseer said to use a cottage for the night. Pointed across the paddock to the toilet. I ran through the sand to it, and as I ran in the whole floor fell through with me. It was a dunny well – 12 feet deep hole. My body went right through the floor. Hung on by my elbows to the rotten timber. The men were three chain away. They couldn't hear me yelling so I didn't waste my breath. I needed it. I swung my legs until I swung up on the bearers. My hands and arms and back were skun. It was frightening. Well that place was bad all round. The doors of the cookhouse hadn't been closed since last shearing, dust was four inches thick on the floor. In the middle of the night a tractor turns up with a furphy of water. That's all the water we had. I said, "Where's the meat?" "Oh, we'll kill a sheep in the morning."

'It took me an hour that first day to clear up the kitchen.

'In 1955 we set off for Muloorula, my baby Stephen was nine months old. The creek was flooded, no bridge, only two planks five inches wide. I said I wouldn't go over that in the car, so I got out and walked over carrying the baby. It ruined my shoes.

'It wasn't easy work, but work never is. Started at 5 a.m. and finished when the last man ate. I didn't mind them coming late, even if I was in bed I'd get up and feed them. I never had a cross word with any of them.

'Some men used to leave their money with me; one man left all his pay for one year with me and didn't ask for it until he was going home and then took it with him.

'Men I cooked for years ago still come and visit me here [her home outside Cavendish, Vic.]. Jackie Goulburn, a shearer, lived at our house for years. He lived life at a great pace, banned from pubs, all sorts of things, but a better bloke you never met.

'If I had my life over again I'd go off with the shearers again. I'd be a shearers' cook. They were the best years of my life.'

41

Damon Runyon Country

Damon Runyon would feel at home out west in Queensland's legendary sheep country. But the names the thriller writer gave his characters of America's prohibition days might pale beside those of the men in the shearing sheds along the outer Barcoo, the Diamantina, the Cooper and the Channel Country.

While Runyon created Last Card Louie, Hot Horse Herbie and Julie the Starker, the lads out west have monikers such as Jim the Ringer and Stumpy McMahon, the Boggi Bull and Wanger McKnight, the Leaning Stump and Snowy Coster. The Corrugated Cow travels with the brothers Shag, Porridge, Humid and Deppy, while back in another town the Nine-Fingered Cassidys get their blades and cutters ready for the next 'run'.

It is claimed that this family of shearing brothers 'lopped a finger off whenever they wanted a bit of quick cash'. Stonk Lewis says one needed money in a hurry for his brother who'd got a girl in trouble, another for a 'sure thing' in the Brisbane Cup, but they say the other one was just careless. 'One day one told me he'd be coming into money soon. I said, "Compo?" and he said, yes. "You going to lop one off mate?" "Yes." Another time one of the younger brothers was on the run and when I asked him why, he said his older brother was after one of his fingers. "He needs dough and I've got ten fingers and he's only got nine." It must have been true because he went like hell when his brother came in the bar. "I'll touch him for compo yet," the brother told me. And another time I was going out to a shed and one told me to tell his brother to come in. I said, "Why would he come in?" and he said, "Why wouldn't he come in when you tell him I've got my finger off." But I saw he didn't have his finger off. "But I will have," he said.

'Dong told me about Dave's turn. It was getting close to Longreach races. 'You'll have to Dave." Dagga the brother stood beside Dave's bed with a block of wood and the axe. Dave up and galloped off. The rest of the nine-fingered Cassidys chased him. "You won't sleep with ten fingers tonight!" they told him.

'Later, Dave told me, "They had bandages and dettol and all that stuff to stop me from getting blood poisoning. They're good chaps. Turnip is a real good chap."

The cooks, the 'Babbling Brooks', have their own handles: the Bagman, the Singer, Topside and the Wood-duck. Some of the sobriquets are purely descriptive of physical traits but others are witty observations on character, such as the Wood-duck – 'snap a twig and he's off!' – a comment on a temperamental cook. 'I went in his galley one day down at Widgiegoara and I said, "By jove it's hot in here." "Oh!" he snapped, whipping off his apron, "If you're not satisfied I'll go then." That's the Wood-duck for you.'

These broad, dry acres that graze one sheep to one and a half hectares (and sometimes one to four) are the heart-land of Australia's legend and lore. The properties are measured in hundreds of thousands of hectares, the place names are tattooed in the folk memory of Australia. The rivers run like magic words: the Barcoo, Paroo, Diamantina, Warrego and Coopers Creek; the towns are redolent of days when Australia rode on the sheep's back: Blackall, Cunnamulla, Charleville, Longreach, Winton, Barcaldine and Julia Creek. You scratch the soul of a people when you talk of the trees that 'Banjo' Paterson and Henry Lawson wrote of, the gidgee, she-oaks, iron barks, sandalwoods, ghost gums, wilgas and borees. Brolgas dance on the outskirts of towns and pelicans sail like galleons on the billabongs after floods that come with the rare rain; emus are a danger to motorists as are kangaroos, both flocking in mobs of hundreds across the plains; dingoes and wild pigs menace the un-shepherded lambs.

Its lingo is as Runyonesque as anything Nathan Detroit ever gritted out in the back alleys of Chicago. 'I grabbed my boggi and I ran her down the whipping side and the Sweat Lover says he'll tell the Cocky if I tomahawk another and I said I'll pull the pin and he says he'll put me down the track.' Which, translated, means that the shearer took up his hand-piece – named after the boggi lizard of inland Australia which it is said to resemble – and ran the clippers down the last side to be shorn while the sheep is on its back. The contractor says he will tell the owner that the sheep is roughly cut. The shearer threatens to cause trouble in the shed by stopping the machines and the contractor in turn threatens to throw him off the property.

The following is heard at the long, scrubbed, mess table. 'The bait-slinger chiakked the blue-tongue but that brownie-gorger called for some roll-me-in-the-gutter for his dodger and sand for his burgoo and the babbler headed for the mulga saying he'd not cook for a sword swallower.' Translation: The shearer's cook rebuked the greedy shearer (blue-tongued lizard with tongue incessantly flicking in and out) but that lover of sweet brown-sugar current loaf called for butter for his bread and sugar for his porridge. At which the cook left, saying he'd not cook for a man who ate off his knife.

'I'm on the board, and a gate-opener is barrowing a bare-belly at smoke-O, and goes for the long blow, when the dreadnought shouts that the Apaches are in and the tomahawks out!' Translation: I'm on the floor where the shearing is done, the wool-roller kid is learning to shear by finishing off an old sheep for a shearer after the bell has gone. ('Barrowed' sheep are counted in the shearer's tally because they have been started by him before the bell went. This is the most common way a lad can get experience with the shears.) He attempts the movement that takes the blade the whole length of the sheep when a man who can shear up to three hundred a day shouts that he is hacking the animal about.

Shearing has a whole vocabulary of its own. The top shearers are known as guns; they shear up to 200 sheep a day. The mighty men who could knock up a tally of 300 a day are dreadnoughts, and the man who tops the shed tally is the ringer.

There are cries of Sheep-O! when more sheep are needed in the pen, and Wool Away! when the fleece piles up around the shearers' feet and the picker-up is busy elsewhere. 'Arming' is the task of taking the wool from the bin to the presser; the finish of each shed is the cut-out.

The western shearers' pubs have names that belong to the century-old shearing routes of the area. At Longreach Mrs Ivy McNally runs the Welcome Home hotel and at Charleville is the School of Arts hotel and the Cattle Camp (changed to Rivette after the owner had a windfall when that horse won the Melbourne Cup but still known only as the Cattle Camp to the locals). The Barcoo is of course in Blackall, 'on the outer Barcoo where churches are few, and men of religion are scanty', as 'Banjo' Paterson wrote.

Much of the language of the shed has lived on since the early days, some perhaps from the sheep clippers who came with the First Fleet in 1788. But time has seen some changes. Today the foot-loose itinerant shearer has almost disappeared. Teams are flown in by air to distant sheds and for those sheds closer to settled areas the men drive out from home of a morning and back at night, some travelling up to eighty miles each way, each day.

Memories of the horsemen who rode hundreds of miles to shear, and those who came by camel, are still in the shearers' and pastoralists' agreement, whereby the owner of a property is obliged to provide a horse trough and hitching rail for the men's horses but no longer insists that 'bull camels are not to be brought onto the property'. (The last of the camel teams to take wool away from these parts plodded off into the sunset in 1934.)

Drought and cattle have reduced the number of sheep in the central-west, and today few teams field more than a dozen shearers in sheds built for fifty or more. Tinnenburra near Cunnamulla had 101 stands at the turn of the century. James Tyson, the multi-property owner built it to be the largest in Australia, but that shed has now been demolished. Isis Downs near Isisford, 'built in 1910 by the architect of the Sydney Harbour bridge' as the locals say, has 52 stands. Today only twelve shearers bend their backs beneath the steel-cantilevered roof. The shearers call this shed 'the iron lung' because of the heat deflected down from the massive semi-circular iron roof. 'It was built to kill shearers,' they say. 'I'd rather shear in a bough shed any day.' A bough shed, used well into this century, with a few still used outback, was simply four posts, roofed with spinifex or boree boughs for shade, the four sides open for any errant breeze. The floor was hard earth, sometimes covered with bullock hides to keep the wool clean.

As men did in John O'Brien's poem 'Around the Boree Log', shearers sit outside the shed at night around a boree log fire, talking not so much of past records but of how many sheep they have been offered 'down the Castlereagh' or 'on the Downs' (Darling Downs) or of who has been 'tramped' (a man must shear a hundred a day to be kept on).

There isn't so much gambling at night as in the past. Perhaps even Last Card Louie no longer plays in Runyon's 'oldest established floating crap game in New York', but he would remember it. These men remember the big tin shed on the banks of the Thompson River where they broke down their cut-out cheque playing local variants of Crap called Yankee Grab, Murrumbidgee and Sevens. But that was back in the days before their faces became weathered and leathered.

42

Swy, Hazards and Yankee Grab

Down by Gin Creek out of Longreach the big Swy school began after World War II and continued there under the trees for ten years until it was eclipsed by The Tin Shed. The Tin Shed on the outskirts of town became the home of hazards and Yankee Grab. 'It's been going since the blacks were bad,' the locals boast.

George Douglas, the town electrician, always knew it. 'Fred, my Dad, he used to keep the fat in the jam tins for them. The shearers would give him 2s 0d to keep the wick alight while they gambled at night. Dad was an all-rounder, a rouseabout, Cobb and Co. driver, navvy, he'd have a go at anything. He saw £2000 on one roll of the dice one night, he told me. And £1000 on one turn of a card. Chaney Dunn would be sitting like a tailor on the table with his lap full of notes.'

Chaney now sits on a box in the corner of the Welcome Home hotel in the main street of Longreach. 'If she's in a good mood she'll give me the dice.' He watches old Ivy McNally, the proprietress. 'I'll teach you Yankee Grab – Highest Score and Murrumbidgee, Highest Alley.

'Box a main!' the ring-keeper calls, and you box the main, get a nine. Throw in a few quid and outbackers bet you 6 to 4. 'Throw to go seven!'

But I throw a seven before the main and must pass the dice and box. 'Took you up the hill,' says Chaney. 'Out like Grogan's eye! Throw a seven!'

'There used to be big gambling schools,' said Jimmy Pugsley. The first draw at any shed was for the boss of the game in those days. But by 1929 the depression cut the really big games out as it cut the grog down. In the 1931 strike the rate of 45s 0d a hundred was cut to 40s 0d. And men were on the move looking for jobs. That slowed the gambling down.' ('You won't be able to get in to Number 3 for sheep shit,' Jim's wife Ellen said of her husband's visit to my room at my motel.)

Stonk Lewis was a gambling man. 'The first job when all the men get to a shed is to call a meeting and vote the rep. in. Then you take bids for the gambling, the right to run the games. The highest bid goes to the local ambulance fund or the Flying Doctor.

'In 1950 I paid £150 for the game. Had the rights for thirteen and a half weeks, cards, dice, two-up and SP [starting price] betting. You had to carry your own games and lights. You came to an agreement with the cook after this. We had biscuits and cake every night.

'On the hour you charged them a shilling an hour. You let the rouseabouts in at £10 a week and shearers £20 a week – you didn't want them finishing the shed with no money. For the last four or five nights you'd give the losers the game so they could try and get their money back.

'At Retreat the men were playing poker, paying £1 for royalty. A young boy wanted to play. "You don't know any poker," they said, but he went on annoying them so they let him in. Seven players, all betting on blind poker. They told him the rules as you must do. "What did you say you were paying for royalty?" says the lad quietly. Hell! Seven hands went into the ring straight off.

'We had the odd game of crown and anchor, we played Unders and Overs, or Sevens, but the person running those games collects so much that there is not much interest in them.

'Always some blokes would come up to you. "Five or six of us want to have a go on the pennies." Once they'd only play the old bike-riders – Queen in front and bike rider on the back (Heads and Tails).

'I always had an empty tobacco tin full of pennies. I'd polish the heads with salt and blacken the tail with carbide lamp black. When they dulled, some would call, "More pennies!"

'It was hard after the war to buy games, packs of cards, anything. My wife used to buy them and fly them out to the shed. Two packs: one man washing them up while the other dealing so there'd be no chance of a dealer setting the game up. It was foolproof handling.

'The Busted Oven was a gambling crank. Some cooks were, others didn't touch it. But few of them minded us using the dining-room for it because whoever owned the game saw that it was cleared up.'

Stonk Lewis heard the calls as he walked over to the big two-up game at Cunnamulla. "Got a grey!" Two tails. "Nobs!" Heads.

'It was a massive game that one. They later dredged out there for a water-hole and got half a ton of pennies. Penny games weren't often played at night. You'd need such a big room the cops would get wind of it. But they did once bring out dice with three heads and three tails and you could play two-up inside with them. We quickly learnt how they loaded the dice with mercury, though. That's what's so great about true two-up: it's the most honest game in the gambling world.

'We played hazards with dice – have outbackers and inbackers. There'd be the call, "Box your dice!" "Boxes his main!" Once you got multiples, e.g. seven, you "passed the box".

'Some blokes were really hooked on gambling. There was a fully qualified dentist, Jim Carey. He had real powerful fingers. He would yank a man's tooth out with a couple of pulls to earn money for gambling. "Give us a look at yer mouth!" and Pow! The tusk was gone.

'He made false teeth, too. In the dust near the two-up game. When it was time to turn the bat he'd leave the teeth in the dust and come back later to them.

'Of course, he wasn't the only odd dentist in the bush. A dentist used to come to Quilpie with a drill to clean up the mess left by bush dentistry. Ted Hansen

The shearers' hut at night

at Ray Station out from Quilpie once pulled a bloke's teeth out with a pair of rusty pliers. Any living in the bush was a gamble in those days.

Classer Jim Carew shakes his head and says: 'The big games in the sheds have gone now. When I was a young man the whole thing was worked out from the man who ran the game, down through the ring-keepers and the lad who kept the carbide lights going.

'You don't see the two-up or the dice come out now. As well as gambling out in the sheds, there were some big games in the towns, fan-tan, hazards, Suy Kai Loo, all sorts of games.

'Gambling made the sheds more difficult to run. You'd have a man lose all his money in a game at night, and by morning he's gone, gone far away where his creditors can't catch up with him – and you're a shearer short.'

At Broken Hill gambling was run on a grand scale, and the shearers' hard-earned cheques scarcely matched that of the men from the mines that were all bonanzas at 'the Hill'. The police would occasionally raid the famous two-up schools and the gaming houses just to show the flag. In March 1903 nine policemen swept into Tattersalls Club in Chloride Street and arrested 63 men. They marched them in handcuffs to the police station through lines of hundreds of locals who hurried up to watch the fun. The gamblers were fined one pound for 'being on Club premises without a lawful excuse'. Those shearers who attended the two-up school at the Block were better protected: spotters were always on duty on top of slag heaps from the mines with field-glasses to watch for police and also for strangers approaching.

The grog on these frontier posts was always suspect. In 1903 Police Sub-Inspector Sawtell made a check on the whisky on sale at Broken Hill hotels. Of 82 samples taken, only two were found to be genuine spirit, the remainder being adulterated.

C. J. Dennis, the creator of the Sentimental Bloke, lived in Broken Hill for a few years and almost died of thirst while walking from Broken Hill to Poolamacca sheep station in February, 1908. The experience didn't encourage him to his usual flippant rhyming. Instead he wrote:

Come, sing us a song of the City of Sand,
Silver, sin and sixpenny ale;
Dumped in a desolate, drought-struck land,
Where the deadbeat pitches his pitiful tale.

There were six hotels in Broken Hill with eleven more in the immediate district. The lock-up facilities at the local police station were minimal so, in 1881, Harry Smith was chained to a log for having become too boisterous when 'blueing' his cheque one night. Next morning Harry fronted up to the bar at De Baurn's hotel with leg irons and chains draped over his arms. His thirst had developed prodigiously overnight, so he tossed down 'a few pints' then went back to the tree and waited until his sentence ended that afternoon.

(The lock-up at that time was a galvanised iron building measuring 12 feet by 10 feet. Prisoners were chained to the flooring joists and, since there was only

'They'd do anything for a laugh,' Meggs Elliott recalls

one cell, whenever there were female prisoners, male malefactors were chained to a log or the station fence. Harry Smith had saved seven shilling of his hard-earned money by being chained to a tree at South Broken Hill, for persons arrested there had three choices: to walk the four miles to the lock-up through the usually above-century temperatures, to be chained to a log and await the cool of the evening to walk the four miles, or, to pay the cab fare of 7s 0d and travel in comfort to the lock-up.)

One of Australia's most famous shearers' hotels is the Conargo. Part of the building is the original pub built in this crossroads of Riverina sheep country in 1856, half a mile down the road from the fording place where bullock teams used to haul the wool waggons across Billabong Creek. The only surviving pub of three in the town and two on the road out from Deniliquin, it saw the wild colonial days, the rush for land, coach staging outposts and the days of wool-boom. It is encircled by the famous Riverina sheep stations. From their vantage point on the wall, pictures of the Falkiner family of nearby Boonoke (NSW) look down on the drinkers. Their famous champion rams also look down from their frames. The fame, scandals and labours of this well-known wool growing family is told and retold around the bar in the way these things are done of someone known to all present. (Rupert Murdoch is the new owner of Boonoke).

The Blue Heeler out from Winton (Qld) is a different pub entirely. Tougher, cruder, but no less friendly, it is the pub of the traveller, the stockmen, the shearer, whereas it was said that the Falkiner sons were 'blooded' at Conargo (and what an incredibly olde-worlde, sowing-of-wild-oats by gay blades of gentlemen farmers feeling that word 'blooded' evokes). The Blue Heeler has the imprint and the character of the real bush pubs that served generations of true bushmen.

Boils were a common ailment in the sheds, and the cure caused other ills. Meggs Elliott saw plenty of both the boils and the remedy. 'You always had a big bottle of metho. It cured the boils. But sent some of the blokes mad. One fellow I had set fire to his hut and the mosquito netting he had over his bunk. The bottles began exploding. I thought someone was shooting. Next day I went over to the chap. "Listen Jack. What's the matter?" He said, "I got that fellow's metho into me. It asked me to drink it."

'Old fellows out here have got cast-iron guts! Some have had a go at petrol. Worst case I ever saw had a go at brake fluid. Sent them mad, these liquors. Yet some can give them away altogether.

'The Australian was the shearers' pub in Winton (Qld). The shearers would stay at the pub until the sheds started. The publican was banker for most of the men. Lots of rooms in big hotels in those days. Eight big pubs. There's still five pubs and four licensed clubs.'

When Jimmy Pugsley was shearing on the Cooper run they'd send in to Charleville (Qld) a request for 'one dozen bottles of cough mixture' and get a dozen rum back on the mail truck. 'Every bar a man walks into in New South Wales and asks for a rum they say, "Hello! Another Queensland shearer".'

Jim Carew says, 'It's true, they would drink anything – a drum of metho with a few bottles of cheap rum in it. Beer with vinegar even, with a grey scum on it. There was a lot of melancholia out there. Station hands mostly, but I've seen several attempted suicides amongst the shearers because of grog. Bottles! You can't imagine the number once motor cars made it easier to bring beer.

'They'd bet on anything, I do know that,' says Stonk Lewis

There was a flood at Mount Margaret once and it washed all the empty bottles from under the shearing shed. When the flood dried away you could see them glistening for miles in the sun. On a run I used to do out round Quilpie, the publican brought a kero fridge down for the fellows. It was in the galley and they were each allowed to keep two or three bottles of beer a day. This publican was also the mailman, so on his round he took the fridge to the next run we had contracted but the cocky didn't agree and reported it to the organisation.'

When shearing time was over in the sheds along the Bree,
We'd raise a keg from somewhere and we'd all have a spree,
We'd sit and sing together till we got that blind and dumb,
That we couldn't find the bung hole of the old keg of rum.

43

Cokenjarri

H. M. Barker, a splendid recorder of times past, wrote of blacks shearing. Perhaps he was the most perceptive of all who saw them at work.

'The shearers were mostly women, big and fat. One would lay her sheep out on the floor and squat beside it . . . and she'd lean over on it shearing away, making small snips with only the tips of the shears. Each one would do about four sheep an hour but they'd make a perfect job of it. The most surprising thing of all was that the sheep did not kick. They lay quiet as if they were enjoying it. If a white man tried to do the same they'd struggle and lash with their hind legs

William Walter Scott took this snapshot in 1901 at Lower Clifton Downs (later Bidgemia station on the Gascoyne river, WA) which covered a million acres. Don Scott tells how 'Ten natives formed the shearing team and some could tally two hundred sheep per day. Paddy (second from right) was a great shearer. The classer was a native woman sorting strong from tender wool. Dirtier wools were later hand-scoured by the natives. 'We were two hundred miles from a port up in the Nor-West.'

1880, South Australia: ten black shearers and sorters line up with the rest of the team

from start to finish. The blacks got their sheep from a pen and sheared them anywhere on a big wide floor. When a sheep was finished it was allowed to get up and find its own way outside. It would do that in a hesitating way, picking its way quietly between the other shearers. The shorn ones would assemble about the door for a while. Occasionally one would remain standing after it was shorn and I'm sure that meant that it had enjoyed the operation and would like it repeated.

'Black children played about amongst the shearers picking up wool but adding to the homely atmosphere. The owner, a man over eighty, sat on a bale of wool supervising. I asked him why the sheep remained so passive throughout such an unorthodox procedure. He said one reason was that "these blacks shear for pastime or pleasure but a white man shears only for money". The blacks, he said, had a friendly approach that fitted the temperament of sheep. He pointed to a gin shearing in front of him, sprawling over the prostrate sheep with her big soft breast holding it down.

'"Gawd, look at that," said the old man, "that would stop any man from kicking, it's the same with a sheep. If that was a white shearer he'd hold that sheep down with his knee and it would be a wooden-headed sheep that couldn't tell the difference and show it".'

We watched her finish that sheep, the points of the blades working only a few inches from her nose.

'"You likeum cokenjarri?" said the old man.

'"Yes, I likeum cokenjarri all right," she said, holding the sheep's face up and pressing with her cheek. (Cokenjarri was their native word for sheep.)

'"Gawd, they're a happy lot," said the old man. "Keep them occupied and they're right".'

At Mundabullangana (WA) native were stationed at the twenty-seven wells on the property 'to raise water by ordinary windlass and bucket' according to 'Bucolic', writing in 1889 in the *West Australian*. 'During my visit shearing was on and I had the good fortune to see the men at work, 315 able-bodied natives . . . one white man, and two chinamen shear sharpening . . . Shearing is done quite as well as by average white men and the sheep get better treatment. The men whilst at work keep up a merry conversation and the utmost good humour prevails. About 1200 sheep are shorn daily, but if it were necessary to make a push a great many more could be got over.'

Thomas 'Fiddy' Fleay (WA) shore with Danny Cooper in the 1930s. 'A blackfellow. He shore 316 sheep in 8 hours 20 minutes. Young Danny, his lad, was good too. There were a few good blackfellows still shearing in my day.'

There is a story in Leonara district (WA) of 'Musso Black', for 'There was always a cloud of black smoke hanging round him from his hand-piece: he kept it out of the wool too long. Then he'd shove it in the wool for a bit and go on shearing but it was as often out as in the wool and so the cloud stayed round

A Victorian shed in the 1890s with two blacks as piece-pickers

him. Musso would grow weary. "Ah, Missus," he'd say, "I think I'll have a little lie down." And he would tell the sheep he was shearing to lie down and have a little sleep too. They would lie side by side and both doze off.'

Jimmy Pugsley was at Buckingham (Qld) in the 1930s; the shearers were told by 'the boss' not to go up to where the local tribe was camped. 'This was the last black tribe on Buckingham . . . a big blue started at the camp and fought with

Shearing time at Mundabullangana (WA), 1898. Aboriginal women penning-up

billies and crow-bars . . . Bill Kelly and I rush up to have a look and there's McConachie up there with gun and headlights . . . threatened them if they wouldn't stop the blue. Well, old Freddie gets hit on the forehead with a crowbar, opened him up six inches, and I wanted to get a needle and sew him up but McConachie wouldn't let me. "If he died you'd have to answer for it," he said.'

When Bill Schuller shore Polterrie station near Lake Alexandria (SA) half the board was black, half white. 'Blacks one side, whites the other. One day, it was raining. A great big black woman climbed up into the shed and in a deep resonant voice said, 'The men are going to have a vote.' It seemed funny at the time.'

Bill was joined by his brothers, Les and Harry. 'The Schuller Boys' rented 'head gear' and took it to small runs with them. 'It was hard going. Tough. Billy Bates, a black overseer, was told to sack the cook one day; the boys couldn't stomach his rubbish any more. And the cook was furious. "Sacked by a black!" he kept yelling. I told him he deserved to get the sack. "But not by a black!"

'That Billy Bates was a good man. Some wool-growers stood by him and helped him. A good man.'

Carunna Downs (WA) 1906 with black women working on the board

Contractor Frank Doecke shore with blacks thirty years ago. 'Simon and George Carpenter were good shearers, real good. If an Aborigine is good he's got fantastic movement in his hands. Lovely to watch him move the hand-piece.'

The most common racist 'story', sometimes told as a 'joke' in shearing towns, is the one about 'the nigger shearer who shore his big toe off mistaking it for a dag'. It is told over and again, many swearing to know the shed where it happened, some swearing to have seen it happen. None tell it as a tall story. One 'sympathetic' telling of it goes: This nigger was shearing, bare-foot of course, and he takes a cut at a dag and off flies his big black toe. Well, the cocky says when the nigger slows up, "You can shear more sheep than that!" and I say, "You'd be the hardest nigger driver I heard of in my life".

The term nigger, or coon, is the lingua franca almost the whole way across Australia north of 20 degrees. 'I gave the barmaid a look that would have killed a coon's dog'; 'If you are going to ride a horse all day you need a backside like the sole of a coon's foot'.

Don Scott, now a life member of the RSL and motel proprietor in Winton (Qld) was born in Western Australia. Some expressions Don uses about the Aborigines (or the blacks, as many Aborigines prefer to be called) may seem harsh, brutal and crude to southerners. But Don has lived in the outback of both Queensland and Western Australia all his life, most of it on sheep stations, therefore his observations must be accepted as pertinent. His brother is on Billabong on the Murchison (WA). His father spent his life in the bush; 'Uncle Jack Scott' rode his horse from the outback to join up with the troops bound for the Boer War, and Don went into the bush when he left school aged 12 years. He served in World War II and when he returned to Australia went back to the outback.

Few Aboriginals have a pen today. Yet everyone who works with them declares them 'Real natural shearers'

'There was no work so I went shearing as did many others.' He stayed on as overseer-manager. 'My father was known as "pick-handle Bill". He built his own home and put down wells, timbered them, a true all-rounder. He knew that blacks are like dogs – if you don't tell them what to do, you hurt your hand punching them all day. So Dad carried a pick-handle.

'Blacks are good musterers. They can track. They taught me. "Look boss! He go this way last night." They can tell when an animal is too poor to waste time on tracking. "Drag feet here Boss." They are all great horse riders and all could get on without saddle or bridle and ride all day.

'At Bidgimie they were wild blacks. You went off with a gun on your belt and handcuffs in your hands and brought your team of shearers in. Teach them like you teach a dog. Kick them in the ribs until they learn. They were very faithful. Followed me on to my own property later.'

Don had Mooloo Downs, he managed Dalgety Downs (304 000 hectares) in Western Australia and in Queensland managed Myrall station near Winton. 'Only 11 400 hectares, a backyard only by standards of this state. So small.' He saw the boom of the 1950s: 'Sold 5000 bales from 14 000 sheep and netted £212 a bale – all in the bank. In the 1960s we got £72 a bale; couldn't afford to put men on.' He has seen the good and the bad. 'Lost 3500 sheep in flood. Myrall. Lost £14 000 in one night. 1955. The bank was foreclosing but I battled on and cleared it in a few years.'

Jack Sparks (Penshurst, Vic.) remembers the strikes of the 1920s. 'They only made use of the blacks . . . employed blackfellows in place of shearers, tommyhawking it off them. Some firms sent experts out who, instead of grinding the shears, began to hack the wool off. What with the blackfellows and the no-hopers the sheep never forgot it. They'd done that in 1922 and when we went back to Tubbo on the Murrumbidgee in 1924 the sheep still remembered the tommyhawking they got. The musterers got the sheep in to within half a mile of the shed, they heard the steam whistle blow and they stopped. They remembered the last time. They wouldn't go on; 5000 sheep ringing – couldn't move them, fifty shearers waiting to get started. They had to cart the sheep up to the shed in drays and things, and as fast as they got them in they tore out again. There were big pens there. You needed two men to catch one sheep. The shearers wasted a lot of time catching them all that season.'

Fred Cotterill also from Penshurst, believes, 'They'd never have got the sheep into the sheds in the good years if it hadn't been for the blacks. In bad years they can trap mobs near water-holes, but good years they depended on large numbers of blacks to find them and bring them in. The pity was, the blacks would come to our gambling school and they'd lose all the little money they had.

'When the shearing was coming to a close they'd start sacking the rouseabouts – 'Out!' No money. Some good shearers were sent off too, for 'misconduct'. "What was it? What did I do?" A man wasn't told. They'd let the blacks they'd kicked out come back to their camps along the river and bring them in to do the work. The shearers objected to the gins being employed sweeping the boards – no pay, just tobacco and grog.'

44

The Shearer's Widow

'Children? I don't know how any shearers' wives got pregnant. He was either too far away, too tired or too drunk.'

– *Shearer's wife, Longford, Tasmania*

Longford, in Tasmania, was once called 'the shearer's widow' because so many of the local men were away 'over the water' or 'over on the mainland' shearing four to eight months of the year. They would leave in March or July and return in November, the day after the Melbourne Cup.

One of the best-known shearing families of the eastern states was the Greig family from Longford. 'Old' Bill Greig and his brother Bob went shearing 'over the water' before the turn of the century, and their sons and nephews followed as they left school. Eventually eight men of the family went to the sheds.

For nine months of the year Bill Greig Senior left Longford and followed the shearing through the mainland states. His daughter Jean remembers, 'Mum was our parent for nine months of the year. Four boys and us two girls. She'd have to make the decisions, worry about us when we were sick or the boys had accidents, and the pranks we got into she had to put up with.' Mrs Greig wasn't alone in this in Longford. Max Greig declares that 'the chief export of Longford district was shearers. There were scores of families, fathers and sons, who headed off every year for the season'. Bill Greig Senior had two brothers who were shearers; there were six shearers in the next generation. 'There was little talk in my family that didn't centre round shearing. Boys grew up with nothing in their heads but shearing and when they left school, aged 14 or so, they'd be ready for off. Some boys had been rouseabout or tar boy during school holidays but anyway, all of us knew what shearing was all about. We'd lived and eaten shearing all our lives.' 'Mum could threaten us all she liked,' Bill Greig Junior recalls. 'She was always saying – as did all the other mothers – that she wouldn't have her sons following the wool track, but as soon as we left school, off we'd go. All my life I'd heard stories of Cordilla Downs, Cunnamulla, Winton, Barcoo. Of course I had to go to see these fantastic places myself. You could as easily have stopped me from breathing as stopped me from going shearing.'

'Young' Bill went off shearing in 1927 when he was in short pants – 'except that Dad put me in long pants so I could get an adult wage. Moleskins. I had to go, you couldn't get work when you left school here.

'I was growing fast but I found out one thing early: shearers come in all sizes and shapes, and often little chaps are the best of all. From the very beginning I enjoyed it. The goat races, riding the camels; the long bicycle ride to Tibooburra and Cordilla Downs – it was enough to say that you'd shorn Cordilla Downs to command respect from lesser mortals!'

Bill started as a 'broomie', sweeping the locks back away from the shearers' range. (Few 'broomies' are seen in the 1980s. The pickers-up do most of their own brooming now.) Then he became a picker-up 'over the other side', then wool rolling. He would start in northern South Australia and work sheds down through New South Wales, or go down south-east to Mount Gambier way, 'a run of sheds, usually finishing at Barunah Plains (Vic.).

'I learnt to shear by barrowing – finishing a sheep for the shearer when the smoko bell has rung. Sometimes a shearer would give me a lesson for five minutes or so before the shed began.' It was always hard to get a pen to start with. 'Sometimes men had to bluff their way into a pen in the depression. In that awful time the contractor was able to put on only the fastest men.'

Later, when he could shear eighty a day, he was taken on as a learner-shearer. Then he got a break. 'Dad was away on a good run of sheds and a telegram came to home for him. He was wanted for a shed out from Hughenden (Qld). I borrowed money and set off for Hughenden, but when I went into the contractor's office he said, "You're not the Bill Greig I know." So I owned up, I told him I was a learner and couldn't shear a hundred yet. He gave me a chance, at Tooleybuck, out from McKinlay, then on to Julia Creek and I got my hundred there so they kept me on. I told them I was "about sixteen" because, according to the award, if a boy was under sixteen he must be with his father.'

'There was no point in us saying they were having us on,' Tasmanian wives declared of their husbands' stories about western Queensland. 'They had snapshots to prove it!'

Little boys with goat carts sold water drawn from the artesian bore dug outside western towns

Ray Watley, 'the Black Prince' they called him, was 'the life of the Tassie party'. 'When we got cars we'd set off with our luggage strapped on the running boards and us packed in all over the place. We'd go over the parched lands, some years through floods, and be forever digging or hauling ourselves out of the black-soil bogs, and then we'd reach the billet we'd been promised. Oh Lord. The boss man would throw a couple of chaff bags at each man, he'd make you sign for them, and you'd see us heading off for a clump of dry grass, of ferns in some areas, filling up our "palliasse".

'Once out west we had to wade through a river of rain-water to get hay for a bed. But then we'd stand round the fire drying our clothes, talking, laughing, telling yarns. Laugh! We'd laugh our heads off.

'I don't know why we were always laughing. We used to soak our hands in Condies Crystals to toughen them against cannonballs and bindi-eyes, the three-cornered-jacks in the wool that ripped the flesh on your wool-soft hands. We shore fly-struck sheep with the maggots crawling over our arms and wrists, and the stink of the poor sheep, when the wool and skin and flesh fell off from where the maggots had burrowed deep into them! Some places you'd have a whole flock struck. Dumb things that they are, they crouch with the pain and make an eerie little noise. And sheep ridden with lice – lousy sheep smell bad. Perhaps if we hadn't laughed so much we wouldn't have gone shearing at all.

'When I first saw the Cordilla flock I said, "Sheep! They're more like bloody mountain goats!" They'd bred these long-legged sheep to walk to distant waterholes. Big buggers they were. "Goodbye Cordilla I'll never see you again!" I'd promise when I left. But I was back again the next year, and the next.

'Oh, everywhere was rough. Kynuna shed (Qld) was a bough roof over a bullock hide floor. We shore 210 000 there one year. It looked a bodgie set up but it was cooler than the big iron sheds.'

For a time Chum Watley lived in Toorak (Vic.). 'We had two beautiful suitcases, and when I'd be setting off on a run of sheds neighbours would say, "Going away for a holiday Mr Watley?" and I'd say, "Yes, off for a few months in the sunshine".'

Lorna Watley, when I asked her what she did when her man went away shearing, said: 'I stayed in Tasmania. I set to work to try to make the 30s 0d he'd left us stretch out until he came back in four months time. I eked it out. Mum was cook at "the big house" [the homestead of a property near Longford]. She gave me a loaf of bread or a bit of a meal occasionally. Then I got a house job. I used to bike four miles over in the morning and four miles back at night.'

The men no longer leave Longford. 'There's no longer a mass exit. The odd man may go. Some go if they "want a tour" but whole communities are not affected by the exodus as they used to be.'

In some ways, those perambulating shearers contributed greatly to the otherwise isolated island state, bringing back home news and stories and artefacts that few had heard of or seen.

Another Longford man, Ray Tapley, 'took his wardrobe' to Cordilla Downs when he cycled there in the 1920s. 'Well, I had two straps for dungarees to tie half-way down the calf of my leg, my "Jackie Howe" singlet with strapping sewn round the left armhole to prevent the little woolly friends from ripping my singlet to bits with their sharp little hoofs, sometimes a grey flannel with a square of material sewn on the left side, and my Morgesons boots made in Tassie with a big flat sole you could grip the floor with, a greenhide sole. Once I was in the shed I'd make a "bag morgeson" – the cap off a wool bale or an old bag sewn up like a moccasin.'

The old men keep their 'tickets', their union membership cards, and *Labor Daily Levy* with its global message: 'A continent to win; a newspaper to win it with.' It is a matter of pride with them, and we are apt to forget today that in their youth the union was the newest, greatest thing the working man had known. As the card announced, a member was '. . . entitled to all the benefits of membership and the protection of the union while loyal to the principles for the year ending . . .'

They keep their references from Westbury (Tas.), Aramac (Qld), Billilla (NSW). 'W. Greig is a good shearer, fast and clean and treats his sheep well. Top tally 109.' Framed on the wall is 'Uncle Bill Greig's' references of 1938. 'Greig is an excellent shearer and man to have in the shed. He never varies in his work and handles his sheep with every care. Greig undertook the shearing of the stud rams which work he carried out in a most satisfactory manner. Conduct excellent.'

The shearing days die hard in a man. The specifications for buildings on the land in Lewis Street, Longford, owned by the Greig family, show 'Residence and shearers' hut'. 'Old' Bill Greig's daughter Jean used to see her father go into 'the hut', with its big brick fireplace. 'He could lie on his bunk there and think he was back on the run.'

Most of the Longford men left between 12–15 July and returned after the Melbourne Cup in November to start shearing in Tasmania. 'They never run the Melbourne Cup until we were ready to go back to Tassie. It was out of the oven into the ice-chest,' Max Greig says. 'Out in the arid lands you'd shear sheep staggering with effects of drought, some of them weighed down with three years growth because they couldn't be walked in the great distances from the outer runs; then back in Tassie we'd see sheep we'd shorn the previous day frozen dead in a blizzard.'

But the years are gone when 'They never started shearing in Tassie until after the Melbourne Cup.' Now they shear all year round. Ewes are usually shorn six to eight weeks before they lamb, as shorn ewes will go to shelter to lamb, instead of lambing in the open where the new-born lamb is apt to die.

Bill, who began shearing in 1927, began contracting later and is still successfully running teams in 1982, shearing some of the best-known flocks in Tasmania at properties such as Quorn Hall and Connorville. He has run up to twelve teams at a time. Instead of the red dust of the great outback he now voyages down the tracks through miles of hawthorn hedges in blossom to the smaller, richer-soiled properties of his own state. At Connorville the beautiful old two-storey shearing shed with its Georgian doorway served for both shearing and men's quarters. After their day's work was done, the shearers had to go down to the river to wash, then return to their quarters above the sheep that were penned ready for the next day's shearing.

Bob Lyons brought this elegant photograph of himself home to Tasmania

'Stink?' Chum Watley remembers, 'Did it stink! And noise! Those bloody sheep bleating all bloody night. A man had to bomb himself out to get any sleep at all. Well, at least the new shed's high enough off the ground to get a front-end loader in to clean out the sheep shit.'

The old shed had only two feet of freeway beneath the shed. 'That was a wet weather job to clean it out.' The living quarters, lined with lathe and plaster, have cedar doors and windows. 'But cedar didn't smother that smell.' Chum gets his last word in.

The old pegs are still there on the wall from the days when a peg was put in its slot for each sheep shorn ('Always a few cut off at the end of every day from every one of them,' Chum swears.) The old shop, where a man could (and did) buy everything from a needle to Friars Balsam, is splendid with its freestone window sills and cedar doors.

Like most contractors Bill Greig is totally loyal to good men in his teams. Seeing Jim Schroeter's name on the wall of Connerville shed he quickly says, 'Jim Schroeter, one of the best experts you could meet. He went away over the other side with us. You couldn't carry his bag off the boat when he'd arrive, he'd have so much gear with him. A wonderful man!' And Joe McInerny: 'One of the best classers in the business. Tremendous fellow!'

On the walls are the names of men who have slept here. Bill knows them all. 'Danny Males 1960 – a quiet little chap. He was an Olympic boxer. And this chap – a real good fellow.' The signatures cover the walls, 'The Mighty Mouse slept here', and this pencilled notice: 'Rouseabouts must have a shower every twelve months whether they need it or not.'

'We heard all these stories,' Jean Greig says of her life in a shearing family. 'They were a part of us, yet we were not a part of them. And of course the life of the women left behind when their men followed the sheds was different. Very different in my mother's day, when we were all little, and Dad and the boys gone and away over the other side.'

Before the glare of dawn I rise
To milk the sleepy cows and shake
The droving dust from tired eyes,
Look round the rabbit traps, then bake
  The children's bread.
There's hay to stook an' beans to hoe,
An' ferns to cut in the scrub below.
Women must work, when men must go
  Shearing from shed to shed.[1]

– Louis Esson

45

O! The Springtime it Brings on the Shearing

O The springtime it brings on the shearing
And it's then you will see them in droves
To the West Country stations come steering, and
seeking them jobs off the 'coves'.

With a ragged old swag on my shoulder, and a
billy quart-pot in my hand,
Well, I'll tell you we'll astonish the new chums, to see
how we travel the land.

And after the shearing's all over, and the wool
season comes to an end,
It's then you will see the flash shearers, making
johnny-cakes round in the bend.

To live in a shearers' town in the days when the great annual migrations of shearers took place was to live by a time clock that chimed in the talk of the town, 'The shearers will be going soon,' and five months later tolled, 'The shearers will be back soon.'

In between times we had only the irregular letter that told little of their every-day lives and the way they filled in their time between sheds or, for that matter, how their domestic life went on during the 'run'.

When men were camped for weeks at a time at the one station (most 'runs' were for three to nine weeks duration) they must make their own entertainment. They told stories around the log fire outside the mess hut, under the stars as they smoked a last pipe before going to bed. Like most bush folk they told and re-told a story and in time it became polished. Sometimes it is hard for an outsider to pick the real from the tall.

The following story was told me at least ten times but said to have come from the local paper at Bourke (1), Port Hedland (2), Leonora (3), Camperdown (4) and a score of other country towns. And each time it was succinctly told and a funny joke against the teller. 'The local rag here in Quilpie made us mad. They reported this accident – "two cars collided in the main street last night injuring three men, a black gin, two dogs and a shearer".'

Some of them mix the real and the fact until some jokes are too true to be funny. Stumpy McMahon of Longreach is a consummate storyteller. In 1950, Stumpy left Longreach for Melbourne with his seventeen-year-old dog, Jerry, at his side. One of the train passengers said 'That's the ugliest pig I ever saw!' 'That', said Stumpy, 'That is no pig! That's my dog.' Says the passenger, 'I was talking to the dog you fool.'

Stumpy declares he made little from shearing. 'After my first nine months work I came home with a union ticket and a pet galah. I went out as hut-keeper once. New South. Carried my push-bike 26 miles over mud to get to the boundary.

'I was in a shed once with a real wild shearer on the board. He could fight like a threshing machine. Caused a lot of trouble. But the owner wasn't game to sack him, so during the lunch break he dismantled this wild shearers' machine. When he came back after lunch and saw his gear dismantled the shearer went straight up to a learner in the end pen. "Move over Sonny. I'll shear enough for both of us," he said.'

Jim Carew was overseer at many western Queensland sheds. 'Peak Downs always had a cut-out ball. Began organising it as soon as the shearers got there. Got the traders from the nearest town to pitch in. Mrs McLean of Peak Downs would . . . cut the ribbon to open the ball. We had a ball at Northampton Downs also. I remember there was never anyone left the next day to clear out the gum boughs and the overseer always had to do it!

'When I was a lad in the sheds, old-timers who had been around at the turn of the century told me that they did a bit of dancing in the hut.

'Sometimes an Irish step-dancer would get going and the young fellows would dance with one another a bit, half jig-half reel. "Anything for a bit of fun," they told me.

'A great pastime out west, of course, was shooting. Most men had guns and most were crack shots. They say there was no trouble training Australians for the first world war because they all could shoot. The best of them out west could shoot through the hole of a bottle, straight down the neck, and blow the bottom out.'

Jack Howe was an athlete and in the interview he gave the Catholic *Advocate* in 1916 he told of an incident at Springsure. A 'cocksure individual from Rockhampton' said he would give any shearer a start of ten yards in the 100 yards for a stake of £10. The then-young Jack Howe was known to his mates as a 'bit of a flyer', and they backed him against the challenger in the race on 'the clay-pan track'. Howe won, 'hands down'. Then they ran level for the same stake money. Again Howe won. Finally Howe gave the challenger two yards for all the money the boaster had left and for the third time the hero of the wool-sheds breasted the finishing tape first.

Howe's race with Frank Rogers on Langlo Downs station gives an indication of his prowess. Rogers was training at Langlo for the big Botany Handicap in Sydney, then the 'pedestrian Derby' of Australia, and Howe, who was shearing there, gave him a trial run at 100 and 150 yards. Although the professional sprinter wore the orthodox spiked running shoes, the shearer ran barefooted (after finishing a hard day's tally of 217 sheep). The two ran a dead heat. Rogers

Shearers gear – including Divil Ryan's bike – arrives at the station

went to Sydney, was put back a yard in the big handicap for 'beating the pistol' and lost by only half a yard in the final.

Taking Rogers's advice, Howe bought spiked shoes. 'I was winning easily in a big race at Warwick's St Patrick's Day Sports and I stopped dead within a few yards of the post. You see, I'd been taking part in other contests during the day and had been meeting too many friends all day long.

'It was at the same sports that I won the Irish and Sailors' Hornpipe Competition.'

It was recorded that he could run a mile shoeless, over grass, after finishing his day's shearing.

Stonk Lewis was in to everything. When bales of jute arrived at the shed they were banded with hoop iron. 'Us shearers used this hoop iron for scores of things – but mainly as handles to fruit tins for use as pannikins – sometimes we would shape them as ladies' legs. We made holders for water-bags out of them – so the bag wouldn't touch the ground because, like a canvas tent, the bags leak if touched.

'At sheds out from Broken Hill iron quoits were popular. Water was poured onto two patches of clay – eighteen inches apart; a peg was hammered down in the centre of each mud patch. Then a shearer would call, "Light her up!" and we pushed a piece of paper on the top of the peg for a marker. We had teams and in turns players tossed an iron quoit at the peg. Now, these quoits were circular, but with one side flat and the other curved and the trick was to get the quoit to stand upright in the mud, and the team with the most near the peg won.'

Stumpy McMahon shore in the 'bough' or 'bag' sheds in western Queensland. 'We'd make quoits out of the band round wool bales, push a stick in the ground and get a game of quoits going. Another game we had was throwing the clips off the wool bales into fruit tins – we'd have money on the game of course. Lots of gambling out there. Lots of poker and two-up.

'We stayed in the bush for weeks at a time – and that was in the thirties, note. If someone had a tin lizzie you might get in to town for a weekend. We played and gambled at everything . . . if someone had an accordion or a guitar, we'd sing around the fire, sitting on logs of wood. Someone would have rum – it's more concentrated than beer, so we drank that.

'A man needed something to do because shearing gives you the shits. I've known the contractor to take the carbide light and hide it on cut-out night so the men wouldn't gamble their pay cheques away.

'I trained Skip, my dog. He was a pure-bred water collie. Had him from six weeks old and he lived for fifteen years. He knew more about shearing than most big guns. He responded to calls such as "Speak!" and he'd speak up, "Eyes damp?" and he'd wail. He'd do anything. I took him thousands of miles on the back of lorries when we were going to sheds. The whole team on the lorry, sitting on our swags, no canopy. Skip loved it.

'If a bloke couldn't get a space on the lorry he'd tie up the mouth of a snake and throw it amongst us – he'd get a seat pretty quick then. Skip got used to this early on. Just stayed where he was and "set" the snake.

'You can see shearers' dogs sitting waiting on the back of their truck by Friday afternoon. They're very knowing. I've seen dogs race out to bring more sheep in when they've seen the pen down to five sheep.'

Snakes have always been good for a joke in Australian bush camps and the shearers have their share of this. Jimmy Pugsley: 'At Jedborough near Windorah, a couple of young fellows killed a snake and we strung it up in place of the pull rope over Hughie T. Smith's stand at Smoko. Boy did he hop when he reached up to pull his machine on!

'The presser, Jim the Ringer, got sick and me and my mate had to press for sixteen shearers. Well, I was "arming" – taking the wool from the bin to the press – and going like a scalded cat when one of the buggers put a snake in the bin and I rushed up, got an armful of wool, toss it up in the air for the press and down slid this bloody great snake straight down my arm.

'Ashton Parry once told a new team in that there was a big snake in the quarters. "When you go to bed tonight take a big log of wood with you. If we don't get him tonight he'll be under the shed in the morning, so watch out!" While the men had tea Ashton crept into the quarters and put a little Joey kangaroo down between the blankets in a bed. God but there was a shindy when the bloke felt the movement when he got into bed.'

'Pizzle guards' are one of the tricks played on young new chums. 'Hop down to the expert and ask for a pizzle guard!' A common reply from the expert will be 'I've only got a left-handed one', or, 'Will the cap of a pen do?'

'We were always up to something. Of course you couldn't or wouldn't purposely hurt a sheep, but sometimes they are damned frustrating. I had this big ram one day and it thrashed around and kicked a treat and I was done in and the boss was watching me, waiting to see if I'd do my block, so I turned the sheep's head round to me and spat in its face. You should have heard the boss laugh.'

The after-hours entertainment was simple, wild, and sometimes downright dangerous. 'We were shearing at Womban in 1939 for Houghton the contractor. We used to sit outside round a fire of logs at night, each on his own log of wood. The expert, Clem McInnes, got peeved because there wasn't any space left for him – men used to go mad and clear off to bed and the bottle if anyone took their habitual possie by the fire – anyway, this night Clem goes away and comes back and throws something on the fire. "Clear out! It's bullets!" Brother did the boys go! And did the bullets fly!'

'There's very little petty thieving in the sheds,' Jim Carew told me. 'They empty their pockets out on the table at night, take the wrist watch off and leave it there and things are rarely touched. Some newcomers haven't believed this and put traps out such as a piece of cotton across their ports and it is never tripped. I saw a few fist fights in the sheds. It's not a thing that happens.'

Jim took snakes calmly enough. 'For the first few days there are plenty of snakes around the shed. But once the shearing starts the sheep disturb them and it's O.K. then. In the quarters the transistors bring them out these days a bit. When the trannies play music the snakes stick their heads up through the gaps between the boards of the floor.

Erudina, Lyndhurst (SA) when Fred Cotterill shore there in 1920

'There were usually some around the dunnies. Many a man's come flying out Scotch-hobbled from a lavatory in the bush because of the snake in the paper box.'

'Out west they piped such hot water from the hot artesian bores that you had to let it cool off in the bath,' said Meggs Elliot. 'It wasn't good to drink but it had a lot of soda in it. Shearers used to lie in it all weekend. Soak all day. The Muckadilla water was really healthy. A cocky out there told us he thought of lowering the shearing rate. "Blokes go overseas and pay the earth for what you blokes are getting for nothing from me."'

Ye sons of sin that tramp and shear in hot and dusty scrubs,
Just keep away from 'heading' 'em, and keep away from pubs,
And keep away from handicaps – for so your sugar scoots –
And you may own a station yet and wear the Boss's boots.

Meggs enjoyed the dances. 'On the Boulia Road at Middleton pub the shearers used to have great dances. We've been so full there they had to pour us back into the bottles on Monday morning.'

Dunkeld and Penshurst were the homes of shearers in Victoria. In winter they had a job to get a football team together in either town because most of the fit young men were away shearing. When I was fourteen years old and living in Penshurst, I was alive to the devil-may-care air of the boys who went shearing. In a town that had stagnated for fifteen years during the Depression they were a whiff of adventure, freedom and excitement; they travelled to far-away places; their speech rolled along on the names that were still legendary outback to us before World War II; they spoke of Cunnamulla, Barcaldine, the One Tree Plain, Old Man Plain, Edeyowie, Yanka, Boggabilla, Hungry Tysons, Toganmain, Alice Downs, Billilla, Cordilla, Eli Elwah, Boonoke, Marble Bar, Coppins Gap, Widgeegoara, Winton and Windorah. Sometimes they left on their long, dusty journey unheralded, almost unnoticed in the lethargy of those hungry, fearful days when there was no work for men in a town that had no industries and depended on the grudging support of the large and wealthy sheep stations surrounding it. Other times, those of us who had brothers or boyfriends breaking out from the zombieland of no-work, no-tucker, no-funds, no-joy and rushing northwards to the sheds – well, we knew, and the heightened excitement rippled round the cross-road of the little blue-stone town.

One Sunday, Mass half-over, when Father Glennon was in the middle of his sermon, the congregation were all half dozing on the hard wooden pews and I was sitting on the high stool at the organ in the choir loft waiting to burst into the Gloria!, there was the roar of three cars out the front of the church door. Then with hoots of laughter and jostling in tramped ten young shearers. I hung over the railing, trying to attract attention while at the same time trying not to let my parents see my excitement. It was the boys on their way north to the sheds at Cunnamulla.

Later, I said, very carefully, to my mother, 'Wasn't it good of those shearers to come to Mass before they left for the shearing.' 'They didn't come because of Mass,' snapped Mum. 'It was the only way they knew they could see you young girls before they left, that's why they came!' Then, she added, 'Anyway, they're away for six months now and the town can settle down and you and the Kennedy girls and the rest of you can forget about them.' But we never did.

Before the shearers left town each year there was a ball, a formal affair, expensive for those times, and when they came back six months later there was another ball. We knew ahead who had rung the sheds, who were the guns, who had blown his cheque and who hadn't, and who had left girls weeping from Hughendon down through the eastern states on the grand shearing route back to Penshurst in the centre of the Western District. For some months they shore in the surrounding areas and sometimes were back in town for the Saturday night dance. Almost always we girls were at the cut-out dance in every shed within a radius of sixty kilometres or so – which was a great distance in those days, when the only cars around the poorer areas were tin-lizzies.

At fifteen I was playing piano and violin in the New Mayfair Dance Band along with a saxophone, banjo, trombone and drums. We were paid to play to 2 a.m. but when we played at a cut-out ball in a shearing shed this was what would happen: a collection would be taken up at 2 a.m. to encourage us to stay longer and then – always – a gun shearer would say, 'Oh, bugger this messing about!', and he'd slap a £1 note on the top of the piano. In next to no time there would be ten or more notes quivering away as we played 'Whispering' *molto forte* in joy. It was a lot of money for those times.

There, in the Western District of Victoria, the old shearers hold memories of the great days when they left home 'for the far north' and worked their way back south to their homes ten months later down 'the shearing route'.

In Dunkeld, two old mates of the 'travelling' days are still the closest of friends. Percy Riddle, who shore for fifty years, belongs to a family with four generations of shearers and contractors. His mate, Tom Duckmanton, who went with him for twenty years as rouseabout, is the son of a blade shearer.

'In 1939 at Golf Hill there were nine Riddles on the board out of 18 shearers. The same year, six of us were at Barwon Head. The nine were Bert, Normie, Archie, Ganger, Agga, Ray, Percy, 'Old Spud' and Andy. Uncle Dan was boss of the board.

'My father, Ted Riddle, was a blade shearer and pushed his bike all over NSW all his life. His father was a blade shearer before him and that man, my grandfather, had seven sons, all shearers. My father was at Mt Sturgeon when machines came in. There were thirty-two stands there and, like my father, every one of the shearers thought the machine would put men out of work.'

Like the men in his family before him Percy holds a union ticket for every year he shore. 'You just keep them, even though you'll never go shearing again. I can't tell you why. I don't know.

'Most of my family were good shearers. In 1927 I saw Uncle Bulgry – Bert Riddle – shear 303 in one day at Ned's Corner, and the next day he did 300. Vince Rainbird of Tasmania was doing 300 a day there too. Ten shearers averaged 240 for the day there. Sugar McLaughlin and I were picking up. We were young, but it near killed us.

'We all wore blue flannels – you never saw a singlet in those days: you'd freeze to death in one of those in a hot shed. Never catch cold in a flannel. At night you'd screw it out, wring it to get the sweat out. No, I don't know how many sheep I've put through my legs. When I was in good nick I could crutch 400–600 a day.

'There weren't many holidays. We worked Good Friday for instance. Things were crude I suppose. The meat was hung up a tree in a sugar bag, or in a Coolgardie safe. At Mt Sturgeon, because it was no distance away, we took beer – when a man could afford it. There was an underground tank there, so we put the beer down it, suspended in a sugar bag. A man ought to go and dive in that tank, I bet he'd find a few!

'At our hut, there was a big cylindrical water-bag hanging up a tree and of course we always had our own water-bag hanging beside us in the shed.

'There was never a dull moment. Musterers would come in off the great dusty plains out from Broken Hill with a water-bag around the horse's neck and a quart pot clanking away and sometimes a Bedourie oven clanking away too.

'Many a time you'd think the shed was about cut out and you'd hear the cry, "Here's the musterers! They've found another 200!" Cripes.

'Sheds were big. Thirty shearers in a line at Lake Victoria (NSW) when I shore there in 1939. You had to walk up eighteen steps to that shed. Hard at the end of the first week. But interesting they were, those old places on the rivers, all worked by steam.'

Percy Riddle and two of his brothers contracted from 1948 as well as shearing in their own teams. 'But it's different now. Owners are different. "Howaah dew youah dew?" they say now. Dressed to kill. Big hats. Many a squatter stalking around today doesn't know which end a sheep eats with.'

Stonk Lewis went into the sheds when he was a lad, 'An old shearer asked me how old I was. I said I was eighteen, "God Struth", said the shearer, "they age quickly up here. You were only sixteen when I gave you a lift from the gate to the shed last week." I'd learned to be quick: you earn more money at eighteen.'

Stonk needed to be quick. In 1980, when he was retired and living in Brisbane, his daughter Sharron came home one day and looked him 'in the eye'. 'I said, "Father, did you ever thieve a train?" and he said, "Yes. I did thieve a train." Sharron Lewis, the licencee of a hotel, had told a drinker that she knew Charleville. 'Lewis?' said the drinker. 'Any relation to Stonk?' 'Yes. Daughter.' And for the first time Sharron heard of her father 'thieving a train'. It was 1954.

'A man was even sillier when he got back from the war,' Stonk explains it. 'We were shearing at Tarago, thirteen miles from Goulburn. It was Saturday night and Con Halfpenny and I had been drinking. Freezing cold. Snowing. We went to the railway station. A train pulled in – steam train. "What time does that train go to Goulburn?" We asked the porter. "In two hours time." "Can't we go now?" "It's not going now."

'We were freezing. The driver and crew had gone for a meal. We could see the porter sitting in by his fire. "Bugger him. We'll go by ourselves."

'We chuffed away to Goulburn. They had sixty police waiting for us. We hadn't fired her up so we couldn't get away, couldn't make a run for it. They took us around to Goulburn gaol and we had to front up on Monday morning.

'He asked me what I had to say. I said there wasn't much I could say. "They caught me with the train. There's nothing I can say." "No," he said, "there isn't." "Would roaring drunk or insanity be an excuse?" "No." "I'd have to be very drunk to do such a thing. Fancy thieving a train belonging to the New South Wales Government! I'd have to get my head examined."

'Well, the magistrate gave me a £30 fine. I paid up. The old police sergeant said when I came back with the money, "I'm not telling you to leave town, but I am telling you that there are a lot of officious young constables in town, all anxious to score points." I only stayed one hour, mostly because I was being ribbed so badly by the other shearers in town.'

Con, who shared Stonk's cell, was a bushman. 'Mad. He'd get a hiding in a fight. "Why can't I ever win a fight?" he'd say. When we thieved that train from Tarago and were sitting in Goulburn gaol, all Con said was, "I hope Mrs Bill Taylor at Corowa doesn't get to hear about this".'

Shearers used to be great readers of westerns. Lewis was not. He read Russian and French novels and biographies. He wrote 'heaps' of poems about shearers and shearing and sheds. 'There was plenty to write about. Bough sheds in 1954 at Oringa between Quilpie and Thargomindah. We liked them, cooler than tin. But the boughs on the spinifex roof was full of snakes that got in for the coolness and before you started the old engine up you had to shake the roof to shake the Joe Blakes down. There was often a hole in the roof. You'd be told to put a bag over it if the sun worried you.

'And men! The Red Onion was outside Mitchell. He couldn't read or write. He used to be chained up at an old ship's tank, the square iron tank, and he had

The team at Ensay, Gippsland (Vic.) November, 1909

a head like a rock. One day he was hit on the back of his head with a lemonade bottle and all he felt of it made him say, "Youse be careful with them bottles. Kids going by can get hit." In the shed he continued piece-picking through smoko. They had a vote to decide whether he should be sacked. There were eleven men, but twelve hands went up in the vote. "Oh Gawd! How can you sack a man who votes to sack himself?" said the rep.

'During the war Mick O'Connor brought his wireless to listen to the war news. "By Jesus," said the Red Onion, "That's the same wireless you had here before. I recognised some of the old tunes on it."

'The whole country seemed to abound with characters [and Stonk would be one of the greatest himself]. Scandalous Graham was asked to help when Bob McCasker was involved in a motor bike accident. "I'm not a medical man,"

Scandalous protested. "But I am a sporting man and I'll bet you ten bob the bastard dies."

'Some of the outback publicans were great people. Warinda was in between pubs, six miles to Hamilton hotel and six miles to Boulia. Tim Howard had one and Maggie had one. Our own money was printed out there up until last war. All notes. We could use them as far as Winton but we couldn't go past there with them. It was a good scheme. We called them shin plasters. We could cash them like real money, good for out there where real money was short but our credit good with a letter from the contractor to say what was owing us. Boulia was the last stop for drovers before they headed into no-man's-land. So it was usually lively. When I was first there in 1940, a young, shy, frightened lad, I wanted to drink with the shearers. Old Ma Haystead kicked me out. When I complained she said, "I tell you what. You go and put the hard word on Maggie and if she says yes, I'll give you a drink."

'Robbo the Lair once wanted to go to the Graziers' Ball. They were charging £5 admittance – expensive, just to keep the likes of us out. Robbo had £5 but they said no, he couldn't go in. But he did go in. He was a beautiful dancer. A grazier's wife said, "You dance divinely Mr Bathurst," and Robbo said, "Lady, you should see me shear!"

'At Warenda races the horses were "sold" through the yard. You could pay, say £20, and race that horse for the day. Jacky Bye the cook bought two horses one day and gave me £5 to ride them. Jacky was swinging the bag [taking bets] and he didn't want either horse to win. He paid me in Boulia Shin Plasters in 10s 0d notes and I had to spend them on beer and cigarettes as we were leaving town next day.

'There were boxers and crims in the teams those days. I had my first "pro" fight when I was aged seventeen. But real crims needing "a rest" came up sometimes. "Melbourne Joe" was mixed up with Squizzy Taylor and John Wren. He scared us.

'They'd get up to anything. Ashton Parry (the contractor) had old Scotty for a rouseabout at Burinda. Scotty called him outside. "No, I haven't got a drink for you Scotty, but I could make you a cup of tea if you feel bad." Parry went out for chips for the fires. When he came back he said, "You've been drinking metho you old bugger!" "Och no!" said Scotty. So Parry held a mirror in front of his face: soot marks, the three prongs of the metho lamp. Scotty followed Parry on to another run. "I never told you to come!" "Nope, but you never told me not to," said Scotty.'

Stonk has shorn though all the eastern states but he claims that north-east Victoria and across the border into southern New South Wales contains the friendliest people in Australia to shearers. 'The squatters there are different. Tremendous people.'

Stumpy McMahon (Qld) says about shearers: 'If they've got any brains they wouldn't be in the industry.' That's his excuse for going to the sheds when aged 18 as opposed to 15, the age most of his mates went on the board. 'I had some sense for three years. Not much after that, though.' It was 1934 and he earnt £3 a week. 'Good money for those times.

'I was at Manneroo. It had twenty-four stands – only twelve now. There's no longer the great number of sheep there were in those days. It took a terrific number of musterers to bring them in, to help keep them up to the shearers.

Not all sheds are like Manneroo. Once, we got to a place where there was supposed to be a shed and there was only a few sticks sticking up. The contractor put wool bales around the sticks and off we set shearing.

'Sinclair and Kerr, Hamilton and Co., they were contractors who would take a whole plant out, press, plant and gear. Take a few bags of straw for us to fill our palliasses, pitch tents, make bag bunks on two sticks, carbide and kerosene lights.

'Sometimes it would be a bough shed with spinifex on top for a roof. Sometimes a bag shed, other times we'd pull up in the middle of scrub and clear a patch and set to work. The owner might have a forcing yard to bring them up, "stake fences", saplings fastened onto stakes.

'There were no floors in our tents, except sometimes we could take a floor with us. You should have seen us on arrival, erecting our tents, rigging up our bunks, shoving straw into our palliasses. And in a couple of weeks the straw would have gone to powder. On real cold nights we'd sleep in Dalgety Tweed – a wool pack. Wrap our blankets round us and crawl into a wool pack.

'We carried a swag or a small port. Didn't take much, after all, we didn't need going-out clothes. A towel, boots, blankets, soap and razor just about carried it. Usually we threw tubs on the lorry for washing. Some of the buggers didn't wash, themselves or their clothes.'

Jim Pugsley has deep memories of his years on the board. 'The depression keeps you drinking. A man's away from his home, can't get out of his dungarees quick enough to get home on a Friday night. I used to get home 2 a.m. Saturday morning and back off to the shed 9 a.m. Sunday. It was just like tying a dog up for five days and then they let him loose, let him go.

'You can see Australia, I'll say that for it. But the wife's alone at home with the kids. The cocky says you don't have to go home every weekend – do they think you're going back to the Stone Age? A man has to live.

'A man has to keep two homes. You pay fifty or so dollars to the shed a week. If there is rain a shearer still has to pay the cook. When you leave the shed, it's the last pay until shearing starts again. No other industry puts this pressure on a man. You can't go straight from one job to another; if you cut out on Wednesday you won't start again until Monday. Owners won't start before a public holiday because they'd have to pay you. You don't get superannuation, no holiday pay, no sick pay. You're only allowed two days off sick in the quarters and if you're still crook, they put another man in your place and you must get out. You get a bit of fencing to keep yourself. You do something, anything.

'In the hopes of making a quid you'd try to get on the Cooper River run (Qld), one of the longest runs. If you wanted to stop and do the whole of the sheds you could stay the whole year on the Cooper run. Leave Quilpie at the end of January and get back to Charleville end of August. Train to Quilpie. Headlines in the newspaper said, "Shearers drink rum like drinking tea in the Quilpie train". Then we'd travel on the tray of Mark Holtz's truck sitting on our swags for two days.

'There were runs of sheds, Retreat with sixteen stands shearing 60 000 sheep, Galway fourteen stands and 40 000 sheep, Jedborough fourteen and 30 000. We didn't come home from the Cooper run. Stayed out.

'The rep. took the cheque list around every Thursday night and you put your name on it for the amount you wanted. There were two store nights, Tuesday and Thursday and we'd buy our tobacco, toothpaste, combs and cutters. The contractor took the stores out, water-bags, towels, things you might need. It would be debited against you.

Jim's wife, Ellen Pugsley, knows the cost. 'You are in real trouble if your man is ill. Trying to bring two boys up... You watch the weather. If it rains, perhaps no shearing, of if they are travelling they can't get across the creek, dry creeks flooding miles wide in a few hours.' Ellen has been very ill but she is a battler. She got 'odd jobs'.

'Today it's harder in a different way. Early in the year they fly in the outsiders, the shearers from New South Wales and New Zealand and they clean up – they flood in, so many of them they get the shearing done in four months, leaving no shearing at all for the locals to do later in the year. They take the money out of town, don't live here and spend here and keep a town floating.'

Shearers ran under so many names they confused even themselves. Jimmy Pugsley put his suit in to the dry cleaners in New South Wales and couldn't remember the name he had given. 'Cook and O'Malley, the Charleville tailors used to sew our names in the back of the coat – our real name – so this girl in the dry cleaner goes looking for mine and she says "I've got Kelly, Hudson, O'Sullivan coats but I haven't got any of these names on my tickets. And you say you're Bradford but you've got Pugsley branded on your clothes. Shearers!"'

Pugsley was with the contractors, the Roly Brothers. 'It was called The Circus. The day shearing began, they called the names. "Kelly!" No-one answered. "Kelly!" Suddenly it hit me. I was Kelly on that run. "Here!" I yells. Later in the day three Demons turns up with Roly. They're watching me. I was shivering and shaking all over. But they'd only come out to watch the shearing.

'We'd have fun of a sort. In our swag we carried a flesh needle and thread for surgery, stitching up cuts in the sheep. "Hey, Freddie. Do you mind if I sew up this black sheep with white cotton?" you'd hear someone say.

'When I was beginning I lied to get a pen. The contractor had his eye on me the next day. 'I thought you said you could shear a hundred?' he snapped. 'Yes, Alex,' I said. 'We were both mistaken weren't we!'

The Pittaways of Gawler (SA) are a well known father and son duo. Steve Pittaway, father of Greg, was a champion shearer in his day and now the son is a top scorer in the sheds and competition. 'We come from country where you're good or you're exhausted. Round Mt Gambier, Naracoorte, most places in South Australia, sheep are big. They send horses in to drag them out of the pens. Heads on them like milk buckets. Any man out of a job will get work there!'

Fred Cotterill keeps his reference from Grazcos (now Farmers Grazcos) after fifty years in the sheds.

'Conduct excellent. As an overseer Mr Cotterill was capable, sober, hardworking, conscientious, loyal and absolutely trustworthy. He had the respect of both station owners and managers and the men working with him. Extreme knowledge of shearing machinery and ability as a bookkeeper made him an excellent dual purpose man and for many years he was regarded as one of our top shed staff men in Australia.' With his reference is a gold watch, given him after fifty years with Grazcos, inscribed 'Boss at 200 sheds'. He knew every state including Tasmania, and there is scarcely a shearer whom he doesn't know about.

Fred went into the sheds when he was 15, and at 74 is still on call in Penshurst to attend to machinery, sharpen shears and advise on conditions. He remembers his days as a shearer. 'We didn't have showers, anything, early on. Carried our own blankets whether we went on horse, bike, or by train or car . . . the men couldn't take beer out because of the weight, so they'd have O. P. Rum and were damned because of it.

'Before my day they did one and a half to two hours shearing before breakfast. But even in my time we still had the old lamps, a slush lamp (a wick stuck in a tin of fat); a bag of straw to sleep on; twenty to a room. We cut a kerosene tin long-wise to wash in and boiled water in a kero tin for hot water. Like most people in the bush we couldn't have managed without kero tins. I went to Moyne Falls when I was 15 and I went back there every season for twenty-four years. There were three men and one woman working for Grazcos when I joined. Until Grazcos came on the scene we were wanderers, most shearers were. In a way, we weren't too different from the old boys of last century, taking pot luck as to whether we'd get a shed. I liked Grazcos. They always abided by the awards. And they employed learners as the award stated. They played the game fair and that made it easier for us to do the same.

'We'd leave Penshurst, ride our bikes to Dunkeld, catch the train up to Mildura, then ride the bikes to wherever we were going. Some Penshurst men drove their horse and cart to Riverina. A station might have thirty to forty horses to run in when the season was over and the shearers ready to drive away. It took two days to drive from Mildura to Deniliquin before the road was bitumenised.

'Divil (Bill) Ryan arrived at a shed once, carrying his bike because the mud had caked his wheels solid. We often carried bikes a mile over the mud. On the way to Arkuna shed (SA) Harold Lewis got a broken crank on his bicycle. Divil found a bit of wire in an old cemetery and towed Harold into Pimba (SA) behind his own bike.

'We'd get off the train at Pimba siding and the station would send a donkey team to transport us. The first night we camped at a little graveyard and the cook would meet us there and turn on a meal. We'd get to the station by the next night. When shearing was over the order was reversed. And the railways would put a carriage into Pimba siding. They'd shunt out at night when we were all asleep. The Port Augusta Bank would open early for us to change our cheques when we got there.

'Further over in South Australia we'd go to Arkuna on the Tea and Sugar train. I remember the boys buying whisky from the chemist on the train.'

His first shed was Blackwood, 'Ritchie's place'. By 1934 he was rated first out of 34 shearers. 'Not too fast. My tally was 176 at Moyne Falls where I shore for 36 years, first as shearer, then overseer-expert.'

He was boss-of-the-board at Boonoke (NSW). 'You could run the Stawell Gift on the board at Boonoke, it was so long. Old Otway Falkiner always said that. There were 8 blades shearers and 30 machines shearing together when I was there.'

In 1942 he became shed overseer-expert in charge of more than 200 Grazcos shearers in Victoria and the Riverina. 'Travelled all over. Kuba at Darlington Point with 100 000 sheep and the owner J. W. Hughes, who also owned Botany Woollen Mills, marching in saying: "Who's the boss here?" and I'd step forward. Sir Samuel McCaughy owned Coree like he owned so much, and Hungry Paterson owned Hartwood and then it was down to Scribling's shed,

Drysdale at Euroa, 25 shearers on the board. The bank manager there could put three fingers on the tally and run them up adding as he went. It took us half an hour!

'Divil Ryan was a fair terror; a good man to have on a team, he was funny, could laugh at himself, and was a great shearer. When he was 70 he could shear 100 sheep a day. He was with me for years. During the shearers' strike Clive Cameron – the politician who used to be a shearer – Clive and Divil and I walked twenty miles to Kalgoorlie. Clive had learnt to shear with me and he appreciated Divil's company.'

Chips Rafferty, the Australian film star, also began shearing with Fred. 'Divil didn't know this. He only knew Chips from the movies, so when Chips came to see me at Moyne Falls, Divil says to him, "Have a go mate," winking at the other shearers. Chips set to and shore the sheep quick and clean. He was a real back country shearer. He'd begun out from Broken Hill.

'Divil could get upset though. He was shearing well one day until the owner came in and sat up on top of the pen. I noticed Divil was being clumsy, seemed to be shearing side on. At smoko I asked him why. "I didn't want that old bugger looking at the backsides out of my seat."

'We shore at Yardi, out Port Lincoln way, down in Tassie at Mt Morrison, Nant, Mona Vale and Merton Vale run by Mrs Foster, and over in Western Australia. Went by train to Perth, took a boat to Geraldton, then on by motor truck with portable gear including a portable press. They took us to back country sheds made of cane grass, bullock hides pegged to the ground for a floor, tents instead of huts and camps. They were very cool sheds with spinifex roofs. They put an iron tank in the ground for the cook to bake bread in. We had a few rows over there. They'd kill old rams for us to eat. You could smell it from the sheds. Stinking old rams. It caused a few rows.

'Shearers were better travelled than almost anyone in the country, but on the first trip away from home you had to keep your eyes on them. Bluey Edlich on his first trip from Penshurst got as far as Geelong and asked, "Is this Melbourne?" I told him to follow me, "And don't let anyone get off with your suitcase." When we got to the Shearers' Hut [Batman Hill hotel, Melbourne] I said, "Where's your suitcase?" "A bloke asked me if he could carry my case for me." "Jesus, didn't I tell you not to let go of your case?" "I gave him a shilling to carry it." He was a real bush boy.'

Fred Cotterill knew the real Jimmy Gibbs. 'There were plenty of tall stories such as "Jimmy Gibbs got the sack on the board and shore four sheep before he could stop." But the real Jimmy Gibbs was pretty good himself. I first shore with him at Yardi. He was one of the fastest shearers round Naracoorte way. But a blackfellow, Simon Carpentier from Tailem Bend, was beating him one day. "How you going?" the blackfellow said. And Jimmy replied, "Something will have to give and I think it will have to be Jimmy."

'Of course there were some great stations to work at. Jerilderie and Goolgumblah would kill a young bullock every year for us. The old-time owner was better to get on with than the younger manager that came up from being a rabbiter. He was always hard – would offer a man a handful of cigarettes each day and a bottle of whisky weekly instead of pay if he could get away with it.'

And of course, some owners were harder than others. 'Old man Millear at Edgarley had five sons. There was always one of them on the board. The men

When Chips Rafferty visited the shed in 1940 Divil Ryan jokingly said 'Have a go mate!', not knowing Rafferty had spent time on the board before becoming the best-known actor in Australia in his day

hated it. It was a mean place. Skim milk only; the cook used to steal cream at night. The boys always got milk, even on a mean place. They used to get a turkey at night, pluck it in the hut. Several times a season we had turkey.

'Lots of things men wouldn't put up with. Old Millear drove a horse in a buggy. Always whipped him hard to start. The men cut the traces on him for that.'

Fred Cotterill believes Ivan Orchard of Winchelsea (Vic.) was the best shearer he ever saw. 'And Cecil Murray, a gun shearer from Adelaide in the 1920s, was a fine shearer too. Jackie Rose of Conargo shore for me. A real good shearer.

'All the Tasmanians were good. You could always pick a Tasmanian. They dropped their haitches at Hamilton and picked them up at Ararat! Whenever you pointed out a pretty bird to them they'd say, "Can you eat it?" Tough men. The Greig family were all good, and good companions.

'Bob and Bill Greig were on the boat going back to Tassie one night. They didn't have bunks. Drunk, asleep in the lounge. I rushed in and shouted, "Fire! Fire on board!" "Bugger the boat. It don't belong to me," said Bill. "Let it burn!" said Bob. They were great days . . .'

Jack Sparks was 81 in 1980 when he spoke of his shearing days. 'I used to ride an old horse, Glenmaggie, up round the mountains when I began shearing at Toongabbie aged 15. It was well before the 1922 strike. That strike was over conditions we were shearing under.

'The clause in the agreement we objected to mostly was, "If a sheep was shorn in a manner not satisfactory to the owner, no pay . . ." Also, the clause that if a man was sacked he got no pay no matter how long he had worked. We lived in a big hut at one end near the cook, slept the other end, chaff bags threaded on poles, three decker bunks touching one another all round. Fifty-one men sleeping, eating, cooking in one room. Some huts were stone, built by convicts. At the station our horses were taken from us until cut-out. If it was a poor year the horses couldn't walk when we got them back. Some men had camels – only one allowed per man – and they fared a bit better than the horses. All our tucker had to be bought from the station store and that was a racket.

'Shed hands got £1 a week plus tucker of boiled mutton, damper and jam. Johnny cakes for lunch with two kerosene tins of tea. The shearers got cakes and things and sometimes the rouseabouts would grab a handful and run off with the good food. We wanted a place to wash our clothes and have a shower. At Neliamba, near Bourke there were about twenty shearers, and Divil Bill Ryan and I used to wash our clothes every night. We told our rep. we wanted buckets and tubs. By Saturday we hadn't got them. Divil said, "We're getting a bit toey." There was a 400-gallon tank kept full out the back with water carried up from the creek. That afternoon Divil carried armloads of dry wood and threw it under the tank. "What are you doing?" asked the rep.

'"Going to boil my washing," said Divil.

'"Oh no! Wait on!"

'Divil went on running backwards and forwards with arms of wood. "I'll give you ten minutes," he called.

'"You won't boil that up!" the rep yelled.

'"By God I will!" said Divil. Everyone knew he would, so they gave us the tubs.

'At Blackwood station, to wash, we had to pull a kero tin of water up a pine tree and trip it over ourselves. We asked for a shower "Oh. They want a shower now, do they?" the manager said. "There's nothing in the agreement about that." He sent word we could shower under the pine trees or go dirty.

'So we got ready. All fixed. And waited for Mrs Grice the manager's wife, along with the governess, who came by in the pony cart every afternoon. We saw them coming "All right! Everybody out!"

'Out we raced in the nude, throwing jam tins of water over each other. Thirty men dancing round, naked as the day they were born, yelling furiously, water everywhere, even on the horse. Old Grice came down and began to complain but he was told he could easily remedy it. "Put a shower in." Two sheets of galvanised iron were brought and a grating put on the ground, and that was the shower and it stood there for years and years.

'Bough sheds were a damned sight cooler than tin,' Stonk Lewis said. 'But a lot of Joe Blakes got up there in the spinifex.'

'Old Ritchie who owned Blackwood married a Canadian. He came on the board one day in leather chaps and a big Canadian hat. "Bet you can't throw a lassoo!" one of the shearers said. At knock-off time he saddled up, tied a rope to girth and around the horn. "Let him out," he yelled and the bullock rushed out, Ritchie swung, missed him, got the lariat round an old stump, horse galloped off hell for leather, rope snapped tight and broke girth and off came saddle, Ritchie and all. With thirty shearers watching, laughing their heads off. Ritchie enjoyed a bit of fun.

'Some of the old squatters considered themselves too good for the workers. None of the Ritchies were like that. The Ritchie boys played football with our boys, had a drink with them. Still do.

'Those of us who shore Thargomindah, Eulo, Charlotte Downs, Widgeegoara and outback places like that would have great talks with the Ritchie boys about people they knew up there in "the wilds" of Queensland.

'We'd start in Queensland late January – early February at Longreach, Quilpie or Winton. Some famous places. I shore for Barney Hall of Charleville, the contractor. He only wanted fast men. The faster we shore the more money he made. He gave us one rouseabout for every shearer. To keep his fast men he lent them money, the only contractor I knew who did that. When I began contracting I never did that.

'I wasn't married then; stayed north while there was shearing to be got; usually finished Christmas Eve at Adelong. Had six years up there at one time and never came down to Victoria. Small cockies could then hire men outside of Graziers' Associations. The cocky, he would pay up to £2 5s 0d a hundred plus tucker. That was very good; we all liked the little cocky. Although some big places wouldn't go through a contractor: Kelly at Barwidgee used to have 14 shearers on the board with 6000 sheep a man. I shore 8000 of them one year at £2 a hundred. That was a good shed. But you didn't know me in those days, did you?'

No. I didn't know Jack Sparkes until I was a young girl and we moved to Penshurst in the late 1930s. He couldn't walk, but in those days of the depression when everything seemed bad, it didn't seem too terrible to see Jack being piggy-backed by his wife. We'd watch her struggle up the hill to the little dole-struck town with the crippled man on her back. The adults told us kids, 'Ivy Sparks is a wonder. She massages his limbs morning, noon and night and swears he will not be a cripple for life as the doctors vow he will be.'

Now, over forty years later, Jack walks and works so hard that he and Ivy have repaired, rebuilt and redecorated a century old bluestone building of nearly twenty rooms that was once an hotel.

'I'd become crippled from the sheds,' the old shearing contractor says. 'I injured my right hand in Queensland at Charlotte Downs, 200 miles out of Cunnamulla. The points that hold the cutters were worn out. It was dangerous, but I went on shearing because the expert didn't have any to replace mine. The machine hit a burr in the wool – they're hard as iron, the burrs up there – I was going up the neck, the machine locked, flew up in the air, fell to the ground. I reached for it as it whipped up in the air, ripped my leg. A couple of men carried me over to the hut when lunch-time came.

'The cook made a couple of crutches out of brooms. There was an artesian bore there, they can be very hot the bores up there. The cook told me that he believed that if I walked up and down in the hot bore it would ease the pain. I did. In a day or so the leg felt easier but the hand was real bad. In a few days I wanted to see a doctor but the station said we'd soon be cutting out, so I waited a week and then they cut out.

'We headed for home. My mate kept pulling up and boiling the billy so he could put hot foments on my hand but it got real bad. Arthritis set in. I couldn't walk, legs and all crippled up. The doctors wanted to cut the sinews behind my knees so at least my legs would straighten, but Ivy wouldn't let them. My wife backed me. She said no. She carried me on her back, lifted me in and out of chairs and bed all those years, massaged my muscles and kept interest in me. When we bought the old hotel people said we were mad to take on such a job.

What would a crippled shearer know about repairing a big old bluestone building? That was the challenge! In time I walked again.'

Jack Hutchins (Hamilton, Vic.) used to go to the Riverina sheds on a push-bike. 'After I graduated I went on a motor bike.' Born in 1897, he started with blades when he was aged 14. 'I took a pen whenever a neighbouring farmer had a bit of crutching or shearing.' He worked around Hamilton for four years, 'then I thought I'd have a go with machines but there weren't many with machines around here.

'So I set off for Yanko, near Balranald. Rode the bike from Hamilton up to New South Wales. There were two brothers and one of the sons drove a gig and two horses up to Yanko from here. We took a couple of blankets and always carried something for water. You'd go along the track until you saw a nice looking place, nice people, and there you'd ask for a bit of tucker. They'd always give it to you. We'd camp anywhere. Sometimes we'd ask if we could camp in a barn or a shed. They always said yes.

'There were forty shearers at Yanko. I was put in pen forty. Last to get paid. I was aged 19 then. We were at Yanko for 8 weeks and 2 days. For one whole fortnight we never went near the shed, it rained that much. Terrible floods.

'They were rough sheep at Yanko. Fifty was my best tally. Earned enough to pay for my tucker and union ticket and came out with £5 after eight weeks.

'After Yanko, I thought, bother the machines, and I bought a pair of blades at Swan Hill. No trouble to get shearing with blades up there.

'We rode to Broken Hill, from there to Buckloo station for a month. In that part of New South Wales were many camels. You'd see fifty in a row yoked up by Afghans. They'd wait until sundown, then they'd pray, facing the sun. And when the sun went down they let us kill a beast for them.

'On our way home we rode south from Mildura and ran into a mice plague fifty miles wide. We rode and camped through them. They ate the bottom off our clothes, leather off our boots. A man can't describe what it's like "on the bike" camping through a mice plague. We rode like hell.

'Later I got a Rudge Multi Motor Bike and rode from Hamilton to Leigh Creek (SA) with Tommy Salter, my mate. It was no road to speak of and we arrived a day after shearing began...

'We rode to Beltana station. There we began shearing two hours before breakfast. There were donkey teams at Beltana, thirty yoked up to a waggon.

'We had it tough up there in the back country. At Mt Shank we struck a good cook: the only thing you could be sure of being decent was the tucker. We used to strike almost every week over accommodation. Tin pannikins had been thrown in a box for twelve months, all rusted, so we cleaned them up with sand, poured tea in them and it poured out the holes. The rep. was sent off to the boss. First thing the boss says is, "I'll have new ones up for you tomorrow."

'Tomorrow wouldn't come. Same with kerosene lanterns: kero leaked all over the table. Rep. goes to the boss. "Tomorrow." Kero everywhere in the food, the water, but tomorrow never came. We'd go out – strike. They called us Bolsheviks.

'I had three titles in the years when I was a rep: Bolshie, Red Ragger, Communist. Whenever a man said, "We need pannikins. We can't drink out of these," he got one title or the other.

'There wasn't much easy those days. If you cut out at 10 a.m. you wouldn't get any dinner. The Jackies would just have to get on their bikes and pedal off. We all had a nose-bag. "Hang on while I get something for my nose-bag." We'd pedal off till we came to a dam and wash our mouths out, didn't drink usually because you couldn't see what was in a dam.

'Burra, at Clare (SA), was so sandy there was as much sand as wool on a sheep's back. And Leigh Creek, it had terrible dust storms. We had to knock off at 2 p.m. one day. Couldn't see.'

Jack went 'private' shearing in 1925. 'The gentleman's way, you might say. I shore for Scott Brothers at New Wilpena Pound. That was like a home from home. Good woolshed, homestead, and good drinking water from a beautiful drain. Wonderful place. Nothing like the tourist place it is today. It was lovely. I shore Barham, over the border from Coondarook; all my five sons shore there, they all learnt there. During World War II I shore there myself, it was that hard to get men. You didn't have to ask for anything. They'd come and say, "Got everything you want?" They'd yarn away with you. Fine people. At night they'd call, "Come on you blokes! Come and have a game of cards!" And we'd troop over to the homestead. They treated us as men.'

There have always been lone shearers, men who contract and shear alone. Some do very well. Herbert Hutchins, son of the shearer Jack Hutchins, shears within a twenty-two mile radius of his home outside Hamilton (Vic.). Aged 46, he has been shearing for thirty-two years, been on his own for the past twenty-two years and has a left hand that is more like a ham than a hand. On this 'rouseabout's hand' the knuckles are massive, flat welts. He can prod them with a pocket knife and not feel it. Over six foot in height and weighing 14 stone, Herbert uses this great hand continuously to drag sheep out of the pen and pull their fleece around on their skin to make shearing easier and quicker.

He is a member of the AWU and the Shearing Competition Federation of Australia.

He began working as a rouseabout in a team with his father, Jack Hutchins, in the Riverina, but after ten years set off alone. 'It took three – four years to build up my own run, but once started it runs well. I've shorn at Nigretta East for 21 years and Ardmore for 15 years.' In this work method a shearer may go to the one shed three times a year for crutching, shearing and lambs. He does up to fifteen sheds in this way. He may be at one shed for four weeks shearing 2000 sheep and crutching 700 in one run; other sheds are only two days' work. Sometimes he is close enough to go home for lunch, other days his wife can bring a meal to him. And every night he is home with his family. The lack of competition in his sheds hasn't slowed his shearing. In a recent Strong Wool competition when four sheep must be shorn within a time span of ten minutes, he shore his in 9.27 minutes, and at the Hamilton Show he shore a sheep in 4 minutes 33 seconds blindfolded – as fast as many shearers when they can see what they are doing. 'It's not clever or unique.' He is a very low-key man. 'Shearers use touch more than they use sight every time they shear.'

The Hon. Mick Young, MP and former shearer tells a hoary old tale. 'It's bloody funny with a beer,' he says. 'Thirty years ago, one of the most famous and fast shearers in Australia was a bloke called Charlie Gibbs. Charlie shore all

Divil Ryan, third from left, 'was a good shearer and a good mate'

the "long runs" between Queensland and New South Wales, shearing 200 a day, day after day, week after week, month after month. It was said of him that he was one of the very few shearers to have shorn 50 000 in a year.

'Well, this story relates to an incident that occured when Charlie arrived back in Bourke after a very, very long run. As he walked into the hotel, the publican caught his eye and said: "Charlie, there's a cocky just out of town who wants to know if you'll shear his sheep. Can you give him a ring?"

'Charlie sauntered off to the telephone and rang the cocky. "Oh Mr Gibbs," said the cocky (cockies always call shearers Mr before they shear the sheep), "I heard your shed had cut out and was wondering if it would be possible for you to come and shear my sheep."

'"Oh I suppose," said Charlie. "How many have you got?"

'"Three hundred and twelve," replied the cocky. There was a silence, then Charlie said, "What are their names?"'

46

A-shearing for the Loot

The bell is set a-ringing,
and the engine gives a toot,
There's five-and-thirty shearers here
a-shearing for the loot.
– A. B. Paterson

For a moment there is quiet, little sound or movement. The odd man on the board may take a last swig from his water-bag or have second thoughts about a cutter he changed only seconds ago: he'll change it again, swiftly, but with the unflurried rhythm that characterises the shearing board. Finished, he stands, as do the others along the board, waiting.

Then the bell goes, and the silence ends, each man flings the gates back, grasps a sheep and drags it backwards to his stand, reaching as he goes for the pull-on cord that will release the clutch on the machine; his motor hums, motors along the board hum, and the first run of the day has begun.

There's noise, smell, shouting, dogs barking, sheep complaining and above all, the whirr of the machines when that bell goes.

The picker-up puts his broom down and stoops for the first fleece, charges along the board, dancing in and out as shearers reach to the pen for another animal – a rousie doesn't get in the way of a shearer if he knows what is good for him. He tosses the fleece high in the air, a skilful throw so that the shorn wool will spread out on the wool-rolling table for the rollers and classers to go to work, but the picker-up sees none of this: he has already run nimbly back through moving men and the upturned legs of sheep and grabbed another fleece, or swept the locks and dags away before a shearer drags another sheep from the pen. His is not an easy task: the shearer concentrates on his work, he is paid by the sheep, not by the week as is the rousie, and it is up to the latter to avoid collisions. He has been taught that shearing hand-pieces are particularly dangerous machines with unguarded teeth and a cutter travelling at approximately 6500 oscillations per minute across the face of a comb. They must be treated with caution at all times, even when not engaged, because the points of the shearing comb can cause a serious wound if fallen on. It is a lethal weapon. The unprotected comb and cutter are very sharp to enable the wool to be cut and many a shearer has scars to prove it. As well as fear of the cutters, the rousie makes sure that he or his broom never hits a sheep; a kicking sheep puts great strain and stress on the shearer and could cause him to lose control of the hand-piece.

Penners-up are whistling instructions to dogs, and dogs bark and run over the backs of sheep, and sheep keep on complaining; the expert adds to the cacophony with his emery wheel, the presser starts up the electric press (these latter-day presses are noisier than the old back-breaker hand presses because the men, who had to use the weight of their body to compress wool, had no energy left even to curse). Everyone is hurrying, and appearing to; everyone except the shearers. They rarely speak. They concentrate so acutely, mentally and physically, that the dropping of a broom by a rousie frightens both shearer and his sheep. If the sheep struggles suddenly and kicks the hand-piece from the shearer's hand, then there is trouble. A hand-piece attached to the down tube and out of control can rotate across an area up to three metres across, dangerous to shearers and shedhands alike. If the machines have to be turned off to recover a maverick hand-piece gyrating round the board, then the shearers really do get cranky. Wool presses have caused broken legs, arms and ribs, particularly so with the box-on-box type when the handle is left to unwind unattended until the upper box reaches the floor again. It swings like a dervish. But by far the most dangerous area of the whole shed is the expert's room. Here, where the 'expert' grinds combs and cutters and carries out maintenance on hand-pieces and down-tubes, are exposed starting handles on engines, flywheel and crank shaft, belts and pulleys and the grinder. All are capable of dragging a man into their motion, but the two solid steel discs of the grinder travelling at 2800 rpm can cut through flesh and bone. (CSIRO did a survey in 1979 on accidents to staff handling sheep. Forty-one percent of injuries were cuts and punctures; hands and arms were the most injured, with feet and legs next.)

Other dangers in a shearer's life are encephalitis and enteritis, and nausea resulting from drinking too much water while working, thus robbing the body of salt. Hydatids are not as prevalent as one might imagine in men working with the fatal combination of sheep and dogs in an arena where personal hygiene is difficult to maintain. Floors are often patched with a piece of tin that rips off the finger-nails of shed hands picking up wool; some floors are rotted and a man can sprain an ankle slipping through a hole. But all in all, few days off occur through illness or injury.

You don't hear them call for tar now. A milk-oil fluid or disinfectant is used in place of the cold tar mixed with water, but in fact there has grown a belief that nature heals all wounds more effectively in country air than any man-made potion. So, when the woollies are rather more than pinked (small cuts rather like a man shaving may suffer) they go off as is; for a severe cut 'Tar!' is called for, and disinfectant supplied. As ever, men use needle and thread to suture severe cuts, and sheep appear none the worse for slips of the cutters, even for nicks that slit the wall and release 'the bag of shit'. No shearer willingly or even carelessly damages a sheep this way. As anyone who has been present at such a time would know, it isn't the prettiest sight or smell, and the resultant clean-up for the shearer makes him curse the often worn-out and rotten-skinned old snagger.

Buddy Wagstaff runs Thornleigh out from Blackall (Qld)

It is an industry where the whole team works to maximum capacity. Many actions appear mechanical or set to a rhythm, due to the high intensity of concentration by all taking part; they can usually anticipate another person's actions. It is this rhythmic pattern that makes watching a shed in action so hypnotic.

Most shearers are expert at preparing their own gear. They must supply their combs and cutters, and at smoko or lunch-time you see a man or two soaking his cutters in the little tin provided on the ledge of the shed, while he prepares to adjust the shear with as little tension as possible to prevent over-heating and secure a lighter grip. The secret of shearing is to hold the shear as lightly as possible and keep the wrist supple, to enable you to get the balance of the hand-piece.

Each man has his own manner of gripping the hand-piece, rather like everyone holds chopsticks his own way. The great shearer of the 1920s and 1930s, Ivan Orchard, held the shear between the thumb and third and fourth fingers with the second finger hardly touching the shear. He used the first finger only to apply pressure on certain blows, such as when opening the neck in 'hard' sheep. He was famous for giving plenty of 'lead' on the cutter: making sure that the comb was far enough forward to drag the wool fibres upright before the cutter harvested them.

Ivan, a gun shearer in his day, shore 242 ewes, lambs and wethers in 7½ hours in 1929 at Sharrow, Bungendore (NSW); in 1931 he shore 286 ewes and lambs in a day at Peccol Plains, Mackerlay (Qld). Winning the Longreach (Qld) shearing championship in May 1936, his score was 98 points and he shore a wether clean in 2 minutes 11 seconds.

There is a subtle break in the rhythms that an onlooker may not notice: sheep differ one from another, age, condition and sex making a difference. Old sheep, rams and ewes in lamb require very careful handling. Old sheep have very tender skin and wool, and if they are gripped roughly in the pen the wool will often 'lift'. If these sheep are in poor condition, they are hard to 'work' with the free hand that does most of the work in shearing; the skin itself may tear away, the gut open.

Rams are another thing again. They have a sulky nature and even holding them in an awkward position for a long period can injure, sometimes kill them. Ewes in lamb are tender. Most properties try to avoid shearing ewes in this condition, but seasonal conditions sometimes demand it. The shearer often has to carry them from the pen across the board to his stand and support them while he shears. When finished he must carefully stand them on their legs and guide them down the chute. It is slow work and tricky, and even with the greatest of care, lambs can be born on the board.

In 1907, at the age of seventeen Ern Barnes headed off for his first shed away from his father's farm. His book, *Easier Shearing*, was published in 1955 and it still has not been bettered as a manual or as a description of a shearer's working life.

'Most of the shearing sheds in Southern Riverina started during late July or early August. I wrote to the Graziers' Association for a list of sheds which were willing to employ learner-shearers. I enclosed references – such as they were – and sent the required deposit of one pound.

In due course I received a list of shearing sheds in Riverina. I chose a sheep station with the largest number of sheep to shear per shearer. This was Urangaline, which employed seventy-five shearers to shear a guaranteed number of one hundred and fifty thousand sheep, or two thousand sheep per shearer. Nearly two hundred thousand were shorn there that year . . .

'Eventually I received notice that a stand had been allotted to me at Urangaline. The shearing was due to start there in early August . . . I left home four days before Urangaline was due to start with the shearing. My bicycle was loaded . . .

'That night I slept in the top bunk of one of the two shearers' huts two hun-

Sir Alex Creswick (standing right of bale) Tupra (NSW) and his shearing team line up at the end of the season as teams have done for a century and more

dred yards or so from the shearing shed. There were forty to fifty bunks to each hut. Half of these bunks were lower ones: the others were above. Wire-netting did for mattresses.

'Being a good and fast shearer depends on a great many things. First and foremost of these is the ability to choose, the moment you put your hand on the swing gate which opens from the shearing board into the catching pen, the sheep which is the easiest to shear and the easiest to catch. This saves valuable seconds which, in the aggregate, may mean several more sheep shorn during the day. As competition is great amongst most shearers, a man soon learns to choose automatically and quickly, if he is to keep his tally of sheep shorn anywhere near to those who are judged to be good shearers . . .

'So unaccustomed was I, and so unsuited was my body, so it seemed, at so early a stage in the shearing season, to the labour entailed in shearing the rams, wethers and hoggets – the hardest sheep to shear and those generally shorn first – that it became a test of strength between my will power on the one hand, and the protests of my body in the shape of pains and aches and general prostration on the other hand, as to whether I would see the finish of the shearing at Urangaline. I would wake in the morning with stiff limbs and with hands swollen from gripping too tightly both the sheep and the hand-piece, with fingers and thumb blistered from holding the hand-piece too tightly, which, although a Wolseley . . . became warmer than usual because of my inability to adjust it properly . . . for the first week I struggled to survive . . . Hoggets and wethers were still being shorn after two weeks . . . Try as I would, I could not shear the fifty a day . . .

'Some days passed, and I was still investigating the problem of why it was that one shearer could shear more sheep with less effort than another. I would stand for a minute or so after I had let a finished sheep down the chute and try, by watching the methods of fast small shearers and slow big ones, to see what I could learn by it . . .

'I learned that to choose a sheep that was handy to the pen gate was far easier

The relaxed, athletic body of a gun shearer – John Harris

and quicker than diving into the middle of the flock in the hope of getting the best one to shear in the pen. I learned that to up-end the sheep and walk backwards at the same time would bring the sheep into an easy position to carry on my knee. I learned to place the sheep always on the same spot on the floor, that spot being exactly the right spot in relation to the shearing gear. I learned to fill the comb, to take long blows with the hand-piece and not to depend on the speed of the stroke alone; in fact to eliminate useless motions of all kinds. For no matter how fast the hand-piece is pushed through the wool, much time and energy is wasted by going over the same ground more than once, or when pulling back the hand-piece to start another short stroke; or again, when the comb is only half full . . .

'I found that shearing sheep was more of a game or competition than work. No cricketer or tennis player found more pleasure in perfecting a new stroke with his bat or racket than I did in trying out and putting into use a new "blow" with my hand-piece; or a new way of holding the sheep that I had discovered while watching first-class shearers whose easy style pleased me . . .

'Eventually there came a day, the longest we had had since the shearing started, with good-shearing sheep and lambs being shorn, when we all gave a good account of ourselves. The team shore a total of over eight thousand sheep that day, and as my tally was one hundred and fifteen, a few sheep above the average shorn by the seventy-five shearers, I was very pleased with myself.'

In a booklet *Hints to Shearers* put out in 1938 by the Graziers' Association of NSW and Dangar, Sedge and Malloch Ltd, the shearer was informed that 'the boss in not greatly concerned with pace. Find your natural pace and stick to it. Trying to get a tally and pushing yourself beyond your usual pace will exhaust you mentally as well as physically, knock the sheep around as well as earn the displeasure of the boss. Excessive speed is the greatest evil in a shed as tempers get frayed, tools abused and bad feeling arises between penmates.'

The tool for the work they do is one of the most efficient machines made. Simple, effective, there has been little overall change since their introduction almost a century ago. Yet, like the blades, 'they are dangerous damned things', Frank Doecke says. 'We were always afraid of lock-jaw, you know, much of the time the cutter is under the wool we can't see it and our hands are out of sight with it. A cut between the thumb and forefinger causing lock-jaw was the fear of many shearers – today also. It was a serious thing. Blades or machines, a cut from either is a cut from an instrument that carries infection.'

Frank prepared his blades (Rhand shears) with string binding, leather knockers and 'driver' – a strap to help push through the wool but which also, in case the sheep got its foot caught in the top of the shears and began kicking this strap, gave the shearer more control over the blades. The beautiful blades from Sheffield, England, are honed sharp enough to shave a man.

Frank Ayres (Vic.) has had several set-to's with the hand-piece. 'My leg, and another time my chest, were gashed. In WA we were shearing sheep that had walked through ti-tree. These sheep hadn't been shorn for three years – you could see the line of each year's growth – and the broken bits of the twigs were concealed by the long straggly wool, and bang! she locks. The hand-piece goes beserk. I saw another chap there have his flannel singlet ripped off by a runaway boggi. They were terribly dangerous.

'If a burr gets into the cutters and the machine locks, all hell breaks loose, and if you aren't quick pulling the power cord the hand-piece whips round wildly and can mince an arm or a leg before a man can clear out.' (The old shearers learnt this 'failing' when they were resisting the introduction of machines in the 1880s and they would purposely let go of the hand-piece; while it thrashed around, it demonstrated to 'the boss' what a dangerous innovation it was.)

Tony Walsh informed me: 'There have been countless thousands spent on research to better the hand-piece, but basically it's the same as it was eighty years ago. Those old designers were unbeatable. But you would think they would come up with something new today. A hand-piece can get to 60 degrees [Celsius] when you are racing. You can cook on that! It's the most neglected thing around the place apart from your missus.'

Concentration is intense even for Australia's most widely experienced shearer, John Harris

Frank Doecke shore Nonning, north of Port Augusta (SA) for many years, and has his notebooks of those days. In 1950 the cost of 48 cutters was £4 8s 0d; six combs cost £2 8s 0d. 'They say the price of everything has gone up, but comparatively speaking we were paying a hell of a price for gear in those days.' From 8 March to 17 April 1944 at Bon Bon Station (SA), Frank shore 3051 sheep for £62 18s 6d. From this was deducted £10 6s 3d for messing fees, and his check was £35. That sum doesn't add up. 'I think I must have had a little draw somewhere.'

In 1944 he shore 4934 for £102 15s 3d. In the same year at Kapunda he shore 595 for £14 2s 9d minus £2 12s 6d tax, leaving him with £11 10s 3d to take home. His daily tallies show steady, even shearing – except for the 29 of the first day: 'Getting the feel of them.' There followed 161, 171, 168, 178, 181, 160, 162, 155, 175, 179, 183, 175, 179, 175, 179, with 82 on a Saturday morning.

It is not the record breakers that the contractors look for. They prefer the steady man who shears his 200 a day every day, is reliable and easy to get on with.

Joe Bliss is such a man. Trever Caldwell from Grazcos says, 'I've known Joe in sheds and districts for years. They call him "old" Joe now because he's in his fifties and there's not many men still shearing at that age, but several times recently he has shorn 300 with the machines just for the hell of it. When the sheep are full of lead [sand] and weighing a ton and the board is pushing the boggi till their arms ache, Joe will still have 40 up on the rest of them when the tally goes up on the board at the end of the day.'

'He's a man who can't bear being unoccupied, he heads straight off if there isn't a stand going for him within a day of cut-out of a shed.'

Conditions differ from state to state. Whereas in Victoria a property with lousy sheep is quarantined until the owner has cleared the infection, out in Queensland lousy sheep can be sold, transported, or shorn in the shed as are clean sheep. For shearers, lousy sheep mean hard work. 'You push your hand-piece all the way because the gear blunts quickly,' Joe Bliss says. The wool is yellow from the lice and Joe is soaking and cleaning his comb and cutters during smoko in a western Queensland shed. 'I'm using four combs and sixteen cutters a run compared with one comb and four cutters at the last "clean" shed.' This would mean that at the end of the run the cutters have got to be ground which takes the metal off and means the shearer is up for new tools more quickly, with combs costing $6.75 and cutters $1.50.

At this shed the ewes were in lamb, another problem for the shearers. 'The wool tightens up on a ewe in lamb. There's a sort of sticky feeling too.' Some of

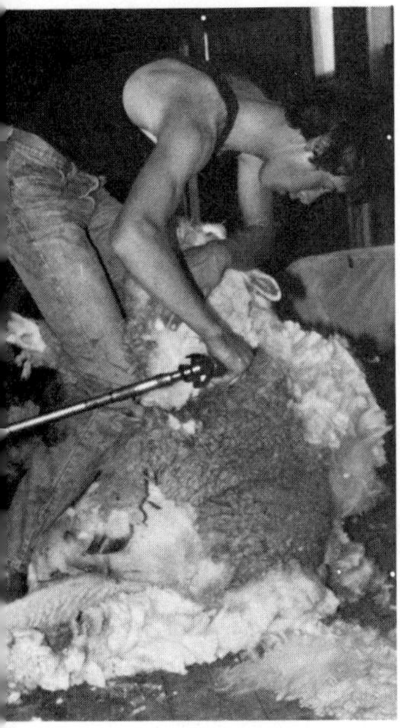

Young Damien Laffin, third generation shearer at Inverlocky (Vic.)

the men are unhappy shearing at this time. 'You're pushing the afterbirth out of the way. The lambs scarcely dry are crying and those that aren't trampled to death in the yards and pens are lucky if they find their mother again. (It is rare for a ewe to feed a lamb other than her own). The lambs born in the pen we push down the chute and they try to scrabble back up to find their mothers. Bleating all the time.'

Other men claim to have used twenty-six combs a run on sandy sheep. 'I've seen a ten-inch wide band of red sand down the backs of a flock in the yards. You do a lot of combs on a mob like that.'

'He's the bane of the contractors and the cocky: unreliable, alcoholic, he'll let you down twice as often as he'll turn up,' say most managers of the big contracting firms of the majority of shearers living in western (Queensland) towns.

To this the shearer born and bred in western towns answers, 'It's all right for those Sweat Extractors [contractors] and for the Parasites [Graziers, "Cocky"]. The cocky inherits the land, sure he works hard, but it's all for himself. Even if seasons are so bad he eventually walks off he's still got the sale-price and he's still got the board-school education his parents before were able to set him up with. What do we get? A broken back, a short working life and whatever we've managed to put by.' ('The cocky hates us, he's never forgiven us for standing up like men and demanding a just rate of pay. As for the contractors, they turn our photo to the wall if we play up.' Jim Pugsley says, 'I've shorn when I'm dog sick but it's the only thing to do.')

In these towns there are the shearers' pubs, the agents' and contractors' pubs, and the clubs. It would be a rare shearer who could make *the* club (as opposed to the RSL and sports clubs) and almost as rare a one would drink in any but the regular shearers' pubs.

Jimmy Pugsley has known The Cattle Camp hotel all his working life. 'I was lead picker-up at the last big shearing at Tinnenburra in 1942. Thirty-two shearers on the board built for a hundred and one. Eight pickers-up, four pressers, one cook and his offsider. This big shed was T shaped. The stem of the T was a neat 100 yards and us young pickers-up used to have races down this during smoko. The shearers would start us off. The stands were in the stem of the T, the steam pressers in the cross-bar. There were big steam engines to run this shed. 'Possum' Larsens [Charles] was the steam-engine-driver-expert: he was known all over the country. He drove at Milo, Thylungra, Ambathargo, the last of the great drivers of steam-powered shearing sheds.

As his shearing time wore on, Jimmy Pugsley took more and more days off. His wife Ellen, with two sons to bring up, had her hands full. 'When Jim was on the grog and didn't want to go out next day he'd try anything. He'd plant his moccasins – "I can't shear without my moccasins" – and then we'd have to find them and shove him on the truck, blind drunk. Many's the time the truck has pulled up with shearers shouting out for him and we had to bundle him out and him protesting all the way.'

'You had to laugh. You'd have cried if you didn't laugh,' Jimmy says. 'You're in an iron tank all day under a broiling sun in a temperature that's gone to 120° F. in some sheds. Often a hundred degrees. Think if, while you're baking and boiling in this shed, you bend down and touch your toes and you stay there for eight

Jack Rose (front of car) with Riverina team setting off for Tasmanian sheds in 1940. The contractor provided the car the team took across Bass Strait on the boat

hours, and all this time you are squeezing a cricket ball in your right hand, flexing, loosening, flexing, loosening. And that hundred pound you're wrestling with your other hand and arm and leg can be lousy, stinking, fly-blown till the wool falls off as the maggots eat into the sheep's flesh – every shearer has seen this. At times a whole back coming away in your hand as you grab the sheep and the maggots falling big and thick all round you. Or like it is when the cocky makes you shear ewes in lamb and the afterbirth comes out over you as you turn the sheep, and she's full of milk, and you're going slow and careful over her udder.

'You can always pick a shearer by the knuckles of his left hand, enlarged, shiny, from working on the sheep, while his right hand gets corns on the ball of the thumb and along the index finger. As for his feet, he has terrible corns along the inside of his foot from swinging down to take the long blow.'

Not so long ago sheds were lit by carbide lights. These gas lamps gave a beautiful white light but were explosive. A much discussed death took place in the 1960s when a shearer was trying to light one and it blew a hole through his forehead. This casualty caused shearers to demand electric light in quarters where it was feasible, but the odd cook can still be seen of a dark morning pouring water onto the 'stone' in his carbide lamp as he lights up.

Cecil Leonard Elliott was called Ned Kelly when he was a kid. Later, he became 'Meggs' from the comic strip kid, Ginger Meggs. His wife Lorelei is known only as Doddy. They live at Winton, the central-western Queensland town where the public conveniences in the main street are marked *Rams* and *Ewes.*

'This is Australia's true sheep country.' Meggs knows. 'We wrote "Waltzing

Matilda" here.' And Meggs and his family have been a part of the industry all their days. When Meggs took Lorelei out cooking in the sheds, 'The town turned out to watch us go, everyone laughing.

'Young Eric was on goat's milk so we had the goat and a poddy calf and the baby's high chair on top of the load on the back of the truck.' They set off from Winton, out on the dirt roads where the flat lands are decorated with six telephone lines – not just wires, but six separate lines, each on its own spindly, thin, knobbly posts, stretched out, party lines to sheep stations.

Doddy is nervous when Meggs talks. 'A good shearing contractor is like a good politician. Can lie like hell.' Meggs turns the crack around and says, 'Yairs. Promise them everything and give them nothing.' Maybe. But when this tough man cannot take on shearers, he tries to get good men a job with other contractors. 'It's only good men look for work,' he believes. 'Stanthorpe men used to send me pumpkins and produce from their farms.

'The sort of fun that stays in your mind is Joey Swain, out at Myrall, getting his hair cut with old, blunt clippers. In the dark. Night-time. Or Snowy Haler, fifteen stone, five-ten high, picking up a ram and hanging it up on the shaft by its horns. It took three men to lift it down.

'Snowy was a gun shearer. He had his own car and would camp a mile from the shed by the boundary to stay fit as gun. Looked after himself. For weights he used to lift two four-gallon drums full of cement with a bar set in them.

'You remember funny things. Shearing at Coralton (Qld) – I wasn't contracting then. They were all rough shearers. Bull-nosing them. The owner said to the boss of the board, "They're too rough." "Snow!" calls the contractor. "Clean them up a bit. They're too rough." Well, Snow picked the contractor up in the air and dumped him down on the board. The contractor got up and staggered away. "Sack him!" shouted the owner. The contractor tried to pass the buck and phoned the firm in town. "Sack him", he was ordered. The contractor yelled into the phone, "Oh no! I'm off. I'm not stopping here. You sack him yourself!" That night, Snowy came to me to tell me about being sacked, "Meggs," he said, "They treated me bad".'

The famous (or infamous, depending on who you are discussing him with) Meggs Elliot was tough. But, 'At Wanalana station the contractor was a big man, Bandy Gill. Had an eighteen-man shed, was trying to finish the shearing but the engine broke down. So, he told the shearers, "You hook in, and I'll crank her." "Faster Bandy!" the shearers yelled. "More speed! More pace!" And Bandy kept them going until cut out.'

'When I started the price was £2 16s 0d a hundred. By 1980 it was $72 a hundred.'

Writing a book like this one, trying to cover a continent as large as Australia, one cannot meet all the interesting men and women involved when one first learns of them – and then, when one gets around to it, often it is too late, because it is the older Australians who know the story. Jack Rose of Conargo (NSW) was such a man. I saw him in 1977 in the famous little pub at Conargo but not until he had finished his beer and left was I told he was one of the greatest of the old shearers, and a great man as well. I wrote to him immediately. Yes, he'd be happy to yarn with me. It was three years before my research again took me to that well-trodden cross-roads of the Riverina. Jack had died two months before.

'You would have got on very well. I don't think you would find anyone better

Young Riverina team during the war years

able to tell you about the shearing industry than Jack and his anecdotes and gift of mimicry – well, I was so looking forward to him meeting you,' his wife wrote. 'He went into the industry aged 16 as a rouseabout and went on to become one of the best and fastest in the game. He went through many hassles in those earlier days, the results of which today's members enjoy the benefits; always very staunch to his union and to his mates and I believe he was deserving of all the good things said of him.' Catherine, his widow, keeps his hand-written 1938 reference from Murroa East, Hamilton (Vic.) with pride. 'The bearer Jack Rose has shorn for me this year with the Graziers' Shearing Co. I found him an excellent shearer in every way. Clean, fast and hard working. He is also quiet and civil in the shed and I can thoroughly recommend him. He shore consistently and reached a tally of 202 in this shed. Eric M. Philip, 14 Nov., 1938.'

Jack shore with forty shearers at Boonoke, 32 machiners and 10 bladesmen. 'The place was a hive of industry. There were two messes, one for the shearers and one for the rouseabouts.' Jack's run started each year in the Riverina on 27 July – 'they all wanted to start on the one day those times' – went on to Victoria and Tasmania, and finished in time to get home to Conargo for Christmas. He was a good fisherman and his life's hobby was fishing in the streams and pools of the billabong near his home.

'In the 'thirties nobody had cars amongst us. The 1933 pay rate was 27s 3d for a hundred sheep. Grazcos provided a seven-seater taxi to travel from one shed to another in some areas. It was a great life in some ways for men,' Catherine admits.

There are women shearing today, union members, but then there were always women shearers. Women members of farming families who shear hundreds of sheep annually have been unsung, but they have been and still are active. Julie Strang of Kurrawong, Byrock (NSW) began shearing her pet lamb, Angelo, when she was aged 7. 'I used scissors then.' She did correspondence school by night, and farm work by day. 'We couldn't afford to put anyone on. Things were bad. I was shearing by the time I was thirteen.' Her father and uncle have been shearers; her grandfather was a carter on the White Cliffs to Broken Hill track.

Now married, living on Meroondah Downs on the road to Jundah, she shears stragglers and rams and goes crutching. Julie is typical of many young country wives. Her idea of a bright time is 'an occasional trip to the Big Smoke, Bourke'. She breaks horses, can ride 50 miles straight, helps farm 16 145 hectares with 12 000 sheep, and the property includes 3240 hectares of stock route through which stock must be escorted. 'In drought it's six and a half miles a day, but – some take a week back and forth!' She has done contract mustering and is, in general, a farm worker. Jackeroos have 'gone out' in all but the big wealthy holdings. 'Less profits, fewer employed. Every penny counts nowadays.'

'Bladeys' shear the rams, Tupra (NSW)

Meggs Elliott's bough shed, Tattoo, out from Winton (Qld) sags with weight of spinifex on its rafters

Also out of Longreach, two other girls in their twenties, the Taylor sisters, Jan and Cindy, do all the farm work on Boree Downs. (They were building stock yards in 1980 when I visited.)

In Queensland the sheep runs are on gidgea, bunya, boree, mulga, gum and sandalwood country on the black-soil plains and the red-soil land, the green valleys down south and the rocky lands north-west. In mid-west Queensland flocks of emus a hundred strong run with the sheep; kangaroos in thousands, wild pigs believed to be in their millions infest the land. Galahs scream and brolgas dance. This is legendary country. Petrified wood lies around, and the red gidjee stones are found only near gidgea trees.

Many of the little towns were born because of sheep and have been kept alive until now by shearers, for the shearers are the industry of the Queensland outback, the people who keep the young men and women in the area.

When the sheep go and the shearers move elsewhere to work, the town fades away, like Isisford did. When Isis Downs shed was built with its fifty-two stands, Isisford was a bustling town. Today, with only twelve stands hooked up at the huge and famous shed, the village is empty.

The shearers call this half-moon-shaped shed 'the iron lung' or 'the corrugated cow'. 'There's not a breath of air in it;' 'It was built to kill shearers'; 'It even blows the sheep out. They can't bear it'; 'We voted the shed out it was so hot. Twice we walked out. Couldn't shear in it. We tried everything. Ran a fan from the motor off the old Vanguard, but nothing gets air into that tin tank.'

There's a lot of wool to be shorn from sheep and rams: hardy rams from East Bungaree, Belcunda (SA)

The fifty-two stands are on the inner semi-circle with chutes behind them, beautifully constructed, beautiful wedge-shaped tables in front, and a steam crane which loaded the wool onto trucks after three pressers had steam pressed it. But nowadays only six stands are used and the electric gear of 1914, when the men first shore there, is in a museum.

The architecture is astounding. The whole 'board' is suspended beneath the dome from a great steel post and steel girders fly from the struts across the rib cage. But the overall impression is of a monument to those days when wool was king.

Buddy Wagstaff remembers the dances held in the Isis Downs shed when she was a girl. 'Lanterns to light the way in, tubs of flowers, big bull ferns in carbide cans, decorations, supper spread on the classers' tables and all of us in full evening dress.'

Buddy owns the 26 730 hectare Thorn Leigh on the Barcoo River. Over the doorway of the beautiful old homestead stretch the great horns of Dusty, the lead bullock of the team that dragged her grandfather's waggon across the plains to found this station before Federation. John Henry Hart came 'when the blacks were still bad' and early spears, shields, nulla-nullas and grinding stones decorate the walls of the wide verandah surrounding the house. Buddy's father was born in this house in 1895. 'Do they ever think of these things, the people who write books about us?' Buddy says.

Buddy has 25 000 sheep and usually runs a twelve-man team, but when I was there it was too wet to muster from the outer paddocks and only six men were on the board.

The shed here is very old, built of boree logs with a gabled roof for coolness. Nearby, the deepest flowing bore in the world was bottomed at 2138 metres. 'Spring Leigh was so deep that grandfather Hart ran out of casing when he was boring it in 1914.'

47

The Call of the North

In recent years shearing in Western Australia has been concentrated mainly in the south and south-eastern areas, but men who have followed shearing for a lifetime were not part of this phenomenon. It is the north, and in particular the north-west, that marked their lives; to have shorn in 'the North' is to Western Australians comparable with an 'Eastern stater' (the Wests' term) having shorn 'down the Castlereagh' or 'up on One Tree Plain'.

Sorenson, the Western Australian poet, captures their spirit as clearly as Paterson or Lawson ever caught the heart-beat of men along the great shearing routes of the east.

'It was 1934, the year we bought the new Bedford truck.' Mark Synnot, wearing white shirt and hat, is near back of lead truck; Arthur Dunbar, his partner, is the tall man, third from right. 'We were on our way from Derby to Noonkambah. The Liveringa team are with the two canopied trucks in the centre, with the Bedford in the front, the Chevrolet at the back. We were having a spell to cool the engines off before tackling the gruelling Fitzroy River crossing to Broome'

The Call of the North

Oh! The western wind is blowing,
  So there's rain and storm in store,
And the teams have long been going
  Down the road to Glindawor:
To where tropic sun is gleaming,
  And the fragrant winds blow free;
I've awakened from my dreaming,
  And the North is calling me.

Oh! the steam is in the boiler,
  In the expert's room below,
While upon the board each toiler
  Waits to hear the whistle blow:
For the shearing is beginning
  And my heart is fancy free,
And the friction wheels are spinning,
  So the North is calling me.

From the Southward to the Nor'ward,
  Where the long brown tracks wind down;
All my mates have hastened forward
  To the wilderness from town:
Gone! by stony hill and hollow
  To where I now fain would be,
Where they lead I needs must follow,
  For the North is calling me.

What's this news I have been hearing?
  Tidings strange to me indeed.
Bidgemia now is shearing,
  With Sawallish* in the lead;
Straining camel teams are swaying
  From the Junction to the sea;
Why so long am I delaying?
  When the North is calling me.

And so Northward I am going,
  For I cannot linger here,
Now the starting whistle's blowing,
  And the 'guns' are into gear;
So to be there I am yearning,
  I will hail the sheds with glee,
For the friction wheels are turning,
  And the North is calling me.

*Robert Sawallish, one of the fastest shearers in the West.

'Shearing in the West has always been different from other states. Most things are different here. We're proud of that.' So says Marcus Synnot, the Grand Old Man of the shearing runs of Western Australia.

The great distances, vast climatic and geographic variations and the history of the state itself make for difference, and the no-man's-land between that state and the rest of the continent has always indicated an individuality that is peculiarly Western Australian.

Marcus Synnot's life as a shearing contractor took him north to runs over 4000 km from his southern runs in a state that is twice the size of Queensland and could swallow three smaller eastern states. Shearers' runs in that state ranged from 33° 45' south to 15° 29' north. Conditions outside the south-west area have always been hard. A survey of deaths in the north-west over the past hundred years shows the number of men under 26 years of age buried north of 26° to treble the total of other deaths. (Coolgardie cemetery on the Eastern Goldfields has almost 1000 men under the age of 26 buried there – and few older.)

To reach those north-western sheds before roads were made, shearers travelled by ship to Port Hedland, taking their trucks on board with them. Once landed, they headed off to the inland stations where there were five blacks to every white man.

It is said that ninety per cent of Western Australian shearers and their families have shorn under the banner of Synnot and Dunbar, contractors for two generations. They were the biggest and oldest contractors in Australia, shearing from a half to one million sheep a year. Marcus Synnot, now retired, had fifty years in the shearing industry 'in the West'.

'We didn't like Eastern staters. They were too union. Too hard over there . . . not welcome in this state. Our shearers didn't get on well with [them].'

His father started the firm with Arthur Dunbar in 1918 when they came home from World War I. 'Young' Marcus came into the firm in 1929 after eighteen months training as a wool classer. 'My father was educated at Geelong Grammar. We always believed a man should be educated, makes it easier to deal with people; the squatters took more notice of us. No, it didn't put us off-side with shearers. On the contrary, they knew it was them we depended on.' Marcus also had a college education. 'In my father's day – even in mine – shearers were rough and tumble boys. A bit of style in the contractor paid off with the men who we were contracting for.

'My employers were, in the main, the best men in the world, but some of them were greedy, grinding bastards, trying to get the last drop of blood out of the men. But one got all types.

'My father and Arthur Dunbar taught me all I knew about handling men. All teams have a man who sticks his neck out. My father told me if a man was worth taking away with a team he was worth persevering with. I only sacked one man in fifty years.

'We had seven or eight teams going at a time. Twenty men in a team. Big teams, up north, in country that is still wild today. We'd be promised "above average conditions". How could they be up there? They weren't honest and you don't like working that way. You had to be not always honest with them to get by.

'I'm talking about the hurricane lamp days, early in the piece, living in tents put up in the bush, no floors, bitterly cold nights, sweltering days.

'Our firm was always respected. We went along with good vehicles always in good repair. We shore the Kimberleys, De Grey, Pilbara, Murchison, from the Kimberleys to Shark Bay. We didn't work the goldfields area much. But we worked an area as big as Queensland. No-one else in Australia had the [same] problems of distance . . .

'When my father started going up north he travelled by ship and then inland by donkey waggon and walked when he had to. We were the first to have trucks in that area. There was a new track eighty miles down from the Fitzroy Crossing to Broome, and trucks could barely get along it when empty so – the men walked; the truck trundled past them; then the truck boiled; the men marched past it; the truck cooled off and trundled past the men; then the truck boiled; the men passed the truck, and so on.

'The Pilbara never carried really big numbers of sheep but there were plenty for us. The spinifex up there would see them through the drought.

'They were big properties up there until the iron ore took over . . . beyond Marble Bar. Warrawagine carried 80 000 sheep in a normal year; on the De Grey River, it was the home station for five properties carrying 200 000 sheep. Beautiful country there, beautiful. Not a sheep on it now.

'The new mine railway from Port Hedland to Mt Goldsworthy went through the De Grey property.

'At Noonkanbah 101 000 sheep were shorn in five weeks in 1954; forty-four men in the shed, twenty shearers on the board. They shore 20 000 one week, 1000 per man per week. That was a world record. We shore over 4400 one day and the learner had tears in his eyes: he had shorn over 200 that day and was still dragging the chain! Of course, he was a good learner!

'With a team of twenty we had four learners and we always made sure we kept our learners' numbers up.

'We shore at Yarrie when Lang Coppins's parents [Owen and Jean] had it. That's real back-country out nor-east of Marble Bar.

'The Pilbara had as many difficulties to throw at a team as any of the northern regions. One year it rained, heavy. The De Grey rose quickly, water everywhere, and there was our team of shearers out helping to build a bridge to get the sheep over the river to the shed. They were anxious to work and knew how long we'd be held up if we didn't get that bridge built.

'Arthur Dunbar and I went out to the normal crossing. It was too deep. We went upstream, going from island to island and we worked out what we thought would be a fairly safe passage for our trucks to get over. But we forgot the tide and in the afternoon when we had the trucks half-way across, down came the tide, as all those northern rivers roll down . . . it took 88 men with a bore cable to drag the two trucks across.

'It's not all beer and skittles getting a team up north I can tell you! As the firm got bigger so did the problems, but there was little we couldn't extricate ourselves from. The work had to be done, and on time. You worked out a schedule and had to keep the teams moving along a run of sheds if they were to get the sort of cheques they wanted.

'All shearers want to go to sheds where the good sheep are. A good shearing sheep is one that is entirely in "evening dress", no points, no belly, no head work. Spinifex country always had good sheep. Men got good tallies in spinifex country. At Myroodah one year we shore 42 000 sheep for two bales of bellies! 387 bales altogether. Today's sheep are really good, thickly covered from hock to head.

'In my days, a team of shearers were usually the same men year after year. They were professional men. Worked eight to ten months of a season and that carried them through the year. They were honest. If they borrowed money they always paid it back. They were scrupulously clean in themselves. In the mess a man had to eat properly or he was sent out to eat on the woodheap. That was their own affair. Their quarters and mess room were their own domain, as was the quality of the cook – if he didn't suit them they could fire him.

Sunday morning, rest day for the team at Shaw River station. Frederick Synnot, (founder of the firm, father of Marcus), leans out 'window' to left in picture

'You couldn't expect a hundred per cent decent blokes. Up at Wallal Downs there were always a lot of natives and as we'd drive through their camp you'd see the odd swag thrown off the back of the truck. Usually we didn't take the man on the board when he caught up with us. It would have been a bad influence on young lads. As for gambling, the only times you interfered were if men were losing too much money, as that created ill-feeling.

'The areas we shore was hard country and a man had to be hard to live or to work in it. The men felt that about the squatter. But in the old days the squatter walked along the board. He could say "Take that wool off", or, "You're sacked". The shearers often had a legitimate grouch against some of the squatters.

'It's different now. The station people can come into the shed any time they like but must go through the contractor if not satisfied with the work. The contractor supplies the shearers, stores, mess, and the squatter supplies the sheep to the outside pens after mustering. Until they are shorn and put down into the counting-out pens the sheep are under control of the contractor. There's no raddling of sheep by squatters now. A contractor . . . is trusted as a go-between. A contractor is a Solomon, has to have all the wisdom in the world and deliver his decision swiftly.

'Booze? Yes, the men were keen to get to a pub up north. "To Broome in a hurry!" They'd turn a town upside down and it was a real battle to get them out. And next day we'd take them a hundred miles out and let them dry off.

'Marble Bar was the place! They'd get in the old Ironclad hotel where the beer was kept cool with strips of wet flannel over the bottles, and there, in Australia's hottest town, they'd decide to dig in. The few stations around Marble Bar were tough, it's tough country, and we sympathised with the boys, but we had to lever them out. And to think those stations are all abandoned now!

'In the early days there was no grog allowed in shearers' quarters – never has been in the sheds. The heavy drinkers were always badly affected when they got into town . . . The union ruled – with the squatters demanding – that no spirituous liquor was allowed on a property, and contractors strictly adhered to this. But it was hard for a shearer to see the station people walking round half-drunk on a weekend and he with his tongue hanging out.

'Now they're allowed a few bottles out on the run and with today's transport shrinking distances men can usually get into hotels at the weekend. But in all our time contracting, grog never caused any bother at work.

'In the old days a nip a night was good for them. If a man said, "I've got a bottle of rum", you'd say to him, "Give everyone a nip", and they never abused this leniency.

The team on the 1934 Bedford, 'our pride and joy,' at Shaw River station

'Today, education has made shearers men of responsibility. A shearer today may have his wife and kids with him. Often there are eight caravans at the shearing shed. The stations are happy and provide caravan facilities for them. It's the ideal set-up. Content with their wives and kids around them.

'Our transport was the most abused in Australia. The roads outback were rough – if there were roads; sometimes we made our own. The 1924 map made of the north had "Motor Tracks" marked on it. But that didn't mean roads as such.

'There were seats running back to back down the centre of the truck, rails on the side for the shearers to put their feet on, swags tied onto the tail board at the back; tucker boxes along the side. Arthur Dunbar always designed our trucks. We had a long way to go and could get little of our needs on the way and he had to make allowances for that. So it wasn't the most comfortable travel.

'Yet the funniest things could happen. The trucks were rumbling over the red tracks through a rocky outcrop to a shed once, each truck taking a turn at leading so they wouldn't over-heat, and there were two shearers bouncing up and down in the back of the lead truck arguing about the Circular Waltz. Each of them claimed to be a better Circular Waltzer than the other. In time the other chaps got tired of the argument and said, "All right! you'd better show us your style!" The truck was stopped and off went the first man, waltzing sedately over and through the rocks. Then someone called him back. "All right Joe! You've had your go. Jack's turn now", and off went the second man, one two three, one two three on his toes through the rocks. The blokes agreed there was nothing between them and called it a draw but Jack wouldn't accept the decision so they had to fight it out. If ever you lose your sense of humour you're finished out there.

'In the old days you must keep politics and religion out of the quarters. Men were sleeping forty to one dormitory.

'Malaria and dengue fever were bad up north at that time. Each man carried his own mosquito net. At Noonkanbah we slept on an old straw mattress and cyclone wire stretchers, sick or well.

'The hurricane lights were out at 9 p.m. "Light out!" the rep. shouted. He'd give them five minutes and over would come a boot. Most lights went out then but some hung on to finish a book and then everyone shouted, "Put the light out!" One bloke even shot one out.

'Men grow very weary shearing day after day. They need sleep like they need good food. They had to put up with a lot. In Kimberly we used to supply gloves to the men because of cork-screw-grass seed. They needed gloves to drag the sheep out of the pen. Those seeds would wind into men's hands and they'd fester and become infected. At Myroodah one year, 12 000 young sheep were lost because the seed hobbled them and the wool became tangled between their legs and they couldn't walk. Sometimes the sheep would be walked through water so that the seed would swell and break away; for all the care, a man's hands were sometimes so full of splinters there wasn't room for any more.

'The old shearing hand-pieces gave you thick calloused hands, always. And always men's wrists grow bigger as each season goes by. Sometimes they blow up with the unaccustomed exercise. Old shearers used to walk around when not shearing with a pair of clippers in their hands stretching the hand all day so both hand and wrist would stay supple. Otherwise muscles can seize up when

Shearing Bidgemia, 1920

you start in a shed. A man has to put a lot of weight on his rouseabout hand, that's why it grows bigger than his shearing hand.

'Backs go. It's never a joke to see a man straighten up with his hand on his back as he finishes a sheep. You see too many and you know the pain. In the old days of poorer machinery men had to bear with the hotbox, and there's many old-timers who still carry the scars of those hand-pieces. Don Young used to carry his finger under the blade of his machine to hold it up a quarter-inch. It gave the most even shearing. But it gave great callouses on Don's finger. The Youngs were famous shearers. At first they were Yings, they changed it to Youngs. There were five of them shearing. All good.

'On some northern properties they had 2–3000 stragglers to be shorn, sometimes more than the actual shearing itself. It was so hard for them to muster. Everything was hard up north. Shearing that country was never easy. At De Grey the shed floor was concrete. A picker-up would walk – or run – twenty miles a day carrying six or seven tons of wool a day. You had to be in good nick.'

Marcus Synnot was only involved in one 'serious' strike. 'In 1931. That strike left a lot of shearers very bitter. It started in the eastern states. It was a great trouble for us getting teams pulled out. We worked in very inaccessible places, you know. We dropped one team at Meekatharra from the Rivers. Then we came to Pindan out of Mullewa to pick up another team of strike-breakers for Shark Bay. The shearers hired cars at Mullewa and tried to overtake us, but police came out from Carnarvon and escorted us.

'Charlie Golding was the organiser in 1931. He was a very good bloke. My firm had very good relationships with the union, even during the strike. A good union man wouldn't break a rule and we wouldn't either.

'I was on the Pastoralists' and Graziers' Association Committee for years and I hoped I helped to break down antipathy to pastoral workers. Any decision they make now they ask the union to send a rep. to make sure they are not breaking union rules – too much anyway.

'We never had as many nicknames over here as they had in the east. We're not so larrikin over here. And men today are more mature, educated... "My name is Bradley, not Brad", a fellow said to me a while ago.

'I'm against eastern staters coming over here to shear. "What are you over here for?" I used to say to them. "You've a mob of shearers over there of your own." And I'm against our men going over east. We never took them back. They were never any good after it. They learnt too many tricks.

Carnarvon team, 1925

'We were always supplied with good men. A few of the criminal element, not many. A few drunks. In the war they opened the gaols to us... Had to overland them.

'During the war of 1939–45 our problems were multiplied. Everything was different. A big mob of RAAF men were camped at Noonkambah. We could hear Jap planes overhead. Broome and Darwin were bombed out. In the event of invasion we were informed that we must go straight overland to Adelaide. Nothing said about how we'd feed our teams on that journey.

'As registered contractors we couldn't pay over the award rate. Being union men we had to pay them union wages. But when the shearers came down south, none of the cockies were members of the Pastoralists' Union. They could offer higher money for their few sheep and our men would go over the fence to them.

'I hated bringing teams from the north down to the south where the cockies waited for them without having done anything for them by way of conditions. Don't speak to me about cocky farmers.

'No big teams go to the north now... The big distances between the places there are being mined out of the ground. That place will end up a desolation. Even the men will go from it – the sheep have already gone. Wild dogs [dingoes] are taking over. The doggers are the only men making money up there from the land.

'Every area had a colony of dogs; when the wool-growers came, the dogs were driven out to the desert; dry seasons have brought them in . . . A lot of the fringe properties on the edge of the desert were abandoned in bad seasons and the dogs are now able to come in unhindered, almost to the coast. Cattle men don't worry about wild dogs, but sheep men, that's different.

'Every industry has its predators in greater or lesser degree. But now, the iron ore has taken over such large areas of the north that all else is finished there. It was the north that was the real shearing land. I can't tell you how exciting it was. You could hear the north calling each year when shearing time came round.'

Thomas 'Fiddy' Fleay shore sheep for fifty years. He started with blades 'doing 100 a day'. His father was a blade shearer and his brothers Scotty, Lardie, Wilfie, Kevie, Alfie and Fid himself all became shearers.

'Most of the men from Toodyay went shearing. You couldn't get no work anywhere else. My wife stayed home with her mother and father.

'There were squatters in the nor-west and cockies in the sou-east – wheat-farmers with a few thousand sheep thrown in – and we'd do them with a [moveable] two-stand plant . . . Then it was up north to places [where] they'd have up to 3000 stragglers. Anything left after mustering and shearing was gathered up on the water points and two men would go back and shear them when the real shearing was over. The station would cook for us then and young black girls would wait on table.

'It was very rough living when I started. Lived in tents, shore in bush sheds with spinifex roof and open walls – Meeka was that way. At Myroodah you could lift the flaps up and let the air in – if there was any. Woodstock was another spinifex shed I shore.'

Fid shore 200 sheep a day, every day, for 48 hours a week. He was a steady, consistent shearer and in the days when men shore for one hour before breakfast he could stay in bed for that time and catch the team up by the night bell.

'The biggest sheep I ever shore were at Jeedamira. I thought they were goats when I first saw them.' Fid shore in sheds 'where the dust was so thick you couldn't tell who was at the other end of the board'. But the trip he remembers best was when he went through the Kimberleys 'just as the Japs finished bombing'.

'They buried a man near the road as we passed. Women were being evacuated, we passed batches of them coming down south as we went up north. We stayed in the Broome pub after the bombing and then we went on to sheds. The authorities gave us tucker to go through.

'Synnot and Dunbar took three teams up, 96 men altogether. On that trip, when we were on the Pilbara, we caught a big groper, 150 pounds, near Port Hedland. We took it to the pub. They didn't charge us to stay there because we brought fish and wild pigs to them.'

Mr Fleay, the centre generation of three generations of the Fleay family to go north to shear, was aged when he talked to me, and tired. 'It's all over', he said. 'It's all gone now . . .'

The shed at Whim Creek with its spinifex roof. 'There were a few brave women up in those latitudes, but only a few.'

Elijah Henry Fleay of Toodyay has been known all his life as 'Lardie'. There were eleven boys in Lardie's family, eight of them fast shearers. And Lardie's sister, Stella, could shear a hundred sheep a day. 'Handy with the boxing gloves too. When it came to work she could hold her own with any man. A great woman.'

Lardie finished his shearing life in 1939 when he went to war as a captain in the commandos, but he went to the Kimberleys in 1926 when he was aged 15. 'We went by boat to Derby . . . it took about two weeks. Another week by truck to Noonkanbah. I was lucky. Only six years before my time the shearers went out by camel team.

'Noonkanbah was a great station in those days. We'd be shearing there for six weeks. In 1939 there were 104 000 sheep shorn there. Now there are none. The blackfellows have it. There were twenty shearers and twenty shed hands in our team. We shore all over the north.

'Our main shearing area was the Pilbara. There's really no shearing there now . . . They've given it back to the blackfellows. We shore 60 000 on De Grey. In 1935 I shore 275 one day there. Now there's nothing to shear. They've eaten them.

'Station people were not very favourable to us. We were shearing 26s 0d a hundred in 1931. Anyway, there was a strike on the Murchison. We set up camps. I was at Yalgoo camp for three months from July to October. Police were protecting the station owners. There were a few little fights. We had some friends. A little telephonist girl was one of our mates and gave us all the news.

'The strike lasted about a year. We lost. Later the price went up to £2 a hundred. Victor Johnstone was around organising us into a union. Later he was a senator. It took until the 1930s to get the unions going. We were very isolated in a strike up there. Hardly any white men . . . few owners, mostly managers; the owners I've found are reasonable enough but managers – well, they're no breed at all. Owners are better altogether. Managers are hard.

'Up north was a land apart. We'd go up in March when the wet season ended. Some men would work their way down south. You couldn't get shearing down south until August. I didn't come down. I'd go on to the Gascoyne and the other areas for nine months, then I'd head home for three months and I'd earned it by then, I needed it. Some years I'd shorn 30 000 sheep by the time I saw Toodyay again.

'And it was never easy. Murchison sheep cut 13–14 lb each. I could only do 150 a day of those but my brother, Scotty, he could shear 200. Those sheep were mostly Peppin blood. Great sheep. Further north in some poor areas you'd find sheep with 7–8 lb, but we've all seen them much less than that too.

'You got a lot of sand on the Ashburton. I've seen three inches thick of sand. And dust . . . At Noonkanbah in two hours you could use two dozen cutters. You'd wear out combs. On the Murchison you'd do ten – twelve sheep with one cutter.

'When we were on the trail the cook would cook enough to keep us going while we were travelling, bread, and stuff that would keep. It was no joke cooking up there. Rancid butter, ropey flour. Meat killed at night and let set and then all cooked the next morning before it went off. The cook would take some started yeast with him from Perth and keep it going with boiled potatoes and hops. It was so hot up there you'd just tip the yeast in an open bowl and watch it work.

'We had a cook, Bob Allett, he used to cook for fifty men and do the lot himself. Good meals. Sweets, everything, even jellies he'd manage. But cooks had a battle. At Malina our cook put his jellies and everything on the window-sill to cool overnight and a donkey team came in and they cleared up everything.

'It would be three months between hitting towns . . . We used to shave our hair off and by the time we hit Broome it had grown again.

'We were transported in open trucks. Shearers on one truck, gear on another. We travelled hundreds of miles . . . We pulled up when we got tired of travelling and it was dark and rolled our swags out, lit a big fire and sat down and yarned.

'You never know what you'd come on up there. In 1931 between Port Hedland and Marble Bar the Shaw River was flooded. Fifteen teams of shearers held up there and we pulled the trucks over the river by ropes.

'We didn't see mail much, about once a month. I was engaged three times but not married. I had a lot of girl friends but they were never there when you got home. You were away too long. You couldn't take a wife on a property in those days. Some said it may make for prostitution.

'Donkey and camel teams carted the wool and tucker up there. Some Afghans, some half-castes, some whites with camel or donkey teams. They'd take a couple of hours to harness them up; have a couple of big strong blackfellows to push them in by the head. About forty donkeys in a team, twenty camels. You'd see them plod off in the direction of the port, their drivers mostly plodding along beside them. Trucks didn't take over until about 1932.

'Half a dozen of the places along the coast couldn't take a ship, the goods had to be lightered out to ships anchored out at sea. Ships could come in to landings at Derby, and Port Hedland, but all the little places had lighters. When it came time to load the wool the station people would send out their blackfellows. The station owners weren't allowed near the blacks' camp and when they wanted them they'd go a hundred yards from the camp and yell.

'There were blacks in the sheds up until recent times. I started shearing up north in 1925. At Noonkanbah black women would come in and sort locks . . . Fifteen to twenty of them sitting there pulling, dragging sheep shit out of the little tufts of wool.

'Wool classers used to have little blackfellows to throw the wool into the various bins for him. Some of Western Australia's best shearers were black. There was one eighty-year-old who shore 200 in a day. He died in the shed that old man. I was shearing with his son.

We had an accordion and mouth organ and we'd sing sometimes and talk a lot and gamble. On Sundays I'd shave fellows for nothing and cut hair for 1s 0d. (my trade was hair-dressing between shearing seasons). And I'd sharpen their cut-throat razors for them.

'There were always boxing gloves hanging up in sheds. Always some boxers in a team. Sometimes you'd get a whole team of boxers and they'd set up a ring in the shed. I was once in a shed with the state heavy-weight champion.

'The quarters were made of tin. As hot in the night as in the day except for some areas, where you get freezing nights inland and then you froze.

'You were sacked if you brought beer onto the property. We'd try to sneak in

The cookhouse at Anna Plains station, like most station buildings in the north, was corrugated iron with domed roof to withstand the cyclones

a few bottles in our cases and then cool it a bit in a bucket of water or cover them with wet bags; we'd stop the bags from drying out in the heat with water dripping from a tin.

'At Liveringa there were two sheds, iron, with domed roofs and wire cables across the walls and fastening between the cement or iron posts to withstand cyclones. I was on De Grey when a cyclone hit. It was a terrific thing. It blew a windmill away but the shed stood: it had been stayed well with cables and powerfully built.

'There were great sheds. At Noonkanbah there were twenty stands, ten each side of the board, 4000 sheep a day were shorn there one year, an average of 200 per man. The expert would set the engine going an hour or so before the bell went, and would we go!

'Les Nolan (a teetotal shearer he was) shore 260 a day at Noonkanbah. Don Young shore 250 and more and that was really going there!

'There was Ellavalla shed that was so hot that if you leant back on an iron post it sizzled your skin. You gained four pounds of a weekend and lost it during the week in the shed.

'I shore at about twelve stations in Kimberley, Mt Anderson, Liveringa, Quanban, Myroodah, all or most of them gone out of sheep now. At Carnarvon area where there were 40–60 000 sheep once there are only 15 000 now.

'Of course a man must not romanticise those days. Conditions were bad. Rations were brought out to us by camel team. Twin Sister tinned butter was oil in that heat; most food was tinned.

'We had dysentery, all sorts. Hygiene very difficult . . . At Ellavalla the water was piped from a mile away and came out boiling. We'd wait until midnight for it to cool enough to wash in . . . In the cooler climates we'd boil the kerosine tin and wash in a tub.

'We had potatoes and onions only for vegetables. No greens. We'd boil milk thistles and that was nearly as good as cabbage; we'd eat the leaves to get iron out of it. It tasted nice; of course we hadn't had green stuff for a long while. We'd boil hops and drink the hot water and it kept the blood in order, and helped with constipation. We took much Epsom salts with us for that.

'Hops was good for yoke boils too. You'd get boils on your legs where you held sheep, especially if you didn't wash your trousers every night. I used to wash [mine]. The yoke comes off easily as lanoline. In that climate pants would dry overnight. I washed my sweat towel every night too. Always wore a flannel I believed that flannel prevented a man getting chills as it absorbed the sweat.

'Well water was bad up there. It gave you scalds in the kidneys from a lot of it. They wouldn't give you fresh. Noonkanbah was all right because the River Fitzroy was fresh, and on De Grey there was fresh water from the De Grey River. Mulga Downs was very bad. Old Hancock's father was an old devil. Lang Hancock was a jackeroo when a lad. You know the rest.'

Lardie Fleay reckoned that it wasn't the wet sheep that hurt a man. 'It's the dry ones that knock a man out. We'd be so tired we'd look forward to the wet! Of course we didn't shear wet. But up north it wasn't the problem it is elsewhere. You have your wet season in January, February up there and that's it. So we didn't have the health problems others complain of from wet sheep.

'A lot went down with malaria in the Kimberleys. Stagnant water in the billabongs bred mosquitos by the squadron. There was a lot of dengue fever

Wool came in from distant stations to jetties on the coast. At Derby, with a tidal variation of nine metres, ships must anchor off at low tide and then it was 'Rush, rush, rush to load as much as possible when the vessel lay alongside at high tide.'

around too. For malaria we took quinine tablets. Lots of men had catarrh. The dust was bad.'

Tommy Fleay of Toodyay is also from one of the best known shearing families in the west. 'I started having a go before I left school. When I left high school I followed in the old man's footsteps. At 17 I shore 184. But Les Nolan could shear 200 when he was 17. I had to wait until 1945 to shear 315 in eight hours at Woodlands station.'

He did 276 grown sheep at Credo, out from Broad Arrow north of Kalgoorlie. 'I shore for twenty years and I wouldn't go shearing if I had my time over. Dad was a great shearer. I felt I had to go. But a kid shouldn't start so young. A boy of 15 isn't set. You work too hard. By the time I was 50 I was done. If I'd taken it steady – but then, you don't. The moment a man goes on the board he's a different bloke.'

At 19, Tommy was averaging 209 a day over a period of five weeks. 'It's a very competitive sport. You set out to shear more than the other blokes.' He shore north to the Murchison, Gascoyne, Kimberleys, over to the Eastern Goldfields and down to the southern properties. 'There was only one thing to be said about it: you met some wonderful men shearing. Shearers, with the exception of one or two contractors I've shorn for, are the finest men you'd ever meet.

'Squatters? You wouldn't pee on them if they were on fire. It's the way the buggers are bred. The big studs from down south, they were too big to say goodday to you. You'd go by with your water-bag in your hand and say Goodday and they wouldn't even see you.

'It made a man go for every little loophole, to get even with them.

'When the niggers went on strike – McLeod's mob – Aubry Hardy and Mr Sainsbury were out there cutting the dags off the sheep themselves. That strike by McLeod's mob of niggers was so well planned that one day they say they are going on strike and the next day they had gone. Those poor buggers would have had to cut the dags off if they were there.

'There's no sheep from Roebourne upwards today – all given back to the niggers that land. I've seen 75 000 sheep at Mt Anderson, 80 000 at De Grey. They can't employ natives nowadays unless they pay them. And they can't get white labour up there. When the old boys were on about a quid a week it was all right, oh yes.

'There was not a horse on Myroodah. Not a horse. All mustering was done by boys on foot. Only one white man on the place. You used to see funny things up there. On Yandeyarra, out from Whim Creek, we saw twenty blacks out mustering on white horses. They (Aborigines) had the odd property, Shelley, Yandeyarra; Don McLeod tried to buy up all the properties along the De Grey for them. But iron ore beat everyone up there.

'I never saw a fight before I went shearing. But in a town, God help any bugger who pushed a shearer. They were a band of brothers. You'd never see a copper around when they hit town. There'd be thirty shearers in a team, two coppers in town, the poor buggers couldn't do much about us!

'Shearers share their money, booze – but in the shed they're racing, terribly competitive. Give anything to beat the other chap by one sheep.

'Contractors are a good mob too, one of the boys. Alf Davis was the best bloke I ever shore for, fair, honest, fought our fights against the squatter if he believed we were in the right but if he didn't he'd press the squatter's view. But off the board he was just one of us. The best.

'Like all contractors he knew that if we slowed down he lost money. Up north conditions were always bad: heat, humidity, dirt, malaria, dengue fever, too far away and too long away from home – well, when he'd see us begin to slow Alf would say, "I'll send for a case of beer". "One case!" we'd snort. "What's the good of that!" "All right. I'll get beer and we'll take the fishing lines and go down to the river for a week."

'A man saw a different life up there. I saw musterers ride by with goannas four to six feet long hanging from their saddle horns – taking it home to share with their nippers for tea. Mustering niggers didn't get paid but got tucker of sorts.

'Men up there had women of course – no white women there. A shearer came to me one night and said had I ever had a black girl? I said no, and he said, "You don't know what you're missing". I reckon more than half the men who went up there were having a bit. Those natives up there were more animal than human. One night one was going to have a baby. She lay down on the ground and had it and caught up with the mob later. You wouldn't see a white woman doing that.

'The buggers were always half wild there. They put on a bloody corroboree one night. I bet you I wasn't the only man needing a change of strides. They were whooping and shouting and stamping round behind us in the spinifex. None of us felt easy about it.

'Up in the Kimberleys you saw blacks all right! Five of them to every white man. I remember King Billy . . . I wouldn't like to tangle with him. You'd see him stalk through the bush, big! and powerful!

'Cooks. Well, we each paid them 8s 6d a week, and mostly they earned it. There was Leo Drummond, the Prince of Cooks he was known as, and Teddy Roberts, real good cooks. But of course there were the others. I remember once, this cook was on the booze a long time. I never ate a hot meal at lunch-time, not after I saw Wilfie, my brother, with a stack of blowflies at the side of his plate.

'The PLB team were a lively lot. They had this chow cook and poor sod, they got at him. They put Double G's [vicious burrs] in his bed, jam in his slippers. The old chow would never retaliate. Paddy, our rep., said to him, "Johnny, we've decided we aren't going to do any more awful things to you." "Why not?" "Because you're a good sport." "You sure? No more Double G's in the bed?" "No." "No more jam in slipper?" "Nothing." "Well, O. K. Paddy," says the chow cook. "I no more piddle in the soup."

'You couldn't blame them for wanting to drink in that climate,' Marcus Synnot says of the shearers who have left the billiard table at the Marble Bar hotel. The tiny town holds the record for the hottest town in Australia

'A lot of poker was played in the sheds. The biggest run I ever had was on the Eastern Goldfields. By the time I got back to Toodyay I'd only spend £1 of what I'd earned. I bought the wife a £10 watch, oh I'd done well! But I've seen poor buggers do their dough in the poker schools. Had to wire the parents for money to get home.

'I saw my first fly-blown sheep up north in 1942. On Myroodah. By Jesus you see them now. They tried everything to eradicate it. I saw a 44-gallon drum cut in half long ways with a dead sheep in it two-thirds covered with arsenic water. For a hundred yards all round the earth was smothered thick with dead blowflies. Underneath the woolly bushes they were piled high on one another. But nothing they did got rid of fly strike. Thirty thousand sheep were dead in one paddock on De Grey. They cleared the sheep out of that country up there.'

Tommy Fleay is an example of the tenet that very few tall men make good shearers. Tommy is 5 foot 7½ inches tall and weighs 64½ kilos. He shore 315 at Goodlands in 1945. In 1942 he shore 276 at Credo out from Broad Arrow. He shore 230 a day of Kimberley sheep, day after day. 'They averaged 6 lb wool a sheep, but get down to Mt Welcome, Roebourne, and you were taking 10 lb of wool off them. I shore right through from Port Hedland, Whim Creek, Yandeyarra in 1942 when Tranquil Star won the Caulfield Cup,' is the way he reckons the periods when he shore best. He shore 98 lambs in a two-hour run and 101 lambs in another two hours.

'Harry Findlay was a great shearer. His family would go with him, son Ted as shed hand, wife Nora as cook. They had a happy acceptance of life and everyone accepted their attitude and gave them credit when they needed it because they always paid when they were in the money. But the government railways. They refused them. They were coming home to Toodyay, been away shearing for months and they always came home the quickest way they could after cut-out. This time they hadn't got their cheques so they got on the railway station and had only enough cash to get them to Clackline – that's a hell of a distance from Toodyay, but they just said to the station master, "Give us tick, we always pay. Everyone knows that." Well, of course railways wouldn't give anyone tick. So they went as far as Clackline and phoned for a taxi to take them home to Toodyay. They booked up the taxi. Now, the cost of that taxi would have been a thing of wonder for most people but when they got home all they said was, "The railways wouldn't give us tick!" We all thought the bloody railways were mad. Everyone knows the Findlays always pay. It was only that they had to work again for a bit to make up for what they lost on the way home.

'Harry Findlay had begun shearing in 1919 when he was aged 16. His best tally was 304 in 8 hours at Edagee Station, Carnarvon, in 1924. I later saw him shear 303. A beautiful shearer.

'Toodyay was the home of shearers and shearers just had to go off. It was an accepted fact of life. Wives mightn't like it, but that's how it was.'

Syd Sinclair, who owns Leatherhead, near Toodyay, is what the Western Australian shearers call 'a cocky', a small farmer. (Eastern statesmen call any sheep man a cocky.) He's seen his friends from nearby Toodyay go off shearing each year for the best part of his life. 'It's a very competitive atmosphere. You see a shearer duck his head and before you know it, half the men on the board are after him and so the tallies roll up.

'They're all adaptable, used to shearing all types of sheep. I once bought sheep from the drylands where the animals were fed on mulga because there was little or no ground feed and they were looking into the air all the time for food. "Stick an axe into a tree," the shearers said, "and they'll come running."

'At Toodyay the locals helped a neighbour out with his shearing recently and so much help was offered that it was possible to get drunk three times in the one day. The keg was flowing. Men were taking over from one another, "Leave your tools." We had a barn dance at night.'

Most of the loners who shear by themselves, non-union, are in for the quick money and early retirement or investment in another life style. Richard Cotterill of Leonora is aged 24 and shears as far as Wiluna to the north-west of his home and Kalgoorlie to the south. He shears stragglers, those sheep not found on the muster for the main shearing, and does crutching when the numbers are too small to bring in shearers from a distance. He is a very handy man to have in a district. 'Food and petrol are usually thrown in so I make as much as if I was working on a team.'

He learned to shear, as most do, finishing a shearer's sheep after the bell (barrowing) and once, when he was jackerooing, 'The boss taught me. I prefer a small board. Just myself and perhaps one other shearer. There is less pressure. In a team there is always competition. You get swept along with it. Working alone you go your own pace. Usually on Fridays I knock off at 3 p.m. I average 150 a day and get $67 per hundred. I've done 200 often but there's no point in it.'

He thinks he'll give it up in five or ten years. 'I don't want to shear all my life. Maybe I'll get a place of my own.'

Like most Western Australian shearers, his rouseabout hand has been punished by the iron-hard grass seeds of the arid lands. 'I got an infection in the rips and tears in my hand once and was off for two weeks.' His right hand is calloused from the machine. With other shearers from his state he believes 'eastern staters are much more militant than we are. We're lazy'.

Alex Bayliss shore at Mundabullangana in 1918, travelling north by ship. Bayliss believed that sheep up in those high latitudes could die after shearing. 'You'd see them go out of the shed, pink all over after shearing. You'd know if they got sunburnt they'd die.' Like other western shearers of the time Bayliss used T.U.S. [Trades Union Shears] or B & B [Bell and Ballantynes]. All men who shore at Mundabullangana recall the donkeys. 'There were 800 to 900 donkeys and mules on Mundabullangana. The owner had brought a jack out from Spain and bred fine animals, none finer in Australia, and they were used for most of the station work. They didn't need the food that must be carried for a horse, so they could be kept out on muster, for instance, weeks at a time.

48

The Guns of Liveringa

Jack Sorenson, the Shearers' Bard of Western Australia (1907–1949) was born on the Kalgoorlie goldfields in 1907. He later became state welterweight champion, some say to win respect as a poet!

As a youth he shore with teams through the Gascoyne, Murchison and Kimberleys. He wrote for his peers, the men who understood rollicking ballads before the day when such verse became unfashionable. They were simple, descriptive and Australian.

Sorenson's poems stir memories in all the old shearers who went north in the days when sheep were thick there. Years after his death, shearers in the countryside of Western Australia quote his lines, 'The Tar Boys and Learners, The Guns of the Liveringa Team'.

Sorenson dedicated 'The Lost Shearing Team' to the shearers that sailed from Port Hedland on the ill-fated *Koombana* in March 1912. He tells of the team setting off from Port Hedland for Derby, from where they would set off overland for Liveringa station.

The driving rain on the waters fell,
As she battled down the stream,
While comrades waved a last farewell
To the Liveringa Team.
From far off Queensland, Lockwood came
(a bladesman gun was he)
To shear more sheep and earn new fame
In Western Kimberley.
And Grand Old Lawrence too was there,
He realised his dream
When he took his sons to learn to shear
With the Liveringa Team.

'A phantom ship glides down the sound . . .'

E.L.Mitchell

The ship struck the 'Nor-West blow' and now, 'Somewhere out where setting suns/Paint the blue with crimson gleam/Lie Tar boys, Learners and The Guns of the Liveringa Team'.

The poet continues with the legend:

Ere dawn when the vapours hover round
The swamps by the Derby pier,
A Phantom ship glides down the Sound
To quietly tie up there.
And ere the silent, sleeping town
Is roused by the sun's first beam
Out east, where the long, brown tracks go down,
Troop the Liveringa Team.'

Most of the old shearers and contractors have shorn at Liveringa. A 'motor track' and the Fitzroy River divide it from the next station, Myroodah. When the Liveringa team was lost, the *Koombana* had left the wharf at Port Hedland and headed out to sea hoping to ride the cyclone out: it is possible to manoeuvre at sea, whereas at a wharf a ship is at the mercy of the willy-willy. Marcus Synnot has been caught in several of the cyclones that regularly lash the north-west.

'One year at Derby we were leaving when a willy-willy came. Our trucks were on board. We were swept away, almost reached the Indies. We ran out of tucker. There were 28 luggers identified, wrecked, dismasted.

'Still, we were used to ships on our runs. A common call in the sheds when the sheep were poor was, "Where's the boat?" And when the tucker was bad, "When's the boat due?"'

49

We Worked all Right!

Frank Welsh has a lifetime of northern living.

'We can see some reasons for sheep leaving the north, but we're told a lot more by people who were never up here. We came under fire from the Agriculture Department for burning at the wrong time of the year. Supposed to burn in November they said.

'Our sheep were mustered by fire in May, June or July . . . we used to change our paddocks round like sheep farmers in many places. Burning was good for regeneration because after three years spinifex became unpalatable to stock. We always had a quarter of our paddocks shut up. Like eucalypt forests and other Australian growth, a fire through spinifex seemed to send up a rich growth that rain couldn't exceed.

'We worked all right! I started at Talga when I was fifteen. Jackaroo? Oh no! Nothing of the sort. I worked my way up to salting meat and I can tell you it wasn't fun on those freezing nights with your hands in that brine. And now the sheep have gone . . .

'I blame management, particularly on the [Aboriginal] Reservations, because they didn't have staff to hunt the dogs and keep them down. At the last we couldn't afford even one man – we used to have six or seven men after dogs once.

'But the hundreds of men on the Reservations could at least have kept the dogs down on their land, instead of which they ate the sheep and let the dogs in and then the same dogs inundated us.'

Talk all over the north-west is of dingoes – 'dogs'. They put strychnine baits out, 'but they are so cunning they can tell what is bait and what isn't. There was a three-legged dingo – saw his tracks often, but do you think we could catch him? Not a chance. They drove sheep into a corner. They don't have to kill lambs, they only have to spread panic for us to lose lambs because if the lamb is separated from its mother they mis-mother and die in a few hours. Lambing percentages are very low up here owing to conditions – weather, feed etc. – and you can lose your whole new flock of lambs in a night. Many settlers have.'

Men like Frank can look back and see clearly how they managed properties. 'You don't work ahead to shearing. From knowing your shearing date you begin to work backwards. It is a campaign with strategy and tactics. Preparing yards, shed and quarters, ordering supplies and getting dipping and other activities arranged. Mustering takes much planning if you've upwards of an area the size of Belgium or Holland to find and bring sheep in from.

'You can only drove sheep so far each day; and you can't get them all into one paddock near the shed at one time. Shearing might take several weeks or months and what are the sheep to live on if they're crowded into one paddock? If you had the men you'd have three blokes still out mustering when the shearers arrived. You'd have sheep moving in from every corner of the property, all planned to arrive at different times. But you must plan to get the first lot in two days before shearers were due and then allocate two days for unforeseen holdups. But it's always a tight schedule.

'Then, after the sheep are shorn, they have to be driven back to the far-out paddocks to clear the way for the sheep that keep coming out from the factory that the shearing shed is. You drive them in carefully planned routes; after all, you can't have them eat out the food the next lot will need on their way in.'

Now living with his goats and a rugged old pony for company near Comet Mine (named after Halley's Comet that flashed across the sky in April 1910), Bill Smith shore for fifty-four of his seventy years. 'Once, there were sheep all the way down from Broome. There's scarcely a station now – except for Pardoo, Boodarie, perhaps another one or two. I shore sheep left over after shearing . . . they'd run the stragglers down later and get them to the nearest well, and we'd blade shear them mostly. Sometimes we had Pineapples, portable machines, and we'd take these and hook them up to the nearest tree where they'd rounded up a few woolly backs.'

Bill was dogging from 1958 to 1968. 'Four dollars a scalp. Better than shearing any time.' Sheep up there were beaten by poor management, economics, dogs and drought. You hear it from all corners. 'And the few hanging on will be beaten too.'

All round the Comet Mine area and Shaw River they used to truck out boatloads of wool, but no longer. 'You can blame dingoes as much as anything. There were so many station hands those days you could keep the dogs down. And blacks on walkabout would go after dingoes. They'd get a bag of flour for their scalps.

'Once I saw blacks bring seventy-six pups in from the edge of the desert. They used to put a little girl down a dingo hole to bring out the pups in her arms – they pulled her out by the legs.'

Les Clayton, typical of many men in the north, has worked as a pearler, shearer, station hand, dogger, commando during World War II, and he now 'lends a hand' at Boodarie, where the Hardies are 'hanging on'. 'There's lots of "hanging on" in these areas,' he reckons. '1980. We had cyclone Amy, Dean and Enid, all within the first two months of the year. Three cyclones in a row. It's unusual, but what do you do?

'Dean left three foot of water in the shearers' quarters, and two foot over much of the land sheep would travel. Moorella on the Ashburton lost the roof off the woolshed and pens, any number of things, and shearing was on and the old man and the expert got it up in twenty-four hours ready for shearing. Their

Loading wool at Mallina. 'There was no easy way to do anything.'

fences had gone, windmill gone to hell in the 160 km winds, lost a lot of their sheep, drowned.'

Hardie lost his shed in Dean. 'That gave me a headache. I had a great problem with the shearers. "Cold water", they complained. I couldn't get at the pipes underground. Next night they complained the water was too hot. "What in hell do you want then?" their contractor told them. We were doing our best. No roof. Our holding pens were left but no holding yards. In twenty-five years we've had two big floods like Dean sent us.'

The Hardies on Boodarie are perhaps the last property to muster by horse and fire: the 'boss' and his 'men' (Hardie and his family) spread across the huge paddocks, 'some more than ten miles square', and when each musterer comes on a mob of sheep, he or she sets fire to the spinifex. The smoke lets one of the others know to come and drive the sheep to the gathering point. 'You can't muster and drive too.'

In 1980 Jenny Hardie watched her husband ride home at the head of a line of eight horsemen and women, mostly family, after the big muster. 'I stood at the kitchen door with tears rolling down my cheeks. I thought, "perhaps this is the last time Australia will see this". They came riding in, dirty, tired, slumped in the saddle from weeks in the spinifex; they'd been in the saddle each morning by 6 a.m. and on it until 4 p.m. They're big paddocks out here. This is the last of the horse and fire mustered properties, and this may be the last time we can do it. So rare. Yet once done all through the spinifex lands.

'Bikes are hopeless. The terrain doesn't suit bikes. But now . . . it seemed so sad to see them ride in after a successful muster and know it was the end of a tradition.'

Further out from Boodari, at Mundabullangana station, Ian McTaggart, a veteran sheepman who also ran properties in South Australia, uses motor bikes (although Ian had introduced push-bikes for mustering in South Australia earlier this century). 'It's the end of the stony country at Munda. There is a ten-stand shed there. Gordon Craig had it built in 1927. Jacobis, the Port Hedland builders, put nine tons of iron nails in it.

'You've got to build to withstand cyclones in this country. And to defeat distance. The shed is in the centre of the station with small holding paddocks and they can bring the sheep in and send them out after shearing without any problem.'

Lighters used to follow ships up and down the coast, ferrying wool out from the properties in the Pilbara. At Mundabullangana the wool was shipped in the grease. It was carted ten kilometres to the coast, then taken by boat to Forrestier Island where it was stored in a large iron building until the wool ship arrived to take it to London. 'Bucolic', writing in 1889, said: 'The schooner *Mira*, formerly used for pearling, is now kept for lightering the wool over.' In that year, the clip exceeded 800 bales. The previous year, Mundabullangana wool had fetched up to 8½ pence per pound, an average for the district.

At Yarrie, out from Marble Bar, the Coppins would muster for six weeks of the year. Jean Welsh, then married to the late Owen Coppin, saw them go off each year. 'Owen would take seven or eight men. It took a day to get the horses in to shoe them as they'd be in rough country. They used three horses per man. In the back country they used to be on camels – those animals aren't afraid of wire, they'd put their feet through any fence.'

Frank Welsh, who managed several properties, mustered often. 'Mid-May it's a wonderful life. Wake at 3.30 a.m. and look for the two pointers of the Southern Cross, one over the other, and call "Horse! Boy!" He'd set fire to a spinifex tussock to let you know he heard. The horses were hobbled and belled. The boy would bring them in and put the hobbles back on while he fed them. We always took a trailer load of feed out for the horses. If you feed them, they'll come back to camp for feed.

'At Talga there'd be eight or nine blacks. Only had to feed them and give them boots and clothing. Sometimes there'd be a gin to cook for you. Other times you'd take a big stew out in the camp-oven.

'Mid-week you'd put a sheep in the camp-oven at midday. Make a fire with good-burning timber, rake out the coal, put sand over the bottom coals and then pack the oven round with hot sand. It cooks beautifully.

'The whites and blacks always had separate camps.

'At Warrawagine fifteen to twenty-five men were out mustering. They always had a cook and camp crew and killed a couple of sheep a night. We took onions and potatoes, jam, flour, tea, sugar, and sometimes we took porridge and tinned milk with us; enamel mugs and quart pots for tea.

'On Talga we made our own dampers. We had corned beef or corned mutton of an evening and breakfast was cold corned mutton or beef plus bread or jam; lunch was the same with damper and jam.

'We carried a bran bag full of common salt to dry salt meat. That meat was like strips of leather. I'd take the slabs out of the oat-bag after tea of a night and wash the salt out of the meat and flex it with my hands in water. Then put the slabs in a kerosene tin of water to boil for 1½ hours. The boss of the muster had to stay up until the salted meat cooked. Everyone else was in bed by 6.30 p.m. in winter, and the next night you repeated the performance and put "another big mob" of meat on to cook.'

'Warrawagine was 506 250 hectares (1 250 000 acres). It took three months to muster. One hundred and sixty kilometres from boundary to boundary it was, with 74 000 sheep between boundaries. Some sheep were being brought in 110 kilometres to be shorn. But when lambs are born during mustering, that blocks the lead, as the lambs hold them back.

'On the Pilbara all mustering was done by fire. Sometimes all black musterers except for the boss. The blacks have fantastic eyesight and they watch the tracks.

'We took a swag each, a piece of canvas and three blankets. At nights we'd put the canvas down so the spinifex prickles would not come through the blankets. A "woman's swag" was one of four blankets and a "married man's swag" was one with a mattress. The easterly wind could be bad of nights, so we'd build a wall a metre high of Bull spinifex on the easterly side of our camp as a breakwall.'

1910, and the camel team that will take the wool away passes through Marble Bar

Frank Welsh has been associated with mustering by fire 'ever since I began to walk. I was a full-grown man before I knew other manners of mustering. Every morning we'd hand each blackfellow a box of matches. "How's your matches?" "Finish boss." You'd give him another box. The blackfellows loved throwing matches into a spinifex clump and watching the beautiful flame fly up.'

Frank believes that mustering by fire gives a completely clean muster. 'You don't miss any this way. Every man had a whip and he'd crack it every four or five minutes. If he couldn't see round a hill or gully he'd crack his whip and the sheep would go like steam. They don't usually hear any man-made sound out there you know.

'Mustering by fire works like this. The boss takes up his stand in the middle of the paddock (the largest paddock I used to muster was twenty by twenty-nine kilometres). Some paddocks take three days to clear up but that big one, oh, you'd get it done in four days if the wind was right.

'Sheep won't go with the wind. Sheep run against the wind. The boss takes the inside posting and men are placed around the fence. Let's say fenceman number one puts up a smoke (lights the spinifex) to let others know he has found a mob of sheep. Those sheep will be handed over to another and across to another, perhaps to four or six men, until they reach the middle or the point from where we can begin to move them to where we want them, such as the shearing shed. By now, other sheep are scared by the smoke and begin to move, more smokes go up as other men send their messages in.

1927, and the little truck is coming in from Bonnie Downs

'The men kept handing their sheep over to drovers and going back to muster more. You can't drove sheep and muster them at the same time. Homestead people could see the smokes across the horizon. "They're in New Paddock today", or "they're a day ahead of themselves. They didn't think they'd get to that paddock before tomorrow". It was like semaphoring when you put up a smoke.

'Everyone I knew as a kid mustered this way. It was unique to the Pilbara. It wouldn't work anywhere else. Later, new methods were used, but horses knocked up in that terrain, men got cranky and there were too many stragglers found after shearing was finished – a few hundred might be excusable but thousands! That's what some had.

'Of course, in these regions it's not possible to know how many head of stock you've got on a place. One sheep-man I knew overstated by 2000 the number he would have for shearing. The contractor was reasonable about it. He understood conditions out here.'

50

The Squatting Company

The history of the north of Western Australia is different from the south, different from the rest of Australia. Apart from the overlanders, some set off by ship from other colonies where the best land had already been taken up. The *Maria Ross* was chartered by eight members of the Portland West Australian Squatting Company Ltd and, on 6 March 1865, with 'all sheep and eight horses', sailed from Portland, Victoria. By 25 March they were outside Fremantle Harbour and day by slow day they sailed north until 2 April. 'Constructed a raft and landed eight horses and waggon. Men belonging to Ship *Tien Tsin* helping us to land...' It had taken one month, a less wearying journey than overlanding would have been, but still a great gamble.

The journal written by Elliot Richardson, one of the eight men at the company's original meeting, lists the runs taken up: 'Block on George River – called Pyramid station; block on Ashburton; two blocks on Fortesque; one block north end of Pyramid.' By the end of 1868 there were forty-nine runs.'[1]

'Aborigines were both troublesome and numeorus,' wrote Battye. 'More treacherous even than those of the South West. Water was scarce except along the rivers and owing to the absence of fences the sheep had to be shepherded. The sheep were generally of a poor class, light fleeces, generally dirty and dusty; the heat in those latitudes was trying to work in.' Any man who persevered with what they encountered deserved to win. They were dependent entirely on small schooners of less than 100 tons which traded more or less irregularly along the coast, forming the only means of communication and supply. Stock losses were enormous.

While the north-west was being settled, an attempt was made to open up the West Kimberley country farther to the north. By 1879, John and Alexander Forrest had explored the Kimberleys and overlanders followed. The flocks increased until the disastrous drought of 1910 began and continued until 1915. Many flocks were practically wiped out. World War I seriously dislocated the wool market.

In the *West Australian* of 6 December 1889, 'Bucolic' wrote of Mundabullangana in the Pilbara, a property of the McKay brothers on the Yule River, 'one of the many instances of showing what may be accomplished from small beginnings by industrious pioneers: This station was selected and stocked in 1872 by Donald McKay and his brother R. L. McKay. [R. L. McKay was lost when the *Ariel* disappeared between Cossack and the pearling grounds in a cyclone in 1886.] . . . Owing to so much new country being settled, sheep were scarce and costly and I remember well their first little flock of 27 ewes, the survivors of 200 shipped from South Australia in 1867. These were added to from time to time until the flock had increased to over 2600 which were pastured on the Maitland when the memorable willy-willy [cyclone] of 1872 burst upon them and carried off 1400 of the flock. This loss was a terrible blow to the brothers . . .

'They moved over to the Yule and settled at Mundabullangana where there is . . . permanent water. As stock increased they added to their runs by selection and purchase until the area reached 1 100 000 acres. About 200 000 acres are now [1889] enclosed with sheep-proof wire fencing and subdivided into ten paddocks, 27 wells.'

Perhaps the small, home-made notebook kept by A. W. Edgar of the De Grey station in the 1880s gives us some feeling for those incredible pioneering days.[2] The outer cover was made from an old water-bag stiffened with several sheets of old English newspaper that had been pasted to the canvas cover with what appears to be flour and water paste. Page 2 gives the numbers of sheep shorn each year from 1879 until the depression years of the 1890s.

1879	27662	1884	44123	1889	54914	
1880	37512	1885	50906	1890	74106	Peak
1881	42347	1886	58600	1891	58039	
1882	30800	1887	62691	1892	33084	Depression
1883	35005	1888	69220	1893	44036	

On page 59 he lists 1085 sheep being shorn as stragglers in 1884, an indication of size of property and difficulty in mustering, and page 60 notes: '16 sheep sold to pearlers'.

One of the Portland men, A. R. Richardson, took up Boodarie, in the 1870s, near what is now Port Hedland. In 1881 his brother-in-law, J. P. Wellard, was managing the property for him. 'We have a spinifex saloon', he wrote, 'to live in in fine weather, and a thatched hut to go in when it rains.' The size of the kingdom was such that three other stations later emerged from this first property: Indee, Wallareenya and Pippingarra, and still Boodarie had 243 000 hectares in 1901 when the Hardie brothers took over.

Today the Hardies are battling. They took three cyclones in two months in 1979. After the third 'blow' their sheep numbers had dropped from 12 000 to 8000, mostly through drowning. They breed merinos and 'mickey rams. It's survival of the fittest here'.

Jennie Hardie, the Port Hedland historian, says that when the Portland men explored the land and took up their runs, 'the coastal plains were covered with salt-bush and bluebush, perennial grasses like weeping grass, woolly butt, bundle bundle and kangaroo grass, all highly nutritious and all producing good stock and high lambings. Plus mile on mile of spinifex, both palatable and unpalatable, of low food value, but providing survival for sheep in poor years.

Mundabullangana polo team, 1898

'By the 1920s much of these nutritious grasses had been eaten out. Lambing figures were low. A grim situation was saved by buffel grass with its high protein value and kapok. Buffel came in with camel fodder. (We were bringing camels into Australia until 1909 and Port Hedland was served for many years by this transport.) Kapok was introduced by the Malays and Koepanjers whose mattresses were stuffed with it. In the 1920s, Mundabullangana had an aerial sewing of the seed, one of the earliest records of planes for that use. Buffel spread quickly and was a great addition to pasture.

'But then came the depression of the 1920s and 1930s and the drought, and the banks had to abandon many leases. Up here they speak of being "stumped up". "The banks stumped up and financed him." Or, "I was stumped up by so-and-so".'

The Pastoralists' Association was formed in Western Australia in 1907. Members were to hold 1000 sheep or 100 'large' stock. In March 1909 the Pastoralists' Association and the Shearers' Union met and fixed shearing rates, the first occasion in the history of the colony or state on which a common understanding had been arrived at between the two bodies. In 1912 another agreement was signed, this time with the AWU.

Figures from the Bureau of Statistics for 1973 show the number of sheep in the West Pilbara to have been 441 571. By 1975 the number was reduced to 434 235. In the East Pilbara in the same years the number had dropped from 167 573 sheep to 126 157. 'Each year we hear of another one or two properties moving out of sheep because the economics and the dogs have beaten them,' Jenny Hardie wrote.

This quite remarkable woman (remarkable for reasons that have no place in this book) wrote in 1976 about the preparations for shearing. 'For these properties the whole pattern of shearing as it was done in the days of the old-timers has completely altered. Gone are the big gangs of horsemen living under tents and taking months to gather the flocks. It is the era of the aeroplane musterer, the motorbike men and the walkie-talkie radio. Mustering is done in weeks now, not months.

'Over the past few years we here at Boodarie have tried the motorbike and the aeroplane, but through various difficulties have once again reverted to what is probably just a remnant of the grandiose methods of the past: an all horse muster.

'Mustering began late this year (1976). When cyclone Joan stopped sending rain, the horsemen and their plant set off. With their oiled gear, washed saddlecloths and girths, their swags, stockholm tar, hobbles, horse bells and big "fly" tent, they moved 40 km southwards to bring down the mob.

'And this year, due to our friend Joan, we have never been so ill-prepared or in such a quandary about sheep numbers. No longer are there mobs of Aboriginal stockmen plus their women in the mustering camp. Just six of them this year. No longer is the camp cook sitting by her open tin-shielded fire, drawing pictures in the sand and watching the "smokes" go up into the blue skies as the men drop matches in the spinifex, sending the sheep homeward.

'Those "yang-yang" horses which were always left in the camp for some new chum stockman to ride do not exist. No passengers in the plant these days. Stock-horses here earn their oats and are carefully pampered to evade saddle sores and girth galls.

Aboriginal polo team, Mundabullangana, 1898

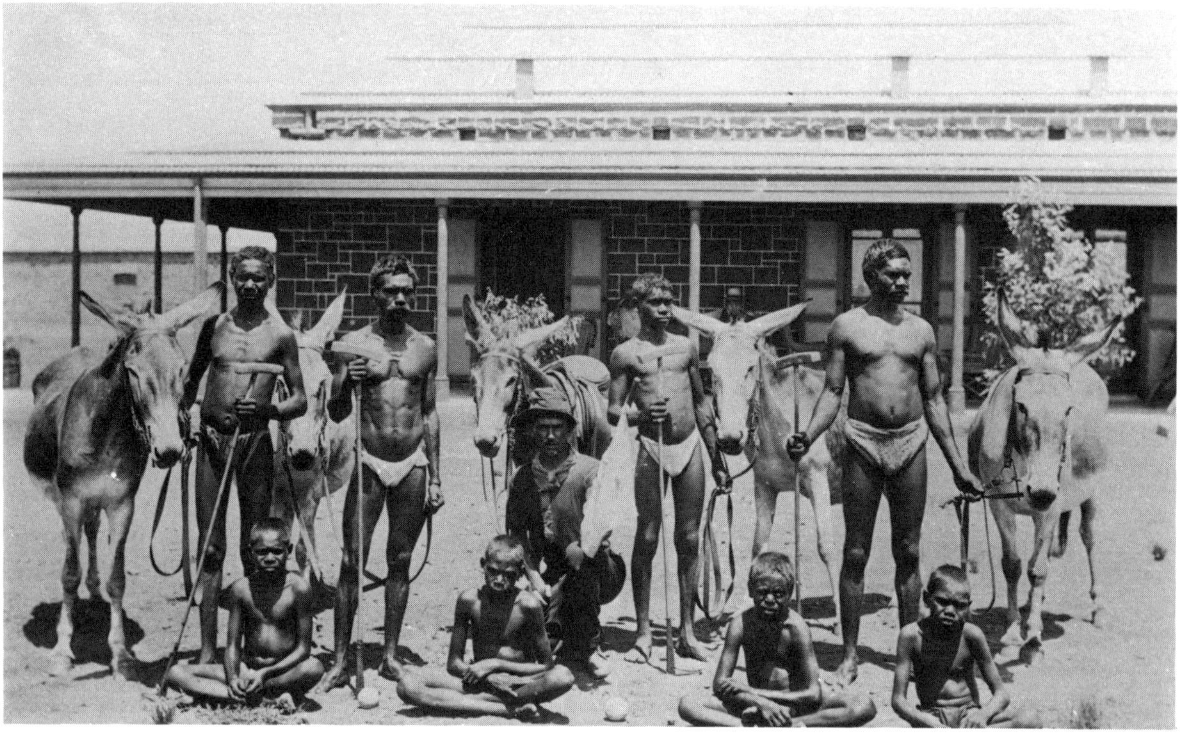

'For the "Mrs" back at the homestead those pre-shearing days, when everyone was out in the camp, are no longer like the calm before the storm. She sees shearing from the kitchen window.

'Daily on my kitchen table, there is a whole sheep, from neck to shank, to be cooked and consumed. Out to the camp it goes in boxes, as stews, roasts and cold legs, with the bread, biscuits and the CWA's "Hard Times" cake. There are the eternal spuds to peel, pots to wallop and meals to plate. There are the empty baking dishes and fire-blackened pots to refill and the saddle cloths to wash and clog up the washing machine. Always for tomorrow.

'As the smokes in the sky get closer to the homestead, the shearers' quarters must be sluiced out, dusty pannikins, plates, cutlery and king-sized boilers have to be washed and set out on scrubbed tables for the shearers' cook. Clean covers must be eased over plumped-up mattresses, rubbish bins, cooks' knives, straw brooms found, and kerosene refrigerators and wood stoves cleaned and coaxed back into their annual two-week life.

'Morning comes early at shearing time. Before the sun rises to fade the stars and turn the sky into pastel-coloured tissue paper. First, the hollow clonk of the night horse's bell stops, next the thunder of hooves as the plant is rounded up into the stock-yards. The cleaver on the meat-house block rings out as the chops for breakfast are cut down. The bell goes, the dogs wail, the galahs screech, the engines roar into life. Another day begins.

'With the shearers' arrival come the familiar smells, noises and scenes of shearing.

'Down in the yards there's drafting and dipping, tailing and tallying. And there's dust. There's that shearing-time smell: of lanolin, sheep dip and ammonia all rolled into something strong, but pleasantly burning to the back of the nose. It wafts through the yards, the sheds and quarters, the homestead and even the bedroom, where dusty, oily shirts and trousers are carefully ostracised from routine wash.

'In the shed when the gong goes, all around is noise, action and sweat. The whole shearing operation is hypnotic in its synchronised teamwork.

'Lined up on the board the six curved shearers in black dungarees, singlets and moccasins drag, stretch and bend the lifeless lumps of woolly sheep. With long rhythmical sweeps of their hand-pieces, they peel back the dusty fleeces to reveal white over translucent pink bodies that, when finished, scramble up into sheep again and leap through the race into the sun.

'The rouseabouts run, the fleece pickers pull, the classer sorts and the presser, a moving mass of wool on legs, piles and pushes mountains of wool into the press. With a whirr and a thump it turns out in heavy oblong bales, stamped and branded, that are rolled out the door.

'But even in the shearing shed there are signs of changing times. Where once one heard the clank, clank of the heavy iron-levered press, now it is the burr of the mobile hydraulic one. The muscle and sweat has gone from the presser's job.

'Once ten shearers lined up on the board; now we have six. Women have moved into the shearing shed. They work the wool table; methodically pick fleeces or sprint down the board with straw brooms and fleeces. "The Expert", who made tin suitcases in his spare time between grinding the combs, has gone. Wives and children follow the team.

'At dusk, from the steps of the kitchen door, the dust goes up and smothers the setting sun as the horsemen push out mobs, now naked cream sheep, bleating and baa-ing down to the river for peace.

'The day is done when, silhouetted against the mother-of-pearl sky, the musterers return. Single file. Slumped, head-hung horses. Slumped shoulders, drooped men.

'We in the Pilbara are watching a dying era.'

Lightering Balla Balla wool out to the ship anchored off shore

'The northern contractors are still really good fellows,' according to the woolgrowers. Jenny Hardie has written of Eric Maidment. 'It is a way of life for the Maidment family. Before becoming a contractor, Eric was a shearer and Joyce his wife, was cook. Their four children grew up with the sounds and smells of sheep and never thought of any other means for a living. The men shear and the women work in the shed.

'The second generation of Maidments all married into the industry and have done well at their trade. They own their own homes in the suburbs of Perth and drive comfortable cars. As a family they look to the industry for a living but not for high profits. "We live hard and work hard", said Eric, "but for us the bonus is a way of life we all enjoy."

'He keeps a sharp eye on economies – he carts his bulk-bought stores himself, saving freight, and to cut down on labour be brings revolving wool tables he has designed himself and a portable automatic press. Before leaving for the North-West, Joyce fills her deep freeze with pre-cooked food and this, too, goes on the truck, with a small engine to keep it going while they are in transit. From their garden they bring cartons of fresh tomatoes, zucchinis, carrots, capsicum and rhubarb which are stored in walk-in chillers that are now on most stations.

'Banter about shearers' cooks . . . could never be levelled at Joyce Maidment or her cooking.

'With shearers now bringing their wives, and having so many of his own family in the team, Eric tries to employ girls as shed hands. He finds them keen, interested, reliable and more fastidious in picking fleeces. He has seen big changes in the shearing operations of the North-West since he began contracting. Once his team started its run in the Kimberleys and finished in the Pilbara. What represented conservatively 100 000 sheep for him to shear annually are no longer there. "I haven't a shed east of De Grey," he said ruefully, as he listed stations like Quanbun, Frazier and Nita Downs, Mandora and Lallal, that were once part of his run and have gone out of sheep.'

All pastoral areas in Western Australia are leasehold (government owned). The lease used to be for ninety-nine years and some have been extended. Nowadays the Pastoral Appraisement Board advises on these matters and wool-growers have representatives on this Board.

So many properties have proved to be not viable over the years that Government, under pressure from the larger, more successful farmers, want to do away with them.

There is a property out from Leonora with 2000 sheep running on 38 475 hectares; the family must find work elsewhere to support the property. The official decision today is for such properties to be either added to larger holdings or to be abandoned. But a man's desire to hold land of his own cannot be so easily abandoned. 'It is in the blood,' they say. And that is different from the unspoken statement of the 'squattocracy' who whispered, 'It is in the pocket'. The distance between the two sheepmen is as great as that between a wharf labourer and the head of the Stock Exchange.

The Glenburgh wool goes down

The Glenburgh Wool Goes Down

The ramp outside the woolshed door,
  Holds yet another load;
So yoke the camel team once more,
  And take the waggon road.
The shafters prop, the leaders pull,
  The wheels creak dismally,
And sixty bales of Glenburgh wool
  Roll westward to the sea.

On down the winding dusty track,
  From dawn till close of day:
The punchers shout, the big whips crack,
  While straining camels sway.
By stony plain, by sandhills brown,
  By wattles o'er the lea,
The hard-won wool goes rolling down,
  From Glenburgh to the sea.

A creek to cross, a hill to climb,
  A stretch of sandy track:
They'll haul it through if given time,
  Though a straw would break each back.
So a morning breaks, a bright sun wanes,
  Till a day, then a week, is gone.
Yet with creaking wheels and clinking chains,
  The Glenburgh wool rolls on.

Cool nights of rest while the camels swell,
  As they munch the mulga near:
While the hobble chain, and the doleful bell,
  Will lull the puncher's ear.
Two more long days from Rocky Pool,
  And then Carnarvon town,
So sixty bales of Glenburgh wool,
  From inland heights go down.

–*Jack Sorenson*

51

Forty-eight by Two

The 'Big Country' people did a TV hour on us living up here out from Marble Bar. Called it "Forty-Eight by Two". 48° Celsius by 2 p.m. up in the Yarrie hills and did they sweat! And they had me making scones. I didn't see it. We don't get TV out here.'

'Out here' is Marble Bar, reputedly the most consistently hot point in Australia with a record of 160 consecutive days of more than 37.8°C. (100° Farenheit) and a highest maximum of 49°C. (120.5° F.). It is an area where flocks of rangy, well-woolled sheep once roamed, but by 1979 the only sheep in the district was the pet lamb belonging to the matron of the local hospital.

Jean Welsh came here as the late Owen Coppin's bride in the days when shearers thronged to the Ironclad hotel in Marble Bar on their way out to the sheds. 'There were some good wheel men over here. Paddy the Pig and Stirrup Iron Jack would ride their push-bikes up from Perth. The squatter those days had much more control over the shearers than he does now. But they got on well. Sometimes if they had wet weather the squatter would go out shooting with the shearers (after he'd put the shed hands to work of course!).

'But I have seen in latter years shearers asking could they go riding, or could they help muster. They're not a bad lot.'

The women up here were always the equal of the men . . . never did complain. Even in the beginning. Eleanor Coppin came to Yarrie with her husband, Christopher, in 1874. He was a police trooper but later took up Yarrie. They travelled 375 kilometres from Cossack to Condon in a small coastal steamer with no cabin and then to Yarrie by bullock waggon. Eleanor didn't visit Swan River for twenty-one years, and in that time spent three years without seeing another white woman.

'Her big excitement was the shearing season. It was like a feast; they boiled a sheep in the copper with vegetables and had melons for sweets.'

Every three months supplies were ordered from Perth. 'They'd come by boat to Condon and then by bullock waggon to Yarrie.' Eleanor was remembered for her medicinal remedies, but nothing could save her stepdaughter, Charlotte.

When this girl died the men of the family took the doors off the house and made a coffin for their sister. 'All the Coppins grew up with the dust of the Pilbara in their blood,' Jean believes. 'Four generations of them have lived here.'

Jean's son, Lang Coppin, now runs Yarrie, out from Marble Bar. The Coppins had this property and ran Warrawagine before Owen Coppin died. When Jean was first a bride, Owen Coppin was away from home mustering for six weeks of the year. 'Once a week he'd come back in for tucker. He'd say to fat Lucy, "You look after Missus". And fat Lucy would follow me all round the place saying, "You all right Missus?" She'd wake us at what she called "first fowl". She said once, "I bin wake you h'early."' Jean can mimic the speech of the blacks in a way few can do, their cadences, rises and falls of pitch, but mostly, her skill is with the tone, a use of muscles and cartilage that Europeans do not use in their speech.

'What heroines these pioneer women were,' Daisy Bates wrote of the north-west. 'British women of all classes set their hand to all and every kind of work without loss of prestige, for there is no social bar other than dishonesty or immorality.' Jane, the young wife of McKenzie Grant, one of the Portland men, was living on De Grey station. In 1875 her baby daughter was so ill she tried to take her south by lugger to Fremantle for help, but the baby died. Jane sewed the tiny body in canvas 'and watched its commital to the deep blue Indian Ocean'. Small children suffered from blight that caused blindness, prickly heat, mosquitos which brought malaria and dengue fever, 'bung eyes' and lips and boils. And the weather at no time helped.

Celia Thompson of Warrawagine was a young bride in 1895 when she was trapped by her first cyclone. When the winds and torrential rain eased the next day and she saw the desolation, 'I thought I could cry for a week.' All her precious boxes she had brought up from the south had been unlocked and opened and the contents now were blown everywhere. 'My Eton jacket and petticoats were found two weeks later a mile down the river eaten by white ants.'

Hardships such as these white pioneers knew are like the wounds an old soldier bears: they don't boast about them, their pride in enduring is deeper than that. But the pride can last only so long as the reason for their suffering is accepted by their peers as being necessary, for the good of their country, or honorable.

Like the survivors of World War I, these landed pioneers have taken a terrible beating in the last few years; all their motives have been questioned and often damned. The station owners are fighting a losing battle. Their income is decided once a year by the tap of a hammer in the great wool sales and yearly they see less profit. Their reactions are sometimes hard to understand for those who didn't share their background.

Jean Welsh has lived her life out past the Black Stump, back o' beyond, the other side of the Never Never, 'out where the dead men lie' in the sand and stone that makes up much of the inland. She lived in the land that no-one takes up unless they are brave in a way few of us ever have to think of in our whole lives; tough in a way few women – or few men either – could, or would want to press themselves to be. The Coppins were synonymous with the unyielding spirit that helped people the outback with Europeans. But Jean hates the blacks with a fervour that is only exceeded by her hatred of Don McLeod, the white man who encouraged the station blacks to strike for wages and leave the stations in 1945.

Oldtimer at Comet Mine with Jean and Frank Welsh, both veterans of the dry stony lands of the Marble Bar region

Jean knows the north, it is a part of her. Her years of strength and youth and vigour were given to it. She is the stuff of which many pioneer women are made if they are to survive. She articulates what perhaps almost everyone, landholder or pioneer, up there feels. She is honest. 'When McLeod came, he took the blacks off when the moon was at a certain stage. McLeod thought we'd go phut! So we brought in jackeroos. These new-style jackeroos thought that all a jackeroo had to do was ride out on a white horse with a guitar. Then the blacks came back. McLeod couldn't stop them. A newspaper had that "all Mr Coppin fed the blacks was damper and treacle". And that was a lie. They had plenty of meat and bread. I varied their tucker. Curry, boiled meat, roast, we killed a sheep a day for them. I'll get that man McLeod yet.

'Owen [the late Mr Coppin] wrote, "For every blackfellow we paid we carried his wife or woman and hangers-on and kids."

'The ones who were working we fed because I thought it would be fine to come home from work and get a good feed. I thought, if your mongrel thing saw the six-pennorth scrag neck of mutton my mother had to keep us on. One of these days I'll come face to face with him.

'All blacks marched down the road. I thought they'd all gone. Saw the shower going. Saw fat Lucy. "Me and Bumble not going Missus. We stop with you."

'At Muckin they put their swags on the goats so they wouldn't have to carry them themselves. But that McLeod, he never did a day's work after he joined the boongs.'

Coons, niggers, prostitutes they are now called, replacing the 'blackies' of the past.

The wool-growing families of the north know they are beaten, not by the blacks, unions, shearing rates and cyclones, not by bad management, lack of suitable pasture, dingoes, the mines and foreign capital, but by time itself. Time is out of sorts for the type of life they once led and the industry they worked. And in their rage and heart-hurt and frustration they lash out wildly. All those years their fathers and grandfathers toiled beyond the ramparts of safety or civilisation, and the mothers and grandmothers buried the little children in the rocky sandy paddocks, children who might have lived had they been near a town or doctor; all the years they lived in their sometimes shack-like homes with loneliness, because they had to send their children south to school; all the times they crouched and tried to shelter as the 'big blows' raged around them, lifting roofs and even houses and tossing them aside. But most of all they are sick deep in the heart for old days that will never come again.

It is true that most people at some time regret the passing of an era, but few lived in an era as rich with self-fulfillment as did these north-western people. There are some places in Australia that are the last bastions where human beings can test themselves against the worst conditions thrown at them by nature, and those who survive and prevail are kings in their own land, poor in money, but superior in the tradition of an area.

Yet now they will go. And their traditions, stories, trials and tales of their courage will go with them.

Walk the streets of Port Hedland today and you will know that must be: this bustling place with huge supermarkets and motels and air-conditioned houses, and rich-from-the-mines families who have come here from afar, is alien to anything the wool-growers knew. The newcomers know nothing of the life they have rolled over and swamped. Indeed, some find it quite distasteful, and speak of the inland areas and the people as southern Americans speak of their 'white trash'.

It is not new, unusual or remarkable; it happens in most places at some time that a whole way of life and livelihood changes, but knowing this cannot help those who are affected. In many, it provokes blind, irrational rage. One feels little sorrow for the absentee landlord wailing down south about the drop in property values, but the greatest pity and understanding is due to those who did it the hard way, lived on their land and gave it all the love they had.

52

The Merry Widow

The standard comb is 2½ inches, with three blades. It is not permitted to bend the comb out a bit to drag in extra wool. This union ruling was also the wish of pastoralists until quite recently, when wide combs were introduced into the south of Western Australia. The shearers call such a comb, 'the Merry Widow'.

This area attracts the greatest number of New Zealanders in Australia, and their use of wide combs with four blades has been an unpopular move with the men who have shorn in the tougher, lonelier, harder areas of the state where less money is made because of time spent travelling and poorer and dirtier sheep. Feeling has run high enough in this state where, in the past, shearers have not been as militant as those in eastern states. No group has yet headed into the south to break combs that are far too wide. Their comment is: 'Many or most of these properties are owned by Americans who take the profit out of the country and the sheep are shorn with illegal equipment by men who come to do their two-year stints, make a packet, and head back home to yet another country. They are not men who have tried to make conditions better over the years or who will be here to have to make a living for the rest of their lives as we are. They work in conditions we fought to achieve and they take away our livelihood in a sense, because we used to have southern shearing to do when we finished the tough north. They don't spend their money in our towns, aren't going to live here to build our state up or bring their kids here as population.'

The sheep owners are divided on the issue. At first they were opposed to the wide combs; they said they would take too many teats and pizzles off, and would leave ridges. They would be too wide and awkward with the thick merino wool. On the other hand, they are an advantage with cross-bred sheep. There are tricks in every chink of this trade. 'I've seen shearers stab their wide hand-piece into the boards and break it when a union rep. comes on the board. "Damned thing's no good!" he mutters. I've seen them push the hand-piece down the chute: it must be in the shearer's hand and be shearing for him to be convicted of using the illegal tool. A fine could cost him $20 and his contractor up to $1000.

'The union has flown a light aircraft over, watching for freshly shorn sheep in areas where no union men are working. But some of the big sheds down south have drovers waiting to move sheep out by night. It's a hard thing to police. Some shear under lights at night. They're up to all sorts. But the union is not strong over here as it is in the east.'

The famous old contractor, Marcus Synnot, believes that apart from the danger of slicing off teats and pizzles 'they even cut the jugular if the combs aren't carefully spread out. I'm against them. New Zealanders with their four-pronged cutters brought it in. New Zealand wool is coarse wool. Big bodied sheep and coarse wool, but Australia has the tight merino fleeces. A shearer is inclined to pick up too much skin with the Merry Widow, the cutter isn't throwing over far enough, leaving ridges.

'It does not make for greater speed in good cutting sheep. If a man filled the Merry Widow all day he'd be a tired man at the end of it.'

53

The Eastern Goldfields

In this great contradiction of a country one can fly out onto the drought lands on the edge of the inland deserts, the Eastern Goldfields, and be unable to land because of water on the air-strip.

In 1980 I set off for Leonora, north of Kalgoorlie, about 860 kilometres from Perth, where the five-year drought had decimated the sheep population, only to look down from the plane at 15 centimetres of water on the tiny strip. It had been there for three weeks but because of the flat land it had nowhere to go. We flew on to Laverton, further out again, and landed on their 'all-weather' strip.

Out here the names are harsh, like the country: Carnage, Siberia, Black Flag, Canegrass, Niagara. Wedge-tailed eagles hang-glide, well fed from the carcasses of animals killed by drought or dingoes; camels, emus and kangaroos are thick on the ground in the acres of bottles that are all that are left of the old goldfields.

'We're not all from the Melbourne Club,' Don Murray says. Don's people managed Mt Eba station (SA). 'Squatters they called us. Well, my people toiled.'

Laverton's only claim to fame in recent years is that it is the satellite town for Windarra, where the nickel boom began in the 1960s and Poseidon shares became part of the most publicised market in Australia since Broken Hill boomed in the 1890s.

The Palace hotel is now the only hotel in Laverton. 'Definitely no dogs allowed' says one notice. Another declares, 'Drink served only to those decently dressed and clean'. Owner, Canadian Elaine Sturch, had nine years in the shearing sheds as picker-up and later as classer. 'Always I had respect from the men. They knew I was there to work. No language. They were mostly boys as old as my sons.'

Sheep properties round here are large but unprosperous. Mt Weld, of 405 000 hectares (1 000 000 acres), has six stands but only four were working when the shearers arrived because the five-year drought had brought them down to 7000 sheep.

I had been invited to stay with Mr and Mrs David Fitzgerald at Nambi. On his 304 000 hectares there are 13 000 sheep, the famous Bungaree strain from South Australia. 'We don't overstock now. You don't want to under-stock either. A family of [dingoes] got 800 lambs in one night recently. We laid 1080 baits but it doesn't make a dent in them.' David, who did his jackerooing days 'up on the Murchison', is the son of the late Irwin Fitzgerald who was killed down a well on the property when the windlass pinned him down. ('Irwin was a great fellow,' Tom Fleay the veteran shearer of Toodyay had told me. 'He'd shear a sheep for you if you just got one in on the smoko bell, or bring you a drink, anything. A real gentleman.')

David followed the plane to Laverton but couldn't take me home to Nambi 'unless you can stay a week or more' because by that evening the homestead would be isolated by water. So he took me instead to Don and Nuala Murray's property, Clover Downs, out of Leonora. 'Got a visitor for you.' And with no more ceremony than that I was fixed up for my stay. Clover Downs and Mertondale stations belong to the Murrays and together they cover 78 325 hectares and run 4000 sheep, roughly one sheep to twenty hectares. There is blue-bush,

'Nana Murray and our family might look romantic on the camels but that was a small part of living in the hard lands.'

salt-bush, spinifex and sago bush. Some plants hold high protein, others not, but in this mean-rainfall area, these sheep will eat anything when they are hungry.

'There was a drought from 1969 to 1973. On Mertondale we lost all our sheep except about 500. We were going bad. So we went off looking for nickel. We took out two mining licences, read all the books we could get and set off looking for green. We walked miles. Week after week. Found nickel. Staked number 2 claim. But you work for it. That's how we got here. Clover Downs.

'Everyone here of the old families knows about scratching for gold. It's the way we've held our land. Now we hold by nickel.'

Some isolated, small properties couldn't press the wool; load on the track from Eucla is bagged

The early squatters came out and pegged the choice land thinking no-one would take it up. They had no interest in using the country, just held it as though it were a prize. But many did follow and take it up. 'We got Clover Downs in 1975. Then big rains drowned 2000 sheep. Sheep were standing in water all over the place, holding their heads out of water. Snakes were on their backs and mice, and worst of all were the lambs that ate the prickly bush, the only thing sticking up out of the water. It tore their lips off and their faces festered for six inches around the mouth.'

In the beginning the gold men excited a whole country with their epic journeys into the scorching unknown north. The pastoralists followed, and now the miners have returned, the circle completed. Clover Downs was named after the goldfields' clover grown here, and Merton's Reward gold mine became the name for Mertondale.

Everything had to be packed in, even building material, for the trees are few and scrubby. At first on men's backs, then by camels, train and now aeroplane.

But air freight is never cheap and houses are still often made of corrugated iron, because it packs into a smaller package than does timber. The spinifex here is poor feed but is used extensively as insulation in walls and ceilings.

In the 31 728 square kilometre shire, the first pastoral lease taken up in 1925 was on Clover Downs. By then the 1901 population of 8000 had dropped to 316. Jack Keenan no longer sold water from his travelling cart and Mr Booden (who found the first gold and called the spot 'The Little Wonder Mine') had gone.

Australians see and hear of the excesses of 'the squatters', know the notorious tales, part legend, part truth, of the greed of the early land-grabbers, of the hundred families who considered themselves an élite, a society-set mafia who controlled the destinies of people and a country, and they don't like them. But while they rumble about the power and political brokers, they lump all wool-growers in with them: 'The Squatter Class'. As Don Murray says on Clover Downs, 'a small percentage got land the easy way, and a small percentage didn't care for the land but only for the money they thought it would bring them. But the great majority got land the hard way, toiled to buy the leasehold, toiled to hold it. We're not all from the Melbourne Club, Geelong and Melbourne Grammar you know.

'One sheep to twenty hectares. Some land can run no stock at all. Not even one to 4000 hectares. You've got to know the land, feel it. You can strip this land very easily, so you must husband it well.

'Visitors see The Breakaways [low craggy hills] nearly fifty kilometres away. "Still your property?" they ask. Their mouths fall away when you nod. They say, "Don't you get lost? How do you find your sheep?" Well, none of it is easy. You spend two to three weeks mustering. Sometimes you have to drive them five miles to get them in a yard at all.

'When we are going to mules sheep we bring them in from the bush and give them a few days rest. After mulesing we leave them in the paddock where the job is done. Give them a spell before starting them off on the track again.

'And flies. You don't get flies in a dry year. When it's wet, flies are more prevalent. They will blow anything. Throw your saddle over the rail, and the smell of sweat and heat, and they'll blow it.

'As for shearers. They're all right. But then you put them in a shearing team and they become most unreasonable. Shearing, like cooking, makes men go mad.'

They had been 'doing a perish' for a few years again. 'Don't know how even the [dingoes] got a feed. Yes, I do. They ate our sheep. Got 800 lambs. You never saw anything so awful. They were tiny. We'd put 800 ewes out to lamb and when we went back in a few days there they were – the lambs born and dead, broken backs on some but most had been used like toys. Dingoes see the little swinging tail and grab it and toss the lamb in the air for fun. We found the culprits . . . a family. Mother, father and pups. The parents had taken the pups out for a training run. There was hardly any blood on any of the little bodies just tails off, or twisted and backs broken. It would break your heart. And they were to have been our next year's flock.'

As for drought: 'David Fitzgerald of Nambi was driving round the low hills day and night, watching the storm cloud, seeing which way it was heading. Rushes home, grabs the phone, "Have you had it?" he yells. Nuala told him

Eventually the road train arrived

she'd phone when we had an inch. And when we got an inch once the phone was out of order.'

And then the rains came.

'Look at the grass,' Nuala kept saying. Then, 'Oh, I mustn't keep saying that! But look at it! Isn't it green!' In front of the house was what appeared to be an ornamental lake with the house and palm trees mirrored in it and the fringe of green grass like a lawn around it. For the past four years there had been nothing but a dust bowl there. The Murrays go to Leonora for supplies. Today, after the rain, the storekeeper jokes, 'Oh God! Here they come! Now it will be the rain has washed out the shearing, the sheep have the scours, footrot next I suppose.'

Well, yes, the shearing had been washed out, and the shearers, too far from home to return, sat disconsolate, without cash in the quarters. They get paid only when they shear. But the sheep men were laughing with the good season ahead to pay their debts and stake them for another year.

54

The Tally Board

Each morning the previous day's tally of each man must be posted on a board in a conspicuous place in the shed. These morning bulletins command interest from everyone from the 'broomie' to the boss, and there have been some remarkable tallies.

At Barenya station, on 19 September 1895, twenty-six men averaged 172 each; on the same day eight men averaged 236 each; and on the twentieth, twenty-six men averaged 175 each. The record for hand-shearing is held by Jack Howe, who shore 321 ewes in 7 hours 40 minutes at Alice Downs (Qld) in October 1892. His tallies for the last eleven days at the shed were 149, 264, 131, 249, 257, 258, 262, 267, 321, and 190 lambs and 30 wethers. On 16 July 1904, he shore 337 sheep in eight hours with the machine.

Back in 1884 at Belalie, on the Warrego, Sid Ross shore 9 lambs in 9 minutes, and at Evesham in 1886, Jimmy Fisher shore 50 lambs in one run before breakfast (about 75 minutes). At Charlotte Plains, Warrego River, in 1885, Alex Miller shore 4362 sheep in three weeks and three days, an average of 203 per day throughout the shed. 'Long' Maloney shore 22 000 in one season in South Australia, Victoria, and New South Wales. In 1876, Angus M'Innes shore against Jack Gunn, at Parratoo (SA) for £50, when the former shore 180 sandback wethers in one day, and but for a timely interference would have exterminated two boundary riders because they had no more sheep in. At Fowlers Bay in 1874, the lengthy Maloney shore 11 big wethers in 11 minutes, using a pair of 'Ward and Payne's 38s'. At a shed in the Paroo in 1884, seven men – Allan M'Callam, Bamphil (the 'Warbler'), Jack Lynch, Jimmy Donaldson ('Maorilander'), 'Warrigal Jack' M'Donald (the 'Barrier Ringer'), 'Long Bob' Hobbs and Jack Reid (the 'Victorian') – shore 1540 sheep in one day, an average of 220 per man. At Parelli in 1885, another Jack M'Donald, who weighed only 6 stone 9 lb, shore 187 full-fleeced wethers in 7 hours 30 minutes.

A shearer who used the 'tongs' all over the Commonwealth during the 'squabble' between pastoralists and shearers in 1902, 'gave the following particulars of his earnings: In 1899 he shore 23 538 sheep, receiving £235 7s 7d; in

1900, 22 976 for £229 15s 2½d; in 1901, 23 142 for £231 9s 5d; and during the first half of 1902 he shore 10 379, his cheque being £103 15s 10d. This man was a long way below being a champion. A fair team of 30 men can average 100 per day. But bad weather, much travelling, and long breaks between sheds, prevent the majority of fairly smart shearers making enough to carry them through the year. Jimmy Power, who was the champion machine shearer and whose record, made at Barenya in 1904, is 315 sheep in 7 hours 40 minutes, shore 40 000 in one season. This record stood until 1909 when Dan Cooper shore 316 sheep in an eight hour day at Bundoran (Qld).

The first hundred full-grown sheep to be put through in a day goes to the credit of Alec Hutchinson at Dunlop (NSW) in 1888, the first shed in the world to put through a complete shearing with machines. The highest tally for the first year of machines, 1888, was Jim Davidson's 173 at Dunlop. Jim was the Grand Old Man of the shearing industry up till his death in 1936.

The first two hundred in a day is credited to S. A. McInnes at Brookong (Qld) in 1889. The first to pass three hundred in a day was Jim Power at Barenya (Qld) in 1893.

Better tallies were made with machines than blades. The tallies are higher in the aggregate, the work is lighter and cleaner, the operator had to find neither shears, stone nor oil, and he never lost any of his rest-time at the grindstone. His only expense, so far as the actual work is concerned, was for combs and cutters.

Harry Livingstone, the Queenslander, shore 223, 225, 237, 237 and 221 in successive days in 1910. The pay was 24s 0d per 100. The same season, 38 men at Cambridge Downs averaged 198½ sheep per man in one day, Harrison, who rung the shed, being top with 265.

Thirty-one shearers at Goonoo Goonoo (NSW) in 1925, averaged 204 sheep per man in 7 hours 20 minutes; the highest tally was 305.

At century-old Darraweit Guim (Vic.) only a shearer's dog moves at smoke-o, 1979

In 1923 at Williamsbury (WA), a team consisting of A. Williams, R. Sawallish, V. Day, F. Leaman, L. Saltmarsh, George Bence, C. Fleming and H. Munro, known on account of their pace as the 'Mad Eight', shore 9167 in the first week and 8770 in the second week (losing one hour through engine trouble). Their highest day's tally was 1806 on 14 September, Sawallish and Day each doing 242, but on another day the same two shearers each reached 250. The sheep shorn averaged 11 lb wool. In 1928 at Dalgety Downs (WA) Bob Sawallish shore 304, and at Three Rivers (WA), Harry Finlay shore 301 in a day.

The Fast Five, at Glenburgh (WA), shore an average of 1102 sheep per day for a week of five days, or a fraction over 220 sheep per man per day. The same team at Dalgety Downs during the same season shore 1173 sheep in eight hours, an average of 234 per man.

Harry Findlay and Norm Richardson knew of other Western Australians with high tallies including:

W. Gray	313	Milly Milly	1928
Harry Findlay	304	Edagee	1924
Snowy Tuckett	304	Noonkanbah	1929
Ray Young	293	(W.A.)	1934
Don Young	290	(W.A.)	1934
Les Nolan	268	Milly Milly	1928
Frank Rowe	266	Milly Milly	1928

Norm Richardson recalled one memorable day in 1932. 'Les and Wally Readey were widely known, notable shearers. On one day the two brothers shore 602 sheep. It was at an outback place . . . no sheds, only yards. So Dad and I said would they give it a go to shear the lot in one day. So they started at 6 a.m.,

Jack Howe's 1892 record stood until 1949; it was bettered yet again in 1950 by W. E. (Ted) Reick (right)

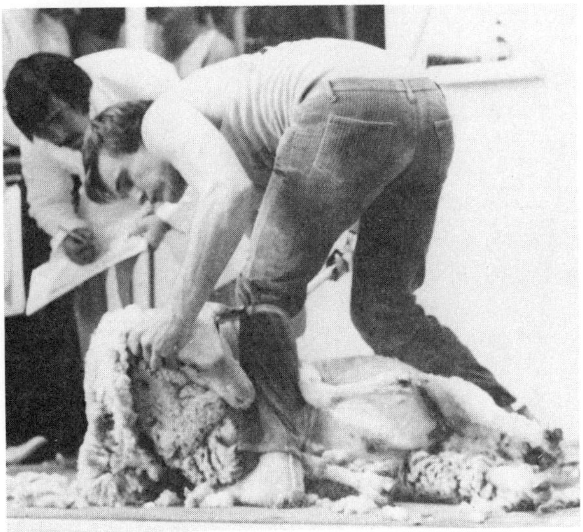

John Conlan

Mark O'Connor

shore to 7.30 a.m. then had breakfast. After that they just broke the day up as they felt like it and by about 6.30 p.m. they had them all done. There was no count-out. We counted them all before starting. No shed. Just a little piece of grating each to stand on was the board.

'Dad and I chucked the wool in an old tin camp, rolled and skirted. The best shearers in the old days included "Ying", Les Nolan, Snowy Tuckett, Boomer Grey and Freddie Byrnes, Roy and Don Young, Joe Herricks. In later days Alby Phillips and Terry Harper were good. The cleanest shearer I knew was "the Singing Shearer", Jim Gibson. He was hard to beat.'

Moffat-Virtue Shears boasted in 1920 of tallies done with their shears:

Murnpeowie station (SA) – 30 shearers shore 20 864 sheep in four days – average 173 sheep per man per day.

Murnpeowie station (SA) – 30 shearers shore 25 688 sheep in six days – average 143 sheep per man per day.

Cordillo station (SA) – 29 shearers shore 21 934 sheep in six days – average 126 sheep per man per day.

Oakwood station (Qld) – 28 shearers shore 20 443 sheep in six days – average 122 sheep per man per day.

Dagworth station (Qld) – 36 shearers shore 26 310 sheep in six days – average 122 sheep per man per day.

Bundoran station (Qld) – shearer, Dan Cooper shore 316 sheep in one day of 8 hrs 40 mins.

Tallies shorn with Moffat-Virtue machines at Curnamona on 9 August 1927 include some of the bicycling Tasmanians. There were only sixteen shearers engaged and two of these were learners who had started at the previous shed.

T. Salisbury	145	F. Jones	164
C. Sims	164	S. Simpson	156
R. Greig	172	J. Heath	241
C. Holloway	178	R. Maroney	252
W. Greig	211	W. Lyon	188
C. Murray	267	W. Stafford	218
D. Marshall	227	A. Greig	153
W. Dordoy	210	E. Jarman	134
		Total:	3080

Brian Morrison turned to contracting in 1965 (as well as producing fine quality wool and sheep skin clothing) after leaving shearing. Before the wool recession of the early 1970s he employed up to 45 men; today he has 12–16 men employed for ten months of the year.

Drysdale was the biggest shed contract he has undertaken – twenty stands. Several years running the 33 000 sheep were cut-out in ten days. Top men enabled this to be achieved. One day the lowest tally shorn out of seven shearers in a row was 224, with several men shearing over 250 per day.

Joe Ives was the best story-teller he met. 'He was quite a good shearer, but a better yarn-spinner, great company, a real character.

'Micky Thomas from Moree (NSW) probably had the biggest reputation in the sheds during my time as a tally shearer. It was claimed he could shear 200 per day in the biggest wethers. People even argued he could do it in Merino rams. He was an extremely big strong man, who dumped the sheep into their smallest position with sheer strength. His style would not have won him many

Longreach team, 1980

Diamond Shears contest, Longreach, 1980

competitions, but it certainly made him a legend in the shearing industry.

'The greatest presser I have ever encountered or for that matter heard about was Bill Chapman, who consistently pressed 60 bales or more on a double-box Ferrier wool press. I saw him press 60 bales until 5 o'clock, come onto the board, shear a sheep for the learner and go down, shower before the shearers had knocked off. He liked to get to a new shed a day early to overhaul the wool press, get the sand bag the exact weight and everything in good working order. He had a perfect system which was followed for each bale.

'In 1963, he pressed for 16 shearers at Bowen Downs (Qld) on his own. Some of the shearers felt two pressers should have been engaged, obviously not realising Bill Chapman's ability. Among themselves they determined to "wool him up", every man shearing the maximum number of sheep possible. Chapman pressed 70 bales, four days running and was not behind. The shearers thereupon acknowledged here was indeed a mighty man, undoubtedly the greatest wool presser ever seen.

'These days, of course, hydraulic power presses are mostly used. The record number of bales is 96 in one day held by Alan "Popeye" Vearing at Drysdale, Victoria in 1969. Popeye now runs a mixed business at Violet Town, having left the shearing industry.'

David Ryan is the new-look shearer. Born at Hamilton in 1946, be began working around the sheds when he was fourteen. He learnt to shear two years later. 'Even then I knew I could shear faster than most others. I knew I could.

'When I first went up north with the shearers, everybody told me to watch out for the ringers because they were the best shearers of all. But I found I could keep up with them. Then I found I could keep ahead of them.' He did that until his thirty-second birthday. 'Then I decided to see how fast I could shear.'

He had a policeman and a Justice of the Peace as witnesses while he went for the record at Spring Valley, David Calvert's property. He shore 120 lambs in the first two hours, 118 in each of the next two 120 minute sessions, and 110 in the last 110 minutes. He had aimed to shear a lamb a minute or 480 for the day but was happy with the record of 466 (the previous record was 421).

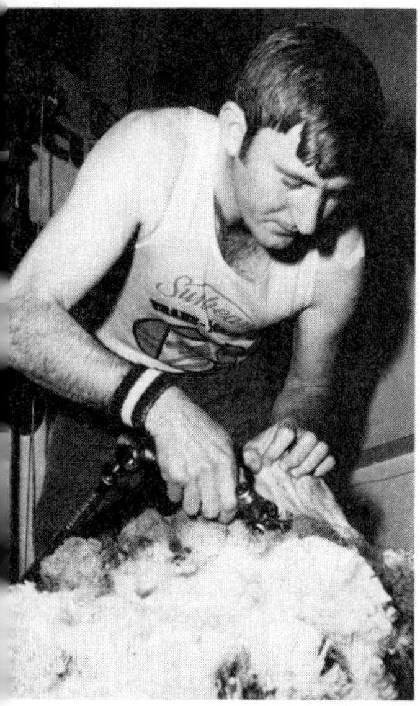

David Ryan, Australian champion

Perhaps no greater demonstration of the difference between blades and machines was seen than at Longreach Diamond Shears Competition in 1981. The machines had whirred all morning, the crowd shouting as each man ran into the pen to pull another sheep out. Then, during lunch-time, the blade shearers came on stage to demonstrate their skill. The audience about to walk off to lunch hesitated, at first unsure what it was that arrested them; then, first one, then another, crept back to their seats and sat in silence to watch the seemingly unhurried bladesmen. The occasional click of the shears only accentuated the vast silence that the sheds used to know before machines shattered the peace.

The bladesmen claim that machines shatter the fibre of the wool. 'Blades cut cleanly.' There are some men who shear only with blades. Peter Clark of Longreach learnt on the show sheep at Terrick Stud. 'Preparing sheep for showing and studs for sale the blades are best by far, you only have to look at the tip of the wool to see that.'

Tasmania has several bladesmen, the Fysh family being the best known. These brothers shore with blades on the 'mainland' for enough years until they saved sufficient money to buy a property for themselves in the midlands of the island state. The Pilley family at Mudgee, father and three sons, have been sought by stud masters for years as has John Hardy, another graduate from Terrick.

Herb. Hutchins from Hamilton shearing blindfolded in 1978. He shore his sheep in 4 minutes 33 seconds. 'A man shears more by touch than by sight.'

'If your back goes you can lay the buggers down with their heads on a bag and you can sit on a kerosene tin and go on with your work!' But there are not many men around who know how to set a sheep up for blades and fewer still who can grind blades. 'As for preparing the knockers, there's hardly anyone about can really get blades ready. The knockers are really only put on to stop the noise of the blades grinding and clicking, and to stop the shears overlapping at the points.' The driver, the leather strap the bladesmen attach to the shears, goes over the hand, and the shearer used it to help push, or drive, the blades through the wool.

Jim Carew saw 'six old bladeys at Denistown, Tasmania in 1929. They were the last of the bladeys to shear an ordinary shed'.

The Hon. E. Chapman, Minister of Agriculture in South Australia (1979) tries this hand at his earlier occupation. He and Kevin Fiebig (now a manager with Sunbeam) both spent their early years on the board

'This fellow might have been a good painter but he was no shearer,' shearers say when viewing Tom Roberts's famous painting, *Shearing the Rams*. 'A man could never have carried a ram out like that. Not even Samson could. By the time you carried a ram out of a pen, and he'd be kicking a bit the way this fellow's holding it, well, you wouldn't feel like shearing after that. See him? He's a full-grown ram. Something like 200 pounds weight!'
Shearing the Rams 1890
Oil on canvas board, 121.9 × 182.6 cm.
Felton Bequest 1932
Reproduced by permission of the National Gallery of Victoria, Melbourne

'Keep him small, keep him under control.'

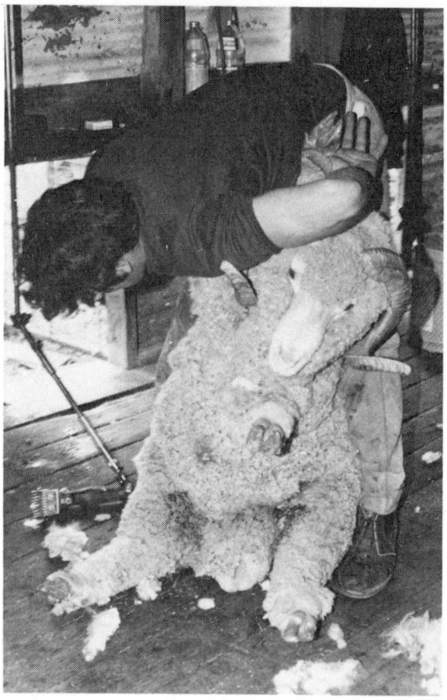

Every movement of the hands, body and feet are stylised to simplify the operation. 'Dump' the animal in as small a space as possible, control him, reach for the machine, pull the starter cord – and you're away

Opening up the belly

The Long Blow

The 'points'

There's so much wool the rouseabout squats near to keep it clear of the shears

The finished job

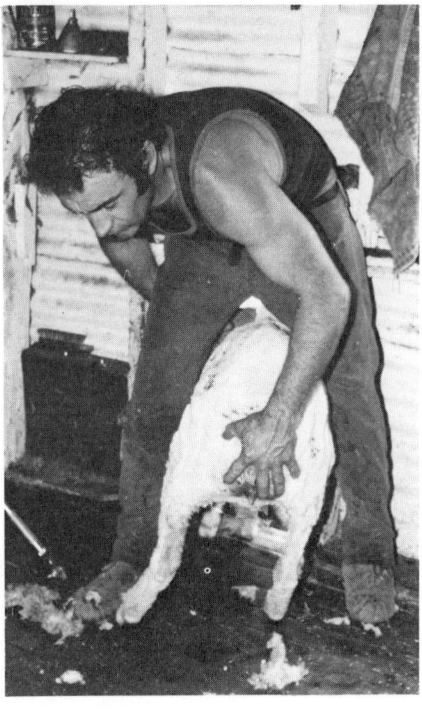

Off the board and down to the counting-out pen

55

Shear Magic!

Australia's fastest growing spectator sport is a surprise, no matter which way you look at it. Who could have expected that shearing, the toughest of outback occupations, would so grab the public interest that they now pack in the onlookers in their hundreds, sometimes thousands; and who would have predicted that women would make up half the audience, or for that matter, that women would pause in their long ball gowns to watch the finals take place in the ball-room, and cheer the shearers as they sweated over the sheep? But then, ninety years ago, who would have imagined shearers turning up dressed for the occasion, well behaved and putting on a show that is exciting and the best of entertainment?

Competitors must be members of AWU and the Australian Federation of Shearing Competitions.

All over Australia shearing competitions are under way. Some are annual events and some are 'once-ers' to raise funds for charity. In many ways it is no surprise at all, for shearing has been going on in this country since the coming of the First Fleet in 1788; and shearers have always lived in a competitive atmosphere and been known as the most colourful of our outback characters.

The whole romance that has grown up around their trade makes for added interest. These men train eight hours a day, five days a week; more intense training than other athletes. So the men who made the legends of the inland take the board today to the cheers of the public.

The highest prize money in the country is the Fabulous Diamond Shears held in central-west Queensland. Here, at Longreach, on the line where the Tropic of Capricorn separates the temperate from the torrid zones, the men with hands like tennis racquets go into action.

Shearers come from all states, as do the judges. Three thousand two hundred spectators pay to enter the large Longreach Town Hall where 102 'guns' compete for prize money of £2500 and an unset diamond. The term 'gun' shearer varies a little in each state, but generally it means a man who can shear over 200 a day, week in, week out. A 'ringer' is one who 'rings' or tops the tally in a shed.

A man was happy enough with a tally of 100 a day once:

> 'I'm out to get my hundred,
> A century or the sack!'

sang the mythical Widgeoara Joe. But then the famous Jack Howe, at Alice Downs station on 10 October 1892, shore 321 sheep in a day using blade shears, a record that has never been broken in the world with blades. In the same year he shore 237 merinos in a day with machines on Barcaldine Downs. (His blade shearing record of 321 was last broken in 1950 with machines when W. E. Reick shore 326 merinos at Brinard (Qld) in an eight hour day.) Today David Ryan, the thirty-four year old Vietnam veteran of Hamilton (Vic.), holds the record of 501 lambs shorn in 7 hours 40 minutes, on 23 August 1979.

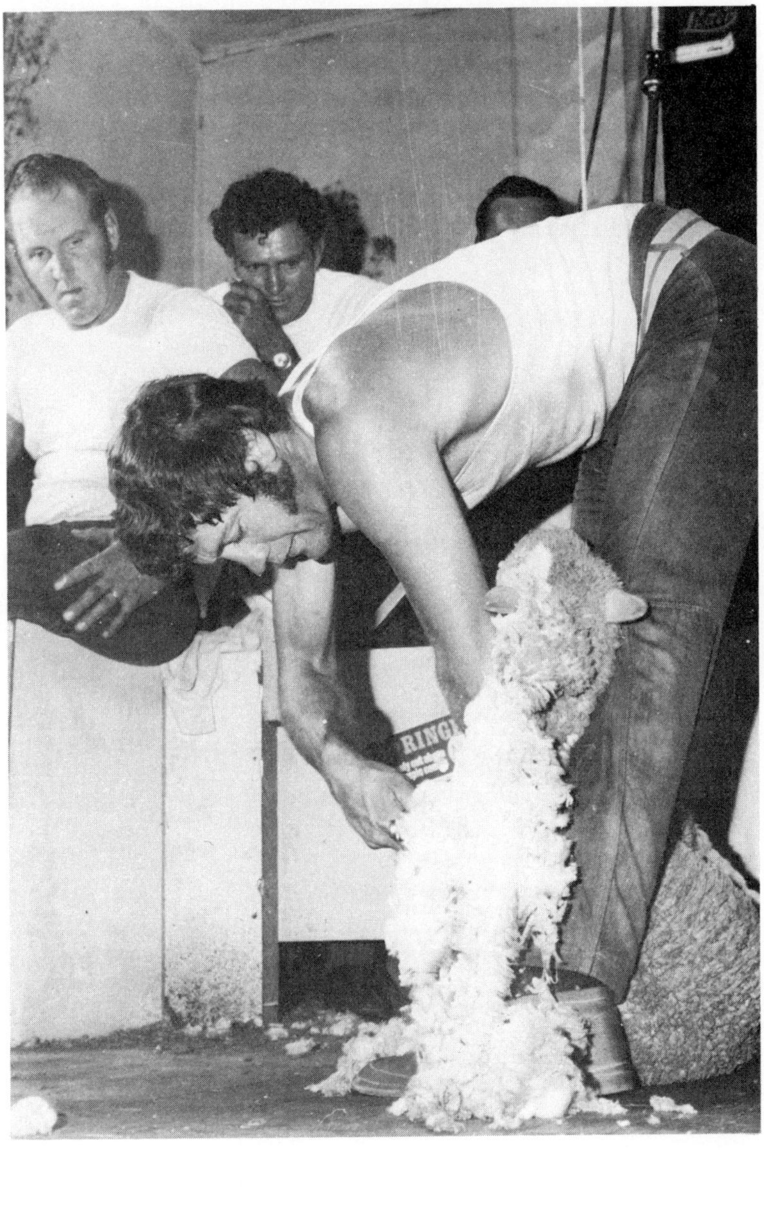

Brian Morrison nearing the end of his record-breaking run

But it is not speed shearing that wins at competitions so much as taking the wool off cleanly and evenly. Each man has eight sheep to shear in 26 minutes. He gains no extra points if he finishes in less time but loses points for every minute he is over time. Few are over time; most have an inbuilt mechanism for timing themselves. They also lose points for skin cuts and second cuts (wool left on sheep requiring a second cut, thus breaking the fleece). In this art, the shearer attempts to shear as near as possible to perfect, he demonstrates his skill as a tradesman and his dedicated attitude to quality shearing. As they say out in the scorching west, 'When it's 120 in the waterbag you soon pick the sheep from the goats'.

For over a hundred years this area has been the heartland of shearing: the Riverina might claim to be more famous for its sheep, but western Queensland is where shearers first banded together. Longreach grew up on wool. Down streets broad enough to turn a camel team, you run into old drovers, shepherds, boundary riders, musterers, blade shearers and teamsters who took up to 120 bales of wool at a time on their camel, donkey or horse teams.

Their stories are a long way from the goings-on in the big Town Hall. All day the heats are run and, at night when the band plays at one end of the hall for the dancers, the finalists shear off at the other end. Each shearer is as nattily dressed as the dancers; some, like David Ryan, in neat-fitting blue trousers with white strap bowyangs below the knees to prevent the trousers from pulling as he bends.

The dancers gather round the stage and cheer the men as they shear. 'Go hard, Dafyd! Push hard!'; 'Go in hard, Bill! You're on borrowed time!'; 'Come on larrikin!; a sort of in-talk. Shearers glance fleetingly at the clock as they grab another sheep from the pen, pull the cord to re-start their machine and make the first cut. To those who are not aficionados there are plenty around to explain when a man is going up the neck, going for the 'long blow', and sliding down the whipping side. In 1980, the winner, Mal Johnstone of Canowindra (NSW), finished his eight sheep in 24 minutes.

The teams took the stand next, four men a team, each shearing four sheep. Some teams are sponsored by local pubs, including the famous Welcome Home Hotel Award by the legendary Ivy McNally. In the winning team, Bill and Terry O'Connor from Euroa (Vic.) come from a family of shearers. Their grandfather, Timothy O'Connor, shore for five years with blades, and their father, Bill O'Connor, shore the eastern states all his young life: now his three sons shear. Timothy won the Australian strong wool shearing in 1976, Terry in 1977 and Bill in 1980. Shearing with the two O'Connor boys this night was the Australian champion, David Ryan, and John Conlan who represented Australia in the World Championship held in New Zealand in 1980. The four Victorians won the event with 739¾ points out of a possible 800 and finished their sixteen sheep in 16.50 minutes. Tests show that, to shear at this speed, these men expend more energy than a man running a four-minute mile. Their bodily fitness and mental discipline, allied with their subtle skill, make these men the super-skilled athletes of any sporting arena.

'How do you feel?' I asked Terry O'Connor.

'Ecstatic. There's no other word. When you finish, and you've done well, you're ecstatic!'

It is a sport not of man against the clock, but of skill against skill. Brian

In India on his tour to demonstrate and train in shearing techniques, left to right, Babu Ram, Kiru Ram, Bramject, Brian Morrison, Peter Curtis, Australian High Commissioner to India, Diwan Chand, Puran Chand

Morrison, twice Australian shearing champion, sums it up. 'Shearing is a cross between an art and a craft, but in truth it is a profession. And here at the competitions the boys set out to prove it.'

The first contest was held at the Sydney Show Ground on Boxing Day, 26 December 1911, the contestants being Dan Cooper (who shore 316 sheep in 8 hours 40 minutes at Bundoran station, Queensland, and Bill Day (who 'cut out' 264 in 8 hours at Murnopeowie station, South Australia). In the contest each man selected the Moffat-Virtue machine, the power being a Davery-Paxman steam engine. Day 'cut out' his lot of 50 sheep in 1 hour 43½ minutes (or at a rate of 253 sheep for a working day), Cooper's time being 1 hour 46¾ minutes (or at the rate of 246 sheep for a working day). The judges, however, declared Cooper the winner by one point.

Possibly the best known shearer in Australia today is Brian Morrison of Euroa (Vic.). Brian went on to the board when he was fourteen years old, following the footsteps of five of his mother's brothers who were all good shearers. He had a goal, and he stuck to shearing until he achieved it. It wasn't the World

Championship record – he got that too – it was a farm. He was determined, and like other shearers who have set a goal for themselves he knew he would never see it out if he didn't like shearing. As all shearers say, 'If you don't like the game, you can't play it well'. From the begining this shearer went well.

'Each time I achieved a better cut or faster, cleaner shear, I got great satisfaction. That makes up for the hard work. I knew it wasn't only important to do a good tally, it became equally important to shear the sheep well. That goes for all top shearers, of course. They all take a particular pride in developing their skills to the maximum. Being top man in a team is important. I found it also ensures a shearer earning a respectable income.'

He achieved both – made history on the board and, as well, bought his farm. He was a good sportsman, played for one season with Richmond Football Team (Australian Rules) but left the 'Tigers' to get on with shearing. 'It's a good life, but I wanted that farm.'

He became the first of the recent batch of champion shearers in the Euroa district. He won the Australian Open Shearing Championship on two occasions, won the Crossbred title together with many state and Country Championships. He represented Australia at the Golden Shears Competition in New Zealand, demonstrated shearing to the Duke of Edinburgh and President Saragat of Italy. He was part of the *Waltzing Matilda Review*, entertaining 3000 delegates at the World Rotary Convention, and he represented Australia at Expo '70 in Japan, where he showed the Japanese – and the world via satellite – the art of the Australian shearer. But his greatest feat is one he swore he would never attempt again under any circumstances. In 1972 he shore 410 sheep in 7 hours 48 minutes, a world record. His average time per sheep, including catching, and changing combs and cutters, was 68.5 seconds. The fastest time for one sheep was 43 seconds.

He is said to be the only shearing record holder to guarantee the quality as well as the quantity of his work. For this attempt at a world record he insisted that the shorn sheep be displayed afterwards for public inspection. Some believe it may be the last public solo shearing record attempt. Morrison had stipulated it was to be his first and last attack on the record; his sole motive was to raise funds for charities in his home town, Euroa, where they had set up catching pens and chutes in the RSL hall for him. The 1500 people who came to watch shearing history challenged, saw this champion in action being monitored by referees and timekeepers constantly checking the quality of his shearing.

Brian retired from competition shearing in 1973, a year after his incredible triumph. His records in shearing sheds were also formidable, with hourly totals of 52, 50, 53, 50, 52, 51, 54, 48; and run tallies of 103, 102, 103, 102.

The Conlan brothers, Mark and John, claimed a new world record for narrow gear on Friday, 14 December, when they shore 852 Perendale lambs in 7¼ hours, exceeding the unofficial South Australian record of 723.

Witnessed by a large crowd, the Conlans shore lamb for lamb all day. The record was set in a shed near their home at Taradale (Vic.). While Mark and John shore, their younger brothers, Matthew and Luke penned up. There were not enough lambs available to shear for eight hours and observers at the Braeside shed believed that a tally of well over 900 would have been achieved in a full working day.

The belief now is that records will be broken only in competition-style shearing in the future. 'There's more wool on sheep now, a man can be cutting 12–14 lb of wool off one animal today, whereas a few years back it used to be 6–7 lb. You still see good tallies, 120 a day in some places, 200 at Winton and Julia Creek where many sheep have bare legs and no top knot. A man can do 325 a day sometimes with machines, but he won't get many more.'

The traditional competition in the shed is going, some say almost gone. 'Occasionally today there is a private duel in a shed. Maybe a bloke that's been cutting another out of a sheep decides to have a go and show him a clean pair of heels. Some, but very few, work to their absolute ability today. Most are quite happy to earn what they do and work steadily.'

'Time rolls on', wrote the young poet-idealist Julian Stuart in his prison cell so long ago. All things change, some things disappear. There are reports of chemical and mechanical shearing devices having been perfected, ready to replace the shearers and the sheds. The sheep, they say, will be trucked to a central depot for treatment. This may well come to pass. Then the poets of the sheds will be silent and the singers songless and a hundred and fifty years of the sight, sound and smells of the life of a country will be ended. The sheds, 'those vast echoing cathedrals of the inland' will crumble, we will not again hear the cries of Wool away! Sheep-O! Rousie! Tar boy, Tar!

Seven Creeks Run, now being commemorated in one of Australia's most ambitious museum complexes

But the spectral figures of the Liveringa team will always tramp, searching endlessly for their shed where the long brown roads go down, and men in Queensland's western lands will hear the horses of those earliest strikers pounding through the night on their wild rides to glory. The memory of a nation will retain the odysseys of the bike-riders pushing across the deserts, tackling The Cobbler; the eternally boyish fun, bawdry, pride, skill, toughness, pain, romance, bravery and endurance of an independent army of men will surely not go unrecorded.

The Pioneer Memorial Gates at Lockhart showgrounds (NSW), with 22 concrete replicas of wool bales branded with station stencils, and the silhouette of the old bullock waggons, salute the days when 'Australia rode on the sheep's back'

Notes

1: THE SQUATTERS

1 G. Ranken (Capricornus). *The Squatting System of Australia.*
2 Ibid.
3 S. H. Roberts. *History of Land Settlement.*
4 Ranken, op. cit.
Other publications referred to for this chapter were:
J. F. Campbell. *Squatting on Crown Lands.*
Angus and Roberta Sykes. *An Outline of Closer Settlement in NSW.*
C. J. King. Various pamphlets published by CSIRO, Melbourne, 1931.

2: THE MASTER – THE SERVANT

1 My source for much information in this chapter was the *Clyde Company Papers*, edited by Phillip Brown.
2 Gigoomgan station records, John Oxley Library Brisbane.
3 Adam Turnbull. Papers. La Trobe Library, State Library of Victoria.

3: LABAN

1 *Clyde Company Papers.*
2 Margaret Kiddle. *Men of Yesterday.*
3 Alexander Mollison. Papers. La Trobe Library, State Library of Victoria.

7: RADDLE

1 William Gough. Droving in the Early Days.

8: THE BREATH OF REVOLT

1 W. G. Spence. *The History of the AWU.*

9: PART OF THE GLORY

1 McKinlay and Drummond. *The ALP. A Short History of the Australian Labor Party.*

11: THE CONSPIRACY TRIAL

1 Pastoralists' Union Report on Conference at Sydney with ASU, 7–8 August 1891.

13: QUEENSLAND ME MOTHER – ARE YE SICK?

1 Telegrams from the Colonial Secretary's files, Public Records Office, Brisbane.

15: THE UNCONQUERABLE SOUL

1 *Barrier Truth*, 10 January 1908, p. 2.
2 *The Worker*, 14 October 1909, p. 13.
3 W. M. Hughes to W. G. Spence. Letter in La Trobe Library, State Library of Victoria.

16: THE BAND OF UNHAPPY AMATEURS

1 *Bulletin*, 18 March 1893.
2 Brian McKinlay. *A Documentary History of the Labor Movement 1850–1975.*

18: THE WHITE WINGS

1 W. G. Spence. *History of the AWU.*

22: LAMBING DOWN

1 'Tally Town' by Henry Lawson was first published in *Freeman's Journal*, 23 December 1893.

26: THE ONE TREE PLAIN

1 Conversation with Mrs Nancy Sheaffe.
2 Hay Historical Society. *Proceedings*. No. 1, September 1962.

30: THE SWEAT EXTRACTOR

1 Docker family of Bontherambo (Vic.). Mss. La Trobe Library, State Library of Victoria.
2 *Stock and Station Journal*, January 1899.

31: CLANCY

1 T. Collins (Joseph Furphey). *Such is Life.*
2 Oscar de Satge. *Pages from the Journal of a Queensland Squatter.*

33: 'WET' CARD OR 'DRY'

1 Pastoralists' Association. Medical Aspects of the Relation of 'Wet Wool' and the Work of Shearing to Health.

35: THE LONGEST, STRONGEST SCAB TRAIN

1 Tom Brislan. A Maverick Among Marxists. 1972. Australian National University Archives.

37: DUCKS ON THE POND

1 'The Shearing Shed' by Henry Lawson. *Bulletin*, 11 December 1897.

38: THE SILLY SEASON

1 Brisbane *Courier Mail*, 22 March 1956.
2 *The Worker*, 26 March 1956.

44: THE SHEARER'S WIDOW

1 'The Shearer's Wife' by Louis Essen, from *The Wide Brown Land* by J. S. and A. Mackaness, Angus & Robertson, Sydney, 1954.

50: THE SQUATTING COMPANY

1 J. S. Battye (ed.). *The History of the North-west of Australia, Embracing Kimberley, Gascoyne and Murchison Districts.*
2 Jennie Hardie of Boodarie station generously gave me this material which she had collected for her history of Port Hedland and district.

Bibliography

Published Material

Allen, James Jnr. *Journal of an experimental Trip by the 'Lady Augusta' on the River Murray*. Platts, Adelaide, 1853.

Amalgamated Shearers' Union of Australasia. *Conference between the A.S.U. and the Pastoralists' Federal Council of Australia, 1891*. Andrew & Taylor, Newport, 1891.

______. *General Rules and By-laws as Revised and Adopted at the Annual Conference of the A.S.U., Bourke, N.S.W. February 1890*. Shearers' Record, Newport, n.d.

______. Manifesto of the A.S.U., 12 July 1890. n.p., n.d.

Anderson, Hugh M. *The Story of Australian Folksong*. Hill of Content, Melbourne, 1955.

______. *The Colonial Minstrel*. Cheshire, Melbourne, 1960.

Argus, 1890–4.

Austin, H.B. *The Merino*; Past, Present and Probable. Grahame Book Company, Sydney, 1943.

Australasian, 1897.

Australasian Pastoralists' Review, 1891.

Australian Star, 1891.

Baker, Sidney John. *The Australian Language*. 2nd edn, Currawong Press, Sydney, 1966.

Barker, H.M. *Droving Days*. Pitman, Melbourne, 1966.

Barnard, Alan (ed.). *The Simple Fleece*. Melbourne University Press with Australian National University, Melbourne, 1962.

Barnes, Ern. *Easier Shearing*. Whitcombe and Tombs Pty Ltd, Melbourne, 1955.

Bassett, Marnie. *The Hentys*. Melbourne University Press, Melbourne, 1962.

Battye, J.S. (ed.). *The History of the North-West of Australia; embracing Kimberley, Gascoyne and Murchison Districts*. Jones, Perth, 1915.

Bean, C.E.W. *Official History of Australia in the War of 1914–18*. 3rd edn, vol. 1, Angus and Robertson, Sydney, 1934.

Bell, E. *Roberts Stewart and Co*. Private Publication, Hobart, 1965.

Beltana Pastoral Company Limited. Private Publication, Adelaide, 1965.

Billis, R.V. and Kenyon, A.S. *Pastoral Pioneers of Port Phillip*. Macmillan, Melbourne, 1932.

Bishop, Isabella Louisa (nee Bird). 'Australia Felix', *The Leisure Hour*, 1877.

Blainey, Geoffrey. *A Land Half Won*. Macmillan, Melbourne, 1980.

Bonwick, James. *Romance of the Wool Trade*. Griffith, Farran, Okeden and Welsh, London, 1887.

Bossence, William Henry. *Numurkah*. Hawthorn Press, Melbourne, 1979.

Braim, Thomas Henry. *A History of New South Wales from its Settlement to the Close of the Year 1844*. Bentley, London, 1846.

Bride, T.F. (ed.). *Letters from Victorian Pioneers*, Government Printer, Melbourne, 1898.

Burfitt, Charles Trimby. *History of the Founding of the Wool Industry in Australia*. Government Printer, Sydney, 1913.

Brisbane Courier, 1891.

Brisbane Truth, 1956.

Brown's Water Holes, History of Lismore, compiled by J. R. Oman, P. S. Lang and J. Ryan. Adams Printers, Geelong, 1961.

Bulletin, 1890–4.

Butlin, Noel George. *Investment in Australian Economic Development 1861–1900*. Cambridge University Press, Cambridge, 1964.

Cannon, Michael. *Who's Master, Who's Man?* Nelson, Melbourne, 1971.

Clark, C.M.H. *A History of Australia*, vols 1 and 2. Melbourne University Press, Melbourne, 1966, 1968.

Clyde Company Papers, P. L. Brown (ed.). Oxford University Press, London, 1941–71.

Collier, James. *The Pastoral Age in Australia*. Whitcombe and Tombs, London, 1911.

Crombie, Andrew. *After Sixty Years*. The author, Brisbane, 1927.

Curr, E.M. *Recollections of Squatting in Victoria*. G. Robertson, Melbourne, 1883.

Daily Telegraph, 1891.

Dengate, R.D. & Carlin, J. (eds). *Songs of the Shearers*. Bush Music Club, Sydney, 1970.

Dubbo Dispatch, 1894.

Ellis, M.H. *John Macarthur*. 2nd edn, Angus and Robertson, Sydney, 1967.

Fetherstonhaugh, Cuthbert. *After Many Days*. Sydney, 1917.

Fitzgerald, John Foster Vesey (formerly John Fitzgerald Leslie Foster). *New Colony of Victoria*. London, 1851.

Folk Lore Council of Australia. *Australian Folk Songs of the Land and its People*. Lowden, Kilmore (Vic.), 1974.
Fraser, Allan Duncan. *This Century of Ours*. Halstead Press, Sydney, 1938.

Glennie, J. *Twigs of Yarran*. Mudall, Nyngan (NSW), 1953.
Gormly, James. *Exploration and Settlement in Australia*. Sydney, 1921.
Graphic, 1891.
Graziers' Association of NSW and Dangar, Sedge and Malloch Ltd. *Hints to Shearers*. Sydney, 1938.
The Great Queensland Shearers' Strike of 1891. Pamphlet issued by United Graziers' Association of Queensland, Black & Co., Brisbane, 1891.
Guyatt, Joy. *Labor History*, no 33, November, 1977.

Hardy, Bobby, *West of the Darling*. Jacaranda Press, Brisbane, 1969.
Harris, W.J.H. *The Bitter Fight*. Queensland University Press, Brisbane, 1970.
Hawdon, Joseph. *Journal of a Journey from New South Wales to Adelaide, 1838*. Georgian House, Melbourne, 1952.
Hay Historical Society. *Proceedings*, no 1, September, 1962.
Henning, Rachel Biddulph. *The Letters of Rachel Henning*. Edited by David Adams. Angus and Robertson, Sydney, 1966.
Herald (Melbourne), 1956.
Hickey, Simon. *Travelled Roads*. Cheshire, Melbourne, 1951.

Ingleton, G.C. *True Patriots All*. Angus and Robertson, Sydney, 1952.
Inglis, K.S. *The Australian Colonists*. Melbourne University Press, Melbourne, 1974.

Kiddell, John. *Choogoowarra*. Macmillan, New York, 1972.
Kiddle, Margaret Loch. *Men of Yesterday*. Melbourne University Press, Melbourne, 1961.

Launceston Examiner, 1891.
Launceston Telegraph, 1952.
Lawson, Henry. *In the Days when the World was Wide, and Other Verses*. Angus and Robertson, Sydney, 1900.
Leader, 1891.

Mackaness, J.S. and A. *The Wide Brown Land*. Angus and Robertson, Sydney, 1954.
McKinlay, Brian. *A Documentary History of the Australian Labor Movement, 1850–1975*. Drummond, Melbourne, 1979.
______. and Drummond, Pat. *The ALP. A Short History of the Australian Labor Party*. Heinemann, Melbourne, 1981.
Mahood, Margaret. *The Loaded Line*. Melbourne University Press, Melbourne, 1973.
Manifold, J.S. *The Penguin Australian Song Book*. Penguin, Melbourne, 1964.
______. *Who Wrote the Ballads*? Australasian Book Society, Sydney, 1964.
Meredith, John and Anderson, Hugh. *Folk Songs of Australia and the Men and Women who Sang Them*. Ure Smith, Sydney, 1968.
Mossman, Samuel. *The Gold Regions of Australia*. Orr, London, 1852.
Murphy, D.J. *T. J. Ryan*. University of Queensland Press, Brisbane, 1975.

New South Wales Employers' Union. *The Labor Crisis*. Sydney, 1890.

Pastoralists' Union of New South Wales. *Brief Statement of Aims and Objectives*. *c*. 1890.
______. 'Report of the Annual General Meeting', in *Australasian Pastoralists' Review*, 1891–5.
Peck, Harry. *Memoirs of a Stockman*. Stockland Press, Melbourne, 1942.
Plomly, N.J.B. (ed.). *Friendly Mission. The Tasmanian Journals of G. A. Robinson*. Tasmanian Historical Research Association, Hobart, 1966.
Punch, Melbourne, 1891.
Pyalong, a Brief History. Back to Pyalong Committee, J. Lowden, Kilmore, 1968.

Queensland. Legislative Assembly. *Debates*. 1891.
Queensland. Legislative Council. *Debates*. 1891.
Queenslander, 1891.

Ranken, George (Capricornus). *The Squatting System of Australia*. Bell and Bradfute, Edinburgh, 1875.
R-s, William. *The Fell Tyrant, or the Suffering Convict*. London, 1836.
Ratcliffe, Francis Noble. *Flying Fox and Drifting Sand*. Chatto and Windus, London, 1938.
Richardson, A.R. *Early Memories of the Great Nor-west*. Perth, 1914.
Roberts, Stephen Henry. *The Squatting Age in Australia 1835–1847*. Melbourne University Press, Melbourne, 1935.
Ronald, Robert Bruce. *The Riverina People and Properties*. Cheshire, Melbourne, 1960.
Ronald, Heather Barry. *Wool Past the Winning Post. A History of the Chirnside Family*. Landvale Enterprises, South Yarra, 1978.
Rudd, Steele (Arthur H. Davis). *On Our Selection*! *Bulletin*, Sydney, 1903.

Scott, Bill. *A Treasury of Australian Folklore*. Ure Smith, Sydney, 1976.
Shackcloth, Irene. *The Call of the Kimberley*. Hallcraft, Melbourne, 1950.
Shaw, M. Turner. *On Mt Emu Creek: A story of a 19th Century Victorian Sheep Station*. Robertson and Mullins, Melbourne, 1970.
Shearers' and General Labourers' Record, 1888–93.
Shearing Wet sheep: Report of Investigation by a Joint Conference of the Pastoralists' Association of N.S.W. and the A.W.U., 1908–1909. Printed by authority of the Government of the Commonwealth of Australia, Melbourne, n.d.
Sheep Shearing Machinery Experts' Association of N.S.W. *Rules and regulations*. Radcliffe, Sydney, 1913.
Singabout (Bush Music Club), 1956–1967.

Sorenson, E.S. *Life in the Australian Backblocks*, Whitcombe and Tombs, London, 1911.
Sorenson, Jack. *The Gun of Glindawor and other Ballads*. R.S.S. Brokensha, Perth, 1932
Spence, William Guthrie. *Australia's Awakening*. Worker Trustees, Sydney, 1909.
Souter, G. *A Peculiar People. Australians in Paraguay*. Angus and Robertson, Sydney, 1968.
Sydney Morning Herald, 1890–4.
Sydney Mail, 1894, 1900, 1902.
Sykes, Angus and Roberta. *An Outline of Closer Settlement in NSW*, 1968.

Town and Country Journal. 1890, 1891, 1894.
Tritton, H.P. ('Duke'). *Time Means Tucker*. Shakespeare Head Press, Sydney, 1964.

Unionist & Gulgong Advertiser, 5 September 1891.

Ward, Russell. *The Australian Legend*. Illus. edn. Oxford University Press, Melbourne, 1978.
Western Herald & Darling River Advocate, 1891, 1894.
Worker (Brisbane), 1956.

UNPUBLISHED MATERIAL

Australian Agricultural Company papers. Australian National University.

Black, Neil. Manuscripts. State Library of Victoria.
Brislan, Tom. A Maverick among Marxists. Original manuscript, Australian National University, 1972.

Drysdale, Anne. Diary. State Library of Victoria.

Gough, W. Droving in the Early Days. *Riverina Press* office, Hay, New South Wales.
Gigoomgan station records. John Oxley Memorial Library, Brisbane.
Goldsborough Mort & Co. papers: Canobar and Meryule, 1881-1900. Australian National University.

Henty Family records and papers. State Library of Victoria.

Mollison, Alexander. Diary. State Library of Victoria.

Piggin, F. S. Pastoralists of N.S.W. and the Strikes of 1890 1891. Thesis. Australian National University.

Turnbull, Adam. Diary. State Library of Victoria.

Uardry station records, Hay, New South Wales.

Van Diemen's Land Company records. State Archives, Hobart.

Queensland State Archives: Colonial Secretary's Office. Correspondence on shearers' strike, 1891–2.

______. Police Department records:
Charge books, 1888–92, 1892–7.
Disbursement book, 1880–7.
Duty and occurrences diaries, 1888–92.
Letterbooks, Isisford station, 6 December 1889–6 January 1895.
Letterbooks, letters to the Inspector of Police, Longreach; a register of offenders and suspects (shearers' strike) and register of outlying correspondence, April 1895–1904.
Memorandum from District Inspector, Longreach to Officer-in-Charge, Aramac, 1889-1903.
Police reports, Adavale station, August 1881–September 1891–1909.
Physical descriptions of certain prisoners, 1890–2.
Register police charges, Adavale station, August 1881–April 1896.
Register persons received in lock-up, Charleville 1889–January 1893.
Register police charges, Cloncurry, 1887–91.
Register persons arrested, Charleville, Adavale, Augathella, Cunnamulla, Eromanga, Eulo, Hungerford, Morven, Nocundra, Wooroorooka, Wyandra, Thargomindah and Upper Warrego, 1901–04.

Glossary of the Sheds

Baitlayer: the cook; babbling brook; silly look; cockatoo; water-burner.
Bale: Pack containing wool already specified by type.
Barcoo buster: A westerly gale in outback Queensland.
Barcoo challenge: In the Barcoo district (Qld), a challenge for the day's best tally was made by a) scraping the points of hand-shears on the wool-shed floor or wall, or b) throwing the belly of a fleece over a shearer's head.
Barcoo rot: Festering sore difficult to cure under inland conditions. It rapidly disappears when sufferer eats fruit and vegetables. Also known as Kennedy Rot or Queensland Sore.
Barcoo spew: A sickness characterised by vomiting after food is taken, also known as Barcoo vomit or Belyando spew.
Bare-bellied-Joe: Or yoe, said to be Irish for ewe. A sheep that has lost its belly wool, i.e. a bare-belly is shorn faster than a full-wool sheep.
Barrowing: The practice of learning to shear e.g. by finishing of the shearers' sheep at smoko.
Basil: Greenhide leather.
Battling for a cut: Looking for shearing, 'cut', shed-work. 'Fine cuts' and 'rough cuts'. In 'fine' sheds the shearer is particular, in 'rough' the sheep are shorn 'anyhow'.
Bell-sheep: Henry Lawson wrote: 'Shearers are not supposed to take another sheep out of the pen when 'smoke-ho', breakfast or dinner bell goes, but some time themselves to get so many sheep out and 'one as the bell goes'.
Benders: Bending fibres – fibres which hold the wool staples together, enabling fleece to be handled as a unit.
Billabong: Low-lying area filled by overflow from river or tributary.
Bins: Space partitioned to hold different grades or classes of wool.
Black-ball: To reject, to declare black, to vote out, to refuse to work with, or for, or on some job.
Black-leg: Non-union labour, particularly a union man who shears non-union. One who exercises freedom of contract.
Blades: Earliest shears, before machines, tongs, scissors, swords, daggers.
Bladesmen: Blade shearers.
Blight: Sand or sandy blight, a form of ophthalmia in which the eye feels full of sand; fly blight, swelling blight 1898: fly-bung and bung-eye.
Blow: Shearing the wool the width of the comb along the contour of the sheep's body. Also to have a rest, take a blow. A strike with the handshears.
Blue-tongue: From the lizard with a flicking blue tongue. applied to shed-hands with big appetites. Also a rouseabout.
Bluey: Swag, matilda, carpet-bag, shiralee.
Board: Strictly speaking, that part of the wool-shed where the sheep are shorn, but application is much wider. Boldrewood, writing in 1878, refers to '100,000 sheep on the board' meaning the number of sheep shorn in the shed in a single season. Also, the whole number of shearers employed in a single season in a woolshed. There is a 'boss over the board' and the woolshed overseer is man-over-the-board; to 'clear the board' is to complete shearing or 'cut out the shed', and to 'ring the board' is to prove the most expert shearer in a shed.
Boggi: Pronounced Bog-eye. A shearing hand-piece, so-called because its shape resembles a boggi lizard peculiar to Australia.
Boss: Shed boss or boss-of-the-board, overseer, head-serang.
Boss's boots: To a shearer, working head down, the first sight of the boss is his boots.
B-Bows: Make of blade shears having a B-shaped curve above the hand grip.
Box: A mixing of sheep (came into use about 1870). By 1880 Boldrewood, in *Robbery Under Arms*, was using the word as the equivalent of confused or muddled.
Boxer: One who holds the stakes being wagered at a game of two-up.
Branding: Stencilling on bales of wool the initials of owner, serial number and type of wool.
Broad gauge: Any comb (on shearing hand-piece) exceeding 2½ inches from the outside of the top tooth to the outside of the bottom tooth. There has been much argument over use of this comb in Australia where it is illegal (and as late as 1982 wide-spread shearers' strikes were called because many pastoralists were demanding its use). It originated in New Zealand where it is legal. When used in Australia by NZ shearers it is known as 'an NZ comb'. Also pulled comb or Merry Widow.
Broken-mouthed: An 8-toothed sheep losing its teeth. A sheep is full-mouthed as long as the teeth remain perfect, but when they decline the sheep becomes failing-mouthed, broken-mouthed, and finally gummy.
Broomie: Woolshed worker. Also sweeper.
Brownie: A sweet currant loaf made with brown sugar.
Brownie gorger: Hungry shed-hands (usually young boys with big appetites).
Brutes: Shearers.

Bullet: Sack. 'Given the bullet.' To be fired or told to hang-up when shearing.
Bowyangs: String or leather thong tied around trousers below knee level to allow freedom of movement while working in a crouched position.
Burgoo: Porridge, rolled oats.
Burry: Wool containing seed pods.
Butt: Package of oddments – wool awaiting additional quantities to make weight.
C.F.A.: Cast for age: a reject old sheep that is past its prime for wool and lamb production under particular conditions.
Call the centre: To solicit any sum of money in the form of a tip to ringmaster and boxer at game of two-up.
Carry the drum: To shear slowly or lazily. A totally different meaning was to carry one's swag or blanket roll.
Catch: Last sheep in pen.
Catching pen: Situated inside shed adjacent to the shearing board, from which shearers catch sheep to shear. Usually two shearers share each pen.
Chaff bandit: Horse. From the days when shearers rode to sheds.
Chinese hundred: Tally of fifty sheep shorn in a day.
Chip (or chop): To gash a sheep, 'remove bootlaces from its hide' or otherwise shear in an inefficient manner.
Cobbler: A rough, wrinkled or difficult to shear sheep. A cobbler need not be the last or any other sheep in a pen but a rather dirty, sticky, wrinkled or matted sheep, one hard to shear. Two men catching out of one pen naturally avoid bad sheep (cobblers) as long as they can; the pen not being filled until the last sheep is out, this sheep is more likely to be a cobbler. Penmates of even speed usually do not select.
Cockatoo: Look-out, sentry at two-up game who 'screeches' a warning if 'John-hops' are approaching.
Cockspur: A stop fastened to the heel of one blade of hand-shears to prevent its crossing the other blade. Also called knocker.
Cocky: Small farmer (who, in early days, broadcast seed by hand only to see it eaten by cockatoos following him). In most states shearers refer to any sheep farmer – large or small – as a 'cocky'. They also call 'grasshoppers' those 'not big enough to be a cocky'.
Cocky's clip: When practically every vestige of wool is removed from the sheep's body (supposedly reflecting the parsimonious nature of cockies).
Coming of the plague: Squatters' term for arrival of shearers preparatory to shearing.
Cots: Matted locks of wool.
Counting-out pen: Situated outside shed to receive shorn sheep.
Crossbred: The progeny of two distinct pure breeds of sheep.
Crutching: The shearing of wool from area of sheep's crutch to minimise fly-strike. Crutching is a specific task representing part of the shearers' livelihood.
Cull: Uneconomic sheep classed out to retain even flock.
Cut: Variation in the manner in which wool comes away from a sheep; the wool can be 'good' or 'bad' cutting. Also relates to shearer's ability i.e. 'high cut' can mean sheep shorn rough.
Cut-out: Original currency during 1840s meant the separating of one or more selected animals from a herd. By the 1880s application had been extended and a wool-shed was said to have cut-out when the season's shearing was completed. By analogy a shed was said to have 'cut-in' when shearing commenced. This use hailed from the shearing term 'cut', sometimes used as a shearer's cut, or stand or pen when he is employed. He also 'cuts-out' a cheque at shearing, and he 'cuts-out' a fellow shearer of, say, thirty sheep a day when he shears thirty more than his companion. 'I'm heading off after cut-out,' the shearers say.
Cut-out horrors: Referred to when shed is about to finish; a mild form of neurosis.
Dag: Consolidated excreta in region of sheep's crutch.
Dagger: Usually employed pre-shearing to clean up daggy sheep.
Darling shower: Dust storm that fills wool on sheep with sand.
Depot shed: A type of wool-shed found in many districts where small graziers who have no shed of their own may send their sheep for contract shearing.
Do a Jimmy Gibbs: Shear a sheep very quickly. (From tall-story character.)
Dook: A fist. Molly Dook, a left-handed shearer.
Dodger: Bread.
Dog-fence: Fence to exclude dingoes from pastoral properties. World's longest fence is in Queensland, area 5632 km long; SA has dog fence of 2172 km. Referring to a poor type of man, 'a dingo', shearers will say, 'the fence must be cut', meaning the dingoes are in.
Donk: Engine.
Double fleece: Two years growth of wool.
Drag the chain: Slow, lazy. Shearer running last in a team is said to be 'on the chain'.
Dreadnought: A tremendously hard-working shearer and/or one who can shear over 300 sheep a day.
Driver: Strap on blade shears through which hand passes to 'drive' the shears through wool while making a 'blow', or strike.
Drummer: Not necessarily the slowest shearer in shed but likely to be laziest. In earlier times the man with the least number of sheep on the tally board was drummed out of the sheds by the rouseabouts.
Ducks on the pond: 'Women approaching the shed' a warning to shearers.
Duds: Clothes. Shearing duds are dungarees etc.
Dump: Wool baling term. Also big, strong gun shearers who 'dump' their sheep to gain speed i.e. putting sheep into its smallest position.
Eaglehawk: Pluck wool from a dead sheep.
Ewe: Female sheep.
Expert: Staff member of shearing team responsible for the care and operation of shearing machinery in the shed and for grinding combs and cutters. (Colloquial: 'Ex'equals unknown quantity, 'spirt' a drip under pressure.)
Eyebrows: Second cuts of wool.
Fadge: Light-weight butt of wool.
Fellmongering: Removing pelt from wool by bacterial action (sweating) or by chemicals.
'Fine day for Travellin': You're sacked. Leave the property.

Flat to the boards: Shearing at top speed; like a blue-arse fly; like a jew-lizard up a gum tree; flat out like a lizard drinking.
Flyblown: Broke, without cash, 'Worse than a fly-blown yoe'.
Fly-strike: Blow-fly maggots on the living sheep.
Forty, a: Gambling shark in huts – one of the Forty Thieves.
Freedom of Contract: The most contentious clause in a contract; the insistence by squatters, particularly during the 1890s, of non-union men's rights to shear beside union men on shearing board.
Freezer: Sheep bred for export as frozen mutton.
Frib: Fribs, Fribby, refer to small, short locks; sometimes shearer leaves on shorn sheep. Also second cuts.
Gaffer: Boss of the board.
Got out among the branches: Leaving wool on sheep on account of shearing 'high' while racing for lead.
Grit: Sugar.
Guard, the: The guns and ringers, the prestigious élite. The 'guard' team is given the best sheds and kept in runs when work is scarce.
Gun: Man who shears 200 a day (although in some regions, such as the Monaro, 80 a day is good going). Top man, high tallies, hard man to beat.
Hang up: Hang up the boggi: to leave, or be sacked.
Hermit: Sheep which becomes attached to a spot and refuses to budge. Also 'placer'. These sheep develop cunning, will not join mob or be worked by sheep dogs.
Hob-nail Express: To walk to shed in earlier days.
Hospital sheep: Sick sheep placed alone. Hospital paddock is put aside for injured or deformed sheep.
Hogget: Sheep 12 months old, as yet unshorn.
Hughie: Mythical figure called upon when rain needed i.e. when run has lasted many weeks and all are weary, shearers have called: 'Play the piano Hughie!' And when work is stopped because sheep are wet, 'Hughie you beauty!'
Hygroscopic power: Power of wool to absorb moisture.
In the wool: Full wool sheep ready for shearing (after shearing 'off shears').
Jingling Johnnies: Hand shears.
July fog: The dead season when no shearing is done.
Jumbuck: Sheep. In 1896 A. Meston (Qld) wrote: 'The word jumbuck for sheep appears in Aboriginal speech as jimba, jimbock, dombock and dumbog. In each case it meant the white mist preceding a shower, to which a flock of sheep bore a strong resemblance. It seemed the only thing to which the Aboriginal mind could compare it.'
Jumbuck barbers: Shearers.
Kemps: Coarse wool around the breech.
Knockers: Fitted to hand-shears to eliminate jarring and control width of blade.
Lamb: With mother, or five months of age.
Lanoline: Product found in all wool.
Learner: Until he has shorn 5000 sheep, a shearer is considered a learner in all states except Queensland where the number is 10 000. Also one unable to earn the equivalent of adult shedhands' wages plus reasonable comb and cutter allowance.
Light on: Team not up to usual standard, being short of men or 'talent'.
Lizards: The hand-piece. 'Load up the lizard.' Used to be applied to musterers who could not keep the sheep up to the shed, also shepherds.
Load up: The process of fitting a comb and cutter to a shearing hand-piece.
Locks: Second cuts or portions of wool from the lower parts of the legs and edges of fleece, greasy staples from under forearm, flank and crutch. Lower quality trimming. Frib wool.
Loppies: Leathernecks, rouseabouts.
Make tracks: Head off, leave the shed.
Merino, a: A Don Juan, a ram (among men).
Monicker: Man's name. 'Put your monicker on the contract.' 'What's your handle?'
Moits: Short pieces of stick and scrub in neck wool.
Moity: Wool which contains vegetable matter other than burr or seed such as ferns, bark, straw etc.
Monkey: Sheep.
Motherers: Shepherds.
Mulesing: Operation performed on the wrinkled breech of a sheep to remove skin to enable bare-skin areas substantially to restrict fly-strike.
Muster: Assemble sheep in a mob to be brought in for shearing, dipping etc.
Noils: Wasty portions of wool.
Nowler: Difficult sheep to shear. A snob.
Overgrown: Fleece of wool more than twelve months growth; Long woolled sheep which missed mustering for a time.
Packsaddle: Overseer. Old woolshed slang. When shearers worked in two rows the outside row was referred to as the packhorse side, possibly because it was under closer observation and had to work harder.
Paddock: This word was in use in sixteenth century England for small field or meadow. In Australia there is no limit to the size of a paddock, it may be one hectare or 60 000 hectares, but it is a paddock so long as it is enclosed by a fence.
Pannikin: Usual drinking utensil in use at shearing huts.
Pannikin boss: Station manager in a small way of business.
Parasite: Property owner.
Paroo sandwich: Beer and wine mixed together.
Pen: Enclosure inside and outside shearing shed for holding sheep. When a shearer is engaged he is offered 'a pen'.
Penner-up: Woolshed worker who pens sheep for shearers.
Peter: The pack-saddle of old shearers.
Piece-picker: Shedhand who sorts fleece skirtings into lines of comparable value – a task rarely done today.
Pink: To shear so cleanly that the pink skin of a healthy hide shows. Top quality shearing. 'Pinking 'em'. The pink skin can reflect in various breed types, usually plain bodied sheep.
Play the piano: To run one's fingers over the backs of sheep to find the easiest to shear.
Plum-jams: Lambs. Usually used of very small lambs.
Pony: Assisted woolclasser in days of big shearing sheds. Responsible for placing individual fleeces in bin line for classer as in a big shed the classer would not have time to walk to the bins.
Puff-de-loon: Johnny cakes of flour and water cooked in fat and eaten with treacle or syrup.
Pull the pin: Hang up the shears; walk off job.

Purse-proud: Cut-out time when the shearer has money in pocket.
Put Kinchela on shears: Sharpen shears.
Put on the track: Sacked. 'Down the road.' 'Tramped.'
Quart pots: Small lambs hardly white with wool. Today seldom shedded for shearing but once a common occurrence as once lambs' tip removed wool commands a higher price.
Rabbit eater: Shearer from hard background. Refers to the constant diet of rabbits by poor during depression of the 1930s.
Raddle: Coloured chalk used to mark bales or sheep.
Raddled: Robbed. From days before union formed, when squatters would mark with red or blue raddle any sheep they said was unsatisfactorily shorn. Whole pen could then be turned out without being paid for.
Scoured wool: Washed wool, usually in a series of bowls, using alkali, soap and water at pre-arranged temperatures.
Scouring: Washing of wool. Also a form of dysentery among sheep pastured on soft, lush feed – hated by shearers.
Second Cuts: Short portions of wool staples resulting from shearer making two 'blows' over same area.
Skirter: Wool shed worker. Skirting requires removing those portions of inferior quality and value from the fleece whilst on the wool table.
Slats: Evenly spaced strips of bevelled timber to allow sheep droppings to pass through grating.
Snagger: A rough, unskilled shearer with little desire to do better.
Spade press: A make-shift press.
Spear or shot: Sacked. 'Speared me.'
Speed Ball: Rissole of meat 'and anything'. Usually served at breakfast or lunch (dinner) and very popular with shearers.
Sweat Extractor: Contractor.
Squatter: Term arrived in this country from America where used for settler having no legal rights to the land he occupied. In Australia the original 'squatters' also took possession of land to which they had no title. Until well into the 1840s, the term was used as one of deepest disrespect or opprobrium and squattage was the holding occupied by a squatter. In 1840 Governor Sir George Gipps of NSW protested that 'squatter' is 'hardly a proper word to describe the occupier of 10 000 acres'. In 1839 the South Australian Company referred to 'the mere squatter, content to lead a savage life in the wilds, remote from the decencies of society, with no company but his fellow dependents'. In 1851 the word was called 'a barbarous appellation' and in 1852 S. Mossman wrote: 'Stockholders erroneously called squatters are in this district (Qld) a very superior class of men and are considered the aristocracy of the colony.'

Usage has changed substantially since those days. Tom Collins wrote in 1903: 'We use the term squatter indifferently to denote a station owner, a managing partner or salaried manager.' Today its use for a large land-owner is well established, although the squatters decided in the 1890s to call themselves pastoralists and, later, graziers.

To shearers, all sheep owners are 'cockies', in 1836 term referring to kindred of the squatters, a farmer on small holdings who sowed seed by broadcasting by hand and cockatoos ate the seed as fast as he scattered it. Shearers use the term in a confused manner, sometimes referring to small farmers only as cockies, but mostly using it for all men they shear for.
Squatterdom: The squatter class in general. Squattocracy (dated from 1846) and squatter aristocracy as well as squatterarchy followed in 1877.
Stained pieces: Urine-stained wool and small portions of yolk-stained wool from the crutch.
Staple: Number of wool fibres which naturally form themselves into clusters or locks.
Station: Distinctively Australian word, in use a little before the 1830s, the earliest written use being as stock-station. Seems to have been used originally for homestead and outbuildings rather than squatting property as a whole.
Stiff, fly-blown: Shearer confessing he is broke, hard-up.
Stir the possums: Liven things up in the hut.
Stir the stumps: Leave, get a move on.
Stopper: Safety device fitted to hand shears to keep blades closed when not in use.
Stragglers: Sheep missed in general muster. Later they are shorn at the straggler-shearing.
Stranger: A stray sheep.
Stumpy: A half-day, as on Saturday.
Super: A station manager was formerly known as a super or cove (both date from before 1850).
Swaggie: Traveller, tramp, old-time shearer between sheds.
Sweat rag: Towel hung beside shearing stand.
Sword swallower: Said of rouseabouts who eat from their knives.
T.S.R.: Travelling Stock Reserves, stock route, public way on which sheep are permitted to travel by permit.
Take a dip: Swim, referring back to the sheep dip.
Tally: Number of sheep shorn by shearer daily, weekly, or for the duration of shearing at a particular shed.
Tar-boy: Woolshed worker. Originally he carried a pot of tar to dab on wounded sheep.
Tick-jammer: Wool-presser.
Tomahawker: Unskilled shearer. 'Shears like a wild black.'
Ton: Tally of a hundred sheep shorn during day.
Topliner: Stylish attractive fleece of wool.
Wagga: Sleeping in or on a wool bale.
Wagga blanket: Two layers of wheat sacks with ti-tree bark filling.
Waiting for a death: Lounging around shed waiting until someone is sacked.
Weaner: 6–9 month old sheep.
White ants in the billy: Attempt to undermine the system.
Whitewashing: Shearing very young lambs.
Wigging: Wool taken from around eyes of sheep.
Wool blind: Wool interfering with sight.
Woolhawks: Shearers, especially if expert.
Woolly: Sheep.
Yakka: The pull-back of a pair of hand-shears. Also hard physical work.
Yolk: Combined secretion of the several glands of sheep. Yolk stain is the stain on shearers' clothing caused by pigment of yolk.

Acknowledgements

The book was written with the help of shearers, contractors, pressers, classers, cooks, rouseabouts, drovers, musterers, teamsters, property owners, managers, wives and sisters of shearers, as well as the people of many small towns where I made my headquarters while researching an area.

The Literature Board of the Australia Council assisted me with a grant.

The men of Farmers Grazcos Co-op Ltd gave me a warm welcome and generous assistance when I arrived in their territories in the four eastern states. Mr Geoffrey Still, managing director, himself an ex-shearer, Mr Frank Sutherland of Melbourne and Mr Les Tout of Longreach guided me during the first two years of the work on this book.

Names of some of the men and women who helped me appear in the text. To these and the others whose names do not appear but whose stories helped make the book I owe a great deal. I took notes of interviews, tape-recorded and collected written material and photographs in all states. I researched in the John Oxley Memorial Library, Brisbane; the Mitchell Library, Sydney; the Tasmanian State Archives, Hobart; the South Australian Archives, Adelaide; the Battye Library, Perth; and the La Trobe Library, Melbourne, where librarians gave generously of their time as did Mrs J. Day and Mr M. Saclier of the Australian National University, Canberra; Ms Julie Whiting of the Australian Wool Commission; Professor Russell Ward, Armidale; Ms Fay Porter, Armidale; Miss M. Turner Shaw, Melbourne; and Mr Warren Henderson, who taught me the geology and botany of the sheep runs in Queensland.

Kind hospitality and assistance was given me in Western Australia by Mr and Mrs D. Murray, Clover Downs; Leonora, Mr and Mrs D. Fitzgerald, Nambi, Leonora; Mr and Mrs C. Turner, Leonora; Mr and Mrs F. Welsh, Marble Bar; Mr and Mrs P. G. Hardie, Boodarie, Port Hedland; Mr and Mrs S. Sinclair, Leatherhead, Toodyay. In Victoria, by Mr and Mrs J. Page, Penshurst; Mr Andrew Lemon; Ms Jean Baker, Creswick. In New South Wales by Mr and Mrs R. McCully, Uardry, Booligal; Mr and Mrs M. Rutledge, Yeadon, Booligal; Mr and Mrs J. Houston, Hay; Mr and Mrs I. Mann, Billilla, Wilcannia; Mrs Catherine Rose, Conargo; Mrs N. Sheaffe, Eurugabah, Booligal; Mr A. Barton, Bourke Historical Society; Mr and Mrs A. Scott, Sydney. In South Australia, Mr K. Fiebig, the Sunbeam Corp., Adelaide. In Queensland, the Bryce family of Colston, Winton; Mr and Mrs W. Scott, Yeerongpilly; Ms Sharron Lewis, Charleville and Brisbane; Ms Ruby Rose, Longreach; Mr S. 'Scowie' Schofield; Mr and Mrs W. Harding, White Hill, Longreach. In Tasmania, the Grieg family and Mr and Mrs R. Jack, Longford. Judy Wookey typed the manuscript, Jenny Carew assisted with research. I will remember with appreciation and pleasure the days spent with Sue Ebury as she skilfully edited the manuscript with patience and understanding.

Mike Charlesworth of Transwest Airlines and TAA assisted me.

Permission to quote from his father's writings was generously given by Donald Stuart, the son of Julian Stuart.

Mrs M. James, Townsville, gave permission to use the photographs of Mundabullangana taken by her father, the late Mr Charles Henry Powell M.I.M.E.; I also thank Miss M. Slater for the photograph of the cookhouse at Anna Plains station; Desmond Dease, of Lafayette Studios, South Perth the photographs of Derby Jetty; and the Department of Community Welfare provided permission to use the photographs of Liveringa shed hands and Moola Boola stockmen. Other individuals who kindly lent photographs or gave me permission to use them were Mr Fred Cotterill, Mr Jim Carew, Mr Len Jenkin, Mr Jim Lowden, Mr Brian Morrison, Mr Don Murray, Mr R. H. Sheaffe and Mr Marcus F. Synnot.

Pictures were supplied or reproduced by kind permission of the following: *Hamilton Spectator* for p. 388; John Oxley Library, Queensland, for pp. 22, 28, 54, 139, 165 and 261; Library Board of Western Australia for pp. 290, 291, 335, 338, 340, 342, 343, 345, 347, 349, 351, 355, 359, 361, 362, 365, 366, 369, 379, 381; Mitchell Library, Sydney for p. 284; National Gallery of Victoria for p. 380 (top left); National Library of Australia for pp. 47, 157, 286; Queensland Newspapers Pty Ltd for pp. 249, 251, 253; State Library of Victoria for pp. 27, 79, 81, 85, 97 (top), 101, 131, 134; State Library of South Australia for pp. 9, 24, 288; *Stock Journal*, Adelaide, p. 391.

Illustrations were also taken from the following publications:

Australasian Sketcher, 30 October 1874, pp. 8, 11, 13, 15; 18 November 1873, p. 224; 21 October 1882, p. 41.
Australia Illustrated, vol. 3, 1892, p. 151.
Bulletin, 1891, p. 121.
Blair, David. *History of Australia*, 1879. pp. 19, 21.
Illustrated Sydney News, 1 March 1869, p. 29.

North Queensland Register, 30 November 1905, p. 265.
Queenslander, 28 March 1891, p. 83; 18 April 1891, p. 107; 2 May 1891, pp. 99, 107 (lower), 109; 28 May 1891, pp. 95, 98; 30 May 1891, pp. 67, 69, 72, 93; 31 March 1894, pp. 136, 137.
Sydney Mail, 27 September 1890, pp. 65, 141.
Town and Country Journal, 15 August 1891, pp. 90, 91; 27 October 1894, p. 132; 19 April 1911, p. 156.
von Stieglitz, K.R. (ed.) *The Port Phillip and Victoria Album of Emma von Stieglitz*. Fullers Bookshop, Hobart. p. 41.

Index

Index